High Performance Cloud Auditing and Applications

Keesook J. Han • Baek-Young Choi • Sejun Song
Editors

High Performance Cloud Auditing and Applications

Springer

Editors
Keesook J. Han
Air Force Research Laboratory
Rome, NY, USA

Baek-Young Choi
School of Computing and Engineering
University of Missouri – Kansas City
Kansas City, MO, USA

Sejun Song
Department of Engineering Technology
The Dwight Look College of Engineering
Texas A&M University
College Station, TX, USA

ISBN 978-1-4939-4435-4 ISBN 978-1-4614-3296-8 (eBook)
DOI 10.1007/978-1-4614-3296-8
Springer New York Heidelberg Dordrecht London

Printed on acid-free paper

Springer is part of Springer Science+Business Media (www.springer.com)

Preface

Introduction

Cloud computing is revolutionizing all aspects of technologies to provide scalability, flexibility and cost-effectiveness. It has become a challenge to ensure the security of cloud computing that supports cross-domain services on a federation of multilevel secure servers. To secure cloud services and resources, cloud auditing and rapid response tasks are very important to meet the Service Level Agreement (SLA) requirements that may specify the levels of availability, serviceability, performance, security, or other attributes of the service. This book mainly focuses on cloud security and high performance computing for cloud auditing. Big cloud audit data sets may consist of client and server audit logs, router logs, etc. Since cloud computing may deploy services in federated cloud environments, audit data are collected and stored in distributed environments. It is necessary to feasibly capture, store, and analyze logs in order to identify threats and prevent attacks. Capturing security relevant information and auditing the results to determine the existence of security threats in the clouds are still challenging problems.

There is a growing demand for cloud computing standards. Establishing cloud computing standards is challenging because it is very complicated to integrate existing standards and new cloud computing related standards to provide reliable cloud services in federated cloud computing environments. Standards organizations (see Table 1) and working groups are documenting the guidelines and specifications to prepare cloud computing standardizations (see Table 2). The National Institute of Standards and Technology (NIST), the Cloud Security Alliance (CSA) and the Distributed Management Task Force (DMTF) Cloud Auditing Data Federation (CADF) Working Group have released essential cloud computing related publications. The NIST cloud computing publications provide comprehensive cloud computing, cloud security and cloud auditing guidelines. The CSA has released cloud security guidelines to establish a stable and secure baseline for cloud operations. The DMTF CADF cloud auditing specifications contain useful information for cloud auditing.

Table 1 Cloud-related standards organizations

Standards organization	
National Institute of Standards and Technology (NIST)	United States
Distributed Management Task Force (DMTF)	International
IEEE Standards Association (IEEE-SA)	International
International Telecommunications Union (ITU)	International
European Telecommunications Standards Institute (ETSI)	European
Organization for the Advancement of Structured Information Standards (OASIS)	International
International Organization for Standardization (ISO)/IEC	International

Table 2 Cloud security and auditing publications

	Publication title
NIST	Challenging Security Requirements for US Government Cloud Computing Adoption
NIST	Cloud Computing Security Reference Architecture
NIST	Guide to Security for Full Virtualization Technologies
CSA	Security Guidance for Critical Areas of Cloud Computing
CSA	Trusted Cloud Initiative (TCI) Reference Guidelines
DMTF	The CADF Data Format and Interface Definitions Specification
CSC	Digital Trust in the Clouds
OACIS	Security, Access and Identity Policy Standards
ITU	Focus Group on Cloud Computing Technical Report (Parts 1–6)

The NIST has collaborated on cloud computing to define and advance standards with United States Government (USG) agencies, federal Chief Information Officers (CIOs), private sector experts, and international bodies to identify and reach consensus on cloud computing technology and standardization priorities. The NIST released a two-volume "USG Cloud Computing Technology Roadmap" document (see NIST Special Publication 500-293) to support secure and effective cloud computing for the purpose of reducing costs and improving federated cloud computing services. These volumes put forth ten NIST strategic and tactical objectives related to cloud computing. The NIST has also established public working groups to achieve the ten requirements by leveraging the expertise of the broad cloud computing stakeholder community. The NIST Cloud Computing Security Working Group (NCC-SWG) is working on six of these requirements which facilitate secure adoption of cloud services.

The goal of the CSA Trusted Computing Initiative (TCI) is to support cloud providers in developing industry-recommended secure and interoperable identity, access and compliance management configurations, and practices. The TCI-Reference Architecture (TCI-RA) has been developed to provide a methodology and a set of tools that enable security architects, enterprise architects and risk management professionals to leverage a common set of solutions to meet the security requirements for a secure and trusted cloud. The NCC-SWG has developed the NIST Cloud Computing Security Reference Architecture

(NCC-SRA) that was derived from the NIST Cloud Computing Reference Architecture (NCC-RA). The NIST leveraged the CSA TCI-RA to identify the set of security components in the NCC-SRA. The set of security components for a particular cloud model is introduced in detail in NCC-SRA. The NCC-SRA security components are carried on the three root-domains (Business Operation Support Service (BOSS), Information Technology Operation Support (ITOS) and Security and Risk Management (S&RM)) and the four service layers. Eighteen security control families are identified in the NIST SP 800-53.

The DMTF CADF Working Group proposed the open standards to meet the cloud customer expectations that cloud providers must provide standard mechanisms for their tenant customers to self-manage and self-audit application security. A cloud providers ability to provide specific audit event, log and report information on a per-tenant and application basis is essential. Therefore, the DMTF CADF Working Group has released the CADF Data Format and Interface Specification to enable information sharing by supporting the federation of normative audit event data in the form of customized reports and logs. This documentation also defines domain specific identifiers, event classification values and tags that can be used to dynamically generate customized logs and reports for cloud subscribers or customers.

Cloud federation is still a new and emerging research area. Federated cloud computing faces challenges relating to policy, technology, guidance, security, and standards. Cloud computing related specifications, standards and implementation technologies are required to establish security, interoperability, and portability to support federated cloud computing. Comprehensive federated computing technologies are critical to ensure cost-effective and secure cloud computing, and to assure mission-critical requirements. Therefore, standards organizations have worked with numerous cloud security and auditing working groups to develop cloud computing standards. In summer 2011, the United States Air Force Research Laboratory (AFRL) CyberBAT Cloud Security and Auditing Team initiated the exploration of the cloud security challenges and future cloud auditing research directions that are covered in this book.

Expected Audience

This book provides cloud security and auditing implementation strategies and research directions to diverse audiences:

- **Academics and students:** This book contains a comprehensive review of cloud security and auditing technologies, secure cloud architectures, programming languages, software and/or hardware based implementation and evaluation strategies for high performance cloud auditing and applications. It also provides introductory course materials for students.

- **Researchers:** This book puts forth future research directions and provides important references to a variety of research areas. Researchers can find this material useful in developing their concepts and strategies.
- **Standard developers and policy makers:** The standard developers need to use the high performance cloud auditing technology to adapt cloud security into the NCC-SRA or other cloud security reference models. Cloud access control and assured information sharing are useful to the policy makers.
- **Cloud vendors and auditors:** This book presents a comprehensive treatment of cloud security and auditing technologies. This book provides useful knowledge for building secure clouds to process and analyze massive audit data sets and to meet the SLA requirements.

Book Overview

The objectives of this book are to present surveys, concepts, algorithms, techniques and components of high performance cloud auditing systems in order to reduce cloud security risks, and to increase availability and performance of cloud computing for surviving in a contested network environment. The book consists of 13 chapters contributed by 40 authors. The book chapters are split into three parts.

Part I: Cloud Architectures and Security Issues

Part I contains surveys of cyber threats and security issues in cloud computing and presents secure cloud architectures. This part is designed to provide introductory materials for cloud auditing technologies.

Chapter "An Overview of Issues and Recent Developments in Cloud Computing and Storage Security" presents an overview of issues and recent developments in cloud computing and storage security. This chapter addresses security and privacy concerns due to lack of data protection transparency and accountability in the cloud. The survey topics include recent security threats, authentication, virtualization, availability, accountability, and privacy and integrity of remote storage and outsourced computation.

Chapter "Moving Target Defense for Cloud Infrastructures: Lessons from Botnets" proposes the novel moving target defense (MTD) for cloud infrastructures. Botnets are fast-moving targets that are difficult to detect with conventional security tools. Therefore, MTD has become a major theme in cyber security researches. The authors comprehensively survey the botnet literature, describe the evolution in botnet technologies, draw lessons from botnets in identifying cloud security challenges, and propose solutions to MTD for cloud infrastructures in order to make the network more resilient against novel and persistent attacks.

Chapter "Secure Mobile Cloud Computing and Security Issues" discusses secure mobile cloud computing and security issues. This chapter provides an overview of the latest mobile computing models and architectures focusing on security properties. A wide range of threats against the availability, privacy and integrity of mobile cloud computing architectures is investigated in order to develop defense mechanisms for secure mobile cloud computing and applications.

Chapter "Information Fusion in a Cloud-Enabled Environment" presents information fusion in a cloud-enabled environment. This chapter describes three aspects of current developments to low/high-level information fusion (LLIF/HLIF) and cloud computing such as agent-based service architectures, ontologies, and metrics for timeliness, confidence and security, and introduces the Cloud-Enabled Bayes Network (CEBN) for wide area motion imagery target tracking and identification.

Part II: Cloud Auditing and Assured Information Sharing

Part II aims to provide in-depth cloud auditing techniques, federated cloud security architectures, cloud access control models, and access assured information sharing technologies. The cloud access control and assured information sharing technologies are in this part because cloud data contain sensitive information which needs to be shared in order to aggregate, analyze and query the data in federated cloud computing environments.

Chapter "Diagnosing Vulnerability Patterns in Cloud Audit Logs" analyzes the diagnosis of vulnerability patterns in cloud audit logs. Existing security standards, protocols and auditing mechanisms can provide audit logs but auditable events of web service compositions in service cloud architectures are not well defined. This chapter specifies the audit log and defines the Vulnerability Diagnostic Trees (VDTs) to determine security vulnerability patterns emerging from Service Orientation Architecture (SOA) communications in conjunction with service composition allocation and cross-cloud communication.

Chapter "Exploiting Timing Side Channel in Secure Cloud Scheduling" exploits a timing side channel in secure cloud scheduling. In multi-tenancy cloud environments, a malicious user can learn about the service usage pattern of an innocent user by a timing based side channel attack. This chapter demonstrates the information leakage of a timing side channel in shared schedulers, discusses timing side channel threats and countermeasures, and introduces the design of secure scheduling policies.

Chapter "Federated Cloud Security Architecture for Secure and Agile Clouds" introduces the federated cloud security architecture for secure and agile clouds. This chapter describes cloud security threats and proposes a novel federated security architecture that consists of a set of seamlessly integrated systematic security mechanisms at the application layer, the network layer and the system layer in federated cloud computing environments.

Chapter "Trust-Based Access Control for Secure Cloud Computing" presents trust-based access control for secure cloud computing. Multi-tenancy, elasticity and dynamicity pose several novel challenges for access control in a cloud environment. This chapter summarizes traditional and modern access control models, explores challenges of cloud access control, identifies various authorization and enforcement requirements and desirable properties of access control models, and introduces graph-theoretic semantics of an access control model.

Chapter "Assured Information Sharing (AIS) Using Private Clouds" introduces the concept of assured information sharing in private clouds. The authors provide basic characteristics of an AIS framework, discuss the state of the art in the realm of AIS, and provide contemporary cloud-based AIS implementation methodologies for Cloud-centric Assured Information Sharing System (CAISS) and CAISS XACML policies (CAISS-X). The CAISS uses a cloud-based framework for both data storage and retrieval as well as policy enforcement and CAISS-X employs a cloud-centric framework to store and query large amounts of data via a non-cloud policy engine that enforces XACML-based policies.

Part III: High Performance Cloud Computing

Part III outlines a wide range of challenges and provides solutions to manage and control very large and complex data sets. It is impractical to process a huge audit data set using existing on-hand database management tools or data processing applications in real-time. CPU-GPU computing, MapReduce and router-based filtering technologies are employed to face the challenges of big data processing.

Chapter "GPGPU Computing for Cloud Auditing" presents GPGPU computing for cloud auditing. There is a growing need for computing platforms that are able to rapidly analyze data-intensive cloud audit data. GPGPU computing can perform data analysis with a high level of parallelism employing tools like Hadoop MapReduce. The chapter contains a broad background on GPGPU computing, architectures, and programming options, illustrated by helpful programming examples.

Chapter "CPU-GPU System Designs for High Performance Cloud Computing" discusses CPU-GPU system designs for high performance cloud computing. This chapter focuses on the improvement of cloud computing performance by combining the powerful scalar processing on CPU with the efficient parallel processing on GPU. The authors also introduce the mainstream and emerging memory hierarchy designs in CPU-GPU systems and optimization techniques of the data allocation and migration between CPU and GPU for performance improvement.

Chapter "MapReduce Performance in Federated Cloud Computing Environments" introduces MapReduce optimization in federated cloud computing environments. The demand for federation among multiple distributed clusters is growing, in order to process data-intensive and compute-intensive applications. The MapReduce framework coupled with cloud computing is emerging as a viable solution for distributed big data processing. The authors describe various cloud based applications over distributed clouds and provide a network aware MapReduce optimization technique.

Chapter "Improving Cloud Performance with Router-Based Filtering" presents improving cloud performance with router-based filtering. The router-based filtering technology has been developed to enhance the availability of cloud computing and performance of cloud auditing. An overview of the specification and generation of filtering rules used by routers, and a theoretical model to find the best locations for hardware routers in a network to block malicious traffic, and experimental results are provided in this chapter.

Rome, NY, USA — Keesook J. Han
Kansas City, MO, USA — Baek-Young Choi
College Station, TX, USA — Sejun Song

"Approved for Public Release; Distribution Unlimited: 88ABW-2013-2591, 31-May-2013"

Acknowledgements

This work was supported by the United States government funds from the Air Force Office of Scientific Research (AFOSR), the AFOSR Summer Faculty Fellowship Program (SFFP), the Air Force Research Laboratory (AFRL) Visiting Faculty Research Program (VFRP), the National Science Foundation (NSF) and the National Institute of Health (NIH). The editors especially appreciated the support of AFOSR/RSL Program Manager Dr. Robert Herklotz; all chapters were partially supported by his AFOSR Information Operations and Security Program funds for extramural projects and AFOSR intramural project (LRIR 11RI01COR). The editors are grateful for the support provided by AFRL Information Institute Director Mr. John Graniero and Mr. Frank Hoke, and also to Elizabeth Schlaegel for helping VFRP and SFFP fellows. The editors are also thankful to Dr. Warren Debany AFRL/RIG, Mr. James Perretta AFRL/RIGA, Mary Allen AF 88ABW/PAX, Sandra Simison AF 88ABW/PAX, and Springer Science+Business Media Editorial Assistant Rebecca Hotowitz for reviewing book chapters. The editors also express their gratitude to Prof. John Kieffer for proofreading the book.

Contents

Part III High Performance Cloud Computing

Contributors

Everaldo Aguiar University of Notre Dame, Notre Dame, IN, USA

Norman Ahmed Air Force Research Laboratory, Rome, NY, USA
Purdue University, West Lafayette, IN, USA

Marina Blanton University of Notre Dame, Notre Dame, IN, USA

Erik Blasch Air Force Research Laboratory, Rome, NY, USA

Tyrone Cadenhead University of Texas at Dallas, Richardson, TX, USA

Heath Carroll University of South Carolina, Columbia, SC, USA

Genshe Chen Intelligent Fusion Tech, Gaithersburg, MD, USA

Yiran Chen University of Pittsburgh, Pittsburgh, PA, USA

Yu Chen SUNY-Binghamton, Binghamton, NY, USA

Chui-Hui Chiu Louisiana State University, Baton Rouge, LA, USA

Rose Gamble University of Tulsa, Tulsa, OK, USA

Qijun Gu Texas State University-San Marcos, San Marcos, TX, USA

Mina Guirguis Texas State University-San Marcos, San Marcos, TX, USA

Jie Guo University of Pittsburgh, Pittsburgh, PA, USA

Chin-Tser Huang University of South Carolina, Columbia, SC, USA

Sachin Kadloor University of Illinois at Urbana-Champaign, Urbana, IL, USA

Murat Kantarcioglu University of Texas at Dallas, Richardson, TX, USA

Vaibhav Khadilkar University of Texas at Dallas, Richardson, TX, USA

Negar Kiyavash University of Illinois at Urbana-Champaign, Urbana, IL, USA

Ralph Kohler Air Force Research Laboratory, Rome, NY, USA

Praveenkumar Kondikoppa Louisiana State University, Baton Rouge, LA, USA

Miriam E. Leeser Northeastern University, Boston, MA, USA

Feng Li Indiana University-Purdue University Indianapolis, Indianapolis, IN, USA

Weiliang Luo University of Texas at San Antonio, San Antonio, TX, USA

Seung-Jong Park Louisiana State University, Baton Rouge, LA, USA

Wei Peng Indiana University-Purdue University Indianapolis, Indianapolis, IN, USA

James Perretta Air Force Research Laboratory, Rome, NY, USA

Indrajit Ray Colorado State University, Fort Collins, CO, USA

Indrakshi Ray Colorado State University, Fort Collins, CO, USA

Virginia W. Ross Air Force Research Laboratory, WPAFB, Ohio, OH, USA

Dan Shen Intelligent Fusion Tech, Gaithersburg, MD, USA

Zhenyu Sun University of Pittsburgh, Pittsburgh, PA, USA

Bhavani Thuraisingham University of Texas at Dallas, Richardson, TX, USA

Rui Xie University of Tulsa, Tulsa, OK, USA

Li Xu University of Texas at San Antonio, San Antonio, TX, USA

Shouhuai Xu University of Texas at San Antonio, San Antonio, TX, USA

Zhenxin Zhan University of Texas at San Antonio, San Antonio, TX, USA

Yihua Zhang University of Notre Dame, Notre Dame, IN, USA

Qingji Zheng University of Texas at San Antonio, San Antonio, TX, USA

Xukai Zou Indiana University-Purdue University Indianapolis, Indianapolis, IN, USA

Acronyms

ABAC	Attribute-Based Access Control
ACID	Atomicity, Consistency, Isolation, Durability
ACL	Access Control List
AIS	Assured Information Sharing
AISL	Assured Information Sharing Lifecycle
AMP	Accelerated Massive Parallelism
AOM	Agile Opportunistic Migration
API	Application Programming Interface
APT	Advanced Persistent Threat
APU	Accelerated Processing Unit
ARP	Address Resolution Protocol
AS	Autonomous System
AWS	Amazon Web Services
BE	Broadband Engine
BFE	Bloom Filter Encoding
BGP	Border Gateway Protocol
BLAS	Basic Linear Algebra Subroutine
BPDU	Bridge Protocol Data Unit
BOSS	Business Operation Support Service
C&C	Command and Control
C2	Command and Control
CADF	Cloud Auditing Data Federation
CAISS	Cloud-Centric Assured Information Sharing System
CASA	Context-Aware Security Architecture
CBAC	Coalition-Based Access Control
CC	Cluster Controller
CCE	Common Configuration Enumeration
CDS	Connected Dominating Set
CEBN	Cloud-enabled Bayes Network
CG	Conjugate Gradient
CGCM	CPU-GPU Communication Manager

CIO	Chief Information Officer
CLC	CLoud Controller
CM	Cloud Manager
C-OODA	Cognitive-Observe-Orient-Decide-Act
CPU	Central processing unit
CRON	Cyberinfrastructure of Reconfigurable Optical Networking
CSA	Cloud Security Alliance
CSP	Cloud Service Provider
CUDA	Compute Unified Device Architecture
CUFFT	CUDA Fast Fourier Transform
DAC	Discretionary Access Control
DBMS	Database management systems
DCGS	Distributed Common Ground System
DDL	Data Definition Language
DDoS	Distributed Denial of Service
DIB	DCGS Integration Backbone
DMA	Direct Memory Access
DMC	Discrete Memoryless Channel
DNS	Domain Name System
DMTF	Distributed Management Task Force
DoS	Denial of Service
DRAM	Dynamic Random-Access Memory
DSCP	Differentiated Services Code Point
DSL	Digital Subscriber Line
DX	DirectX
EC2	Elastic Compute Cloud
ETSI	European Telecommunications Standards Institute
FAT	File Allocation Table
FCL	Fusion Control Link
FFT	Fast Fourier Transform
FIFO	First In, First Out
FPGA	Field-Programmable Gate Array
GDDR	Graphics Double Data Rate
GFS	Google File System
GIG	Global Information Grid
GMC	Graphics Memory Controller
GPGPU	General-purpose Programming on GPU
GPS	Global Positioning System
GPU	Graphics Processing Unit
HDD	Hard Disk Drive
HDFS	Hadoop Distributed File System
HFS	Hierarchical File System
HLIF	High-Level Information Fusion
HPCC	Hybrid P2P-based C&C
HSF	Hard-Soft Fusion

HSI	Human Systems Integration
HTTP	Hypertext Transfer Protocol
HVMR	Heterogeneous VM Replication
I/O	Input/Output
IaaS	Infrastructure as a Service
IEC	International Electrotechnical Commission
IEEE-SA	IEEE Standards Association
ICMP	Internet Control Message Protocol
IFS	Information Fusion System
IFSA	Information Fusion Situation Assessment
IM	Information Management
IP	Internet Protocol
ISO	International Organization for Standardization
ISP	Internet Service Provider
ISR	Intelligence, Surveillance and Reconnaissance
ITOS	Information Technology Operation Support
ITU	International Telecommunications Union
JASYPT	Java Simplified Encryption
JDL	Joint Director of the Laboratories
JTAG	Joint Test Access Group
JVM	Java Virtual Machine
KVM	Kernel-Based Virtual Machine
LAIR	Large Area Image Recorder
LBS	Location-Based Service
LLC	Last Level Cache
LLIF	Low-Level Information Fusion
LTL	Linear Temporal Logic
MAC	Mandatory Access Control
MAC	Media Access Control
MBR	Master Boot Record
MIO	Managed Information Object
MLC	Multi-Level Cell
MLS	Multi-Level Security
MOS	Mobile Operating System
MPI	Message Passing Interface
MPLS	Multi-Protocol Label Switching
MSASP	Microsoft Active Server Page
MTBDD	Multi-terminal binary decision diagram
MTD	Moving Target Defense
MTJ	Magnetic Tunnel Junction
NAT	Network Address Translation
NCC-RA	NIST Cloud Computing Reference Architecture
NCC-SRA	NIST Cloud Computing Security Reference Architecture
NCC-SWG	NIST Cloud Computing Security Working Group
NIC	Network Interface Card

NIDPS	Network Intrusion Detection and Prevention System
NIST	National Institute of Standards and Technology
NLP	Natural Language Processing
NLRI	Network Layer Reachability Information
NTFS	NT File System
OASIS	Organization for the Advancement of Structured Information Standards
OGC	Open Geospatial Consortium
OMTP	Open Mobile Terminal Platform
OODA	Observe-Orient-Decide-Act
OpenCL	Open Computing Language
OpenGL	Open Graphics Library
OWL	Web Ontology Language
PaaS	Platform as a Service
PAP	Policy Administration Point
PCA	Proof-Carrying Authorization
PCI	Peripheral Component Interconnect
PCM	Phase Change Memory
PCT	Mathworks Parallel Computing Toolbox
PDP	Policy Decision Point
PEP	Policy Enforcement Point
PGI	Portland Group, Inc
PIN	Personal Identification Number
PIP	Policy Information Point
PLA	Parallel Linear Algebra
PPBT	Poisson Pareto Burst Traffic
PRA	Permission-Role Assignment
QoS	Quality of Service
RBAC	Role Based Access Control
RDF	Resource Description Framework
RDFS	Resource Description Framework Schema
RFI	Receive Requests for Information
RH	Role Hierarchy
RISC	Reduced Instruction Set Computer
RM	Resource Management
RPC	Remote Procedure Call
RR	Round-Robin
RTT	Round-Trip Time
S3	Simple Storage Service
S&RM	Security and Risk Management
SC	Storage Controller
SCT	Security Context Token
SIM	Subscriber Identity Module
SIMD	Single Instruction Multiple Data
SIMT	Single Instruction Multiple Thread

SIS	Secure Information Sharing
SJF	Shortest-Job-First
SLA	Service Level Agreement
SM	Session Manager
SM	Streaming Multiprocessors
SMH	Session Manager History
SMS	Short Message Service
SMT	Simultaneous Multi-Thread
SoA	Service-Oriented Architecture
SOAP	Simple Object Access Protocol
SoD	Separation of Duty
SOI	Silicon on Insulator
SOP	Same Origin Policy
SaaS	Software as a Service
SOP	Standard Operating Procedure
SPMD	Single program multiple data
SQL	Structured Query Language
SP	Streaming Processor
SRAM	Static Random-Access Memory
SSD	Solid State Drive
SSH	Secure Shell
SSL	Secure Sockets Layer
STID	Simultaneous Tracking and Identification
STIG	Security Technical Implementation Guide
STS	Security Token Service
TBAC	Trust-Based Access Control
TCB	Trusted Computing Base
TCI	Trusted Cloud Initiative
TCI-RA	Trusted Cloud Initiative Reference Architecture
TCP	Transmission Control Protocol
TCSEC	Trusted Computer System Evaluation Criteria
TDMA	Time Division Multiple Access
TLS	Transport Layer Security
TMAC	Team-Based Access Control
TOR	The Onion Router
ToS	Type of Service
TPM	Trusted Platform Module
TRBAC	Temporal RBAC
TTL	Time To Live
UDP	User Datagram Protocol
UFCC	URL Flux-based C&C
UFS	Unix File System
UI	User interface
UMA	Unified Memory Architecture
URA	User-Role Assignment

URL	Uniform Resource Locator
USG	United States Government
UTS	Upper Tier Server
VLAN	Virtual Local Area Network
VRF	Virtual Routing and Forwarding
VM	Virtual Machine
VMI	Virtual Machine Introspection
VMM	Virtual Machine Manager
VMM	Virtual Machine Monitor
WAMI	Wide Area Motion Imagery
WAP	Wireless Access Point
WLAN	Wireless Local Area Network
WMIR	Millimeter-wave infrared
WSDL	Web Service Definition Language
XACML	EXtensible Access Control Markup Language
XML	Extensible Markup Language
XSS	Cross Site Scripting

Part I
Cloud Architectures and Security Issues

An Overview of Issues and Recent Developments in Cloud Computing and Storage Security

Everaldo Aguiar, Yihua Zhang, and Marina Blanton

Abstract The recent rapid growth in the availability and popularity of cloud services allows for convenient on demand remote storage and computation. Security and privacy concerns, however, are among the top impediments standing in the way of wider adoption of cloud technologies. That is, in addition to the new security threats that emerge with the adoption of new cloud technology, a lack of direct control over one's data or computation demands new techniques for service provider's transparency and accountability. The goal of this chapter is to provide a broad overview of recent literature covering various aspects of cloud security. We describe recently discovered attacks on cloud providers and their countermeasures, as well as protection mechanisms that aim at improving privacy and integrity of client's data and computations. The topics covered in this survey include authentication, virtualization, availability, accountability, and privacy and integrity of remote storage and computation.

1 Introduction

Cloud computing enables on-demand access to computing and data storage resources that can be configured to meet unique constraints of the clients with minimal management overhead. The recent rise in the availability of cloud services makes them attractive and economically sensible for clients with limited computing or storage resources who are unwilling or unable to procure and maintain their own computing infrastructure. The ever increasing need for computing power and

E. Aguiar (✉) • Y. Zhang • M. Blanton
University of Notre Dame, Notre Dame, IN, USA
e-mail: eaguiar@nd.edu; yzhang16@nd.edu; mblanton@nd.edu

K.J. Han et al. (eds.), *High Performance Cloud Auditing and Applications*,
DOI 10.1007/978-1-4614-3296-8_1, © Springer Science+Business Media New York 2014

storage accounts for the steady growth in popularity of companies offering cloud services. Clients can easily outsource large amounts of data and computation to remote locations, as well as run applications directly from the cloud.

As an example, Chow et al. [21] point out that some of the world's largest software companies today operate entirely from the cloud or at least have a major portion of their services outsourced to a cloud environment. With all the convenience being offered for a relatively low price, one would expect that most, if not all, companies would migrate to the cloud. There are, however, impediments to this transition, with security commonly cited as the number one concern by both the private sector and the government [33,72]. Public clouds have not yet reached a level of security that enables private users and corporations to entrust them with sensitive data. As a concept that was created from a combination of several building blocks, cloud computing naturally inherited the security flaws present in each component. In addition, new issues that did not previously exist arose with this new technology trend. The evidence for this problematic side of cloud computing can be seen in the considerable literature [1, 20, 21, 42, 49, 58, 60, 62, 75] that surveys several of the general and specific security issues that affect clouds. Cloud security is also the focus of this work.

Unlike prior surveys of cloud security issues, our ultimate goal is to provide a much more complete and thorough coverage of the research literature related to this topic. We give a broad overview of publications in the fields of cloud computing security and security of remote storage and computation. In particular, the topics covered in this work include:

- *Client authentication and authorization*: We cover the current body of work on methods for disrupting and exploiting the interface between a cloud provider and its clients, usually carried out via a web browser.
- *Security shortcomings of hardware virtualization*: We describe the problems that have surfaced along with the massive use of hardware virtualization by cloud providers. We indicate how virtualization can be exploited to obtain unauthorized information from vulnerable users, and also indicate mitigation techniques that can be employed. In addition, we also address vulnerabilities related to the usage and sharing of virtual machine (VM) images.
- *Flooding attacks and denial of service (DoS)*: Because cloud computing systems are designed to scale according to the demand for resources, an attacker may use that characteristic to maliciously centralize large portions of the cloud's computing power, lowering the quality of service that the cloud provides to other concurrent users. We discuss different types of attacks on cloud availability and their potential consequences.
- *Cloud accountability, or its ability to capture and expose wrongful activity*: We discuss capabilities that an accountable system should have and solutions for achieving these capabilities.
- *Challenges and solutions for remote storage protection*: We describe several techniques that can be employed by cloud clients to verify integrity of their outsourced data.

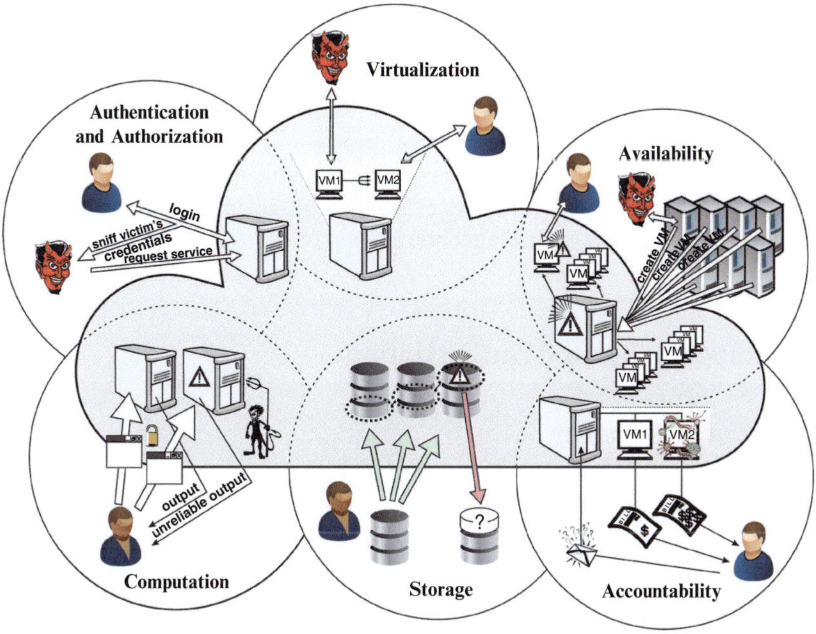

Fig. 1 Overview of the topics covered in this article

- *Protection of outsourced computation*: Finally, we give an overview of current approaches for assuring privacy and integrity of outsourced computations.

Figure 1 provides a summary of the topics covered in this work.

Before proceeding with the coverage of the above mentioned topics, we define some of the common terminology used in the context of clouds. Although the members of the cloud community in its early days did not all agree on which features should actually be part of the broad concept of cloud computing, the definition that was later provided by the U.S. National Institute of Standards and Technology (NIST) [58] precisely subdivided cloud computing into three distinct models, which offer differing capabilities to the consumer.

- *Software as a Service (SaaS)*: Following this model, the cloud service provider makes both the software and the cloud infrastructure to run it available to the client, while it retains complete control over the underlying physical settings of the cloud (i.e., the operating system, network, storage, etc.) and the individual application capabilities. Thin client interfaces such as web browsers are often used to allow access to these applications.
- *Platform as a Service (PaaS)*: Similar to the previous model, PaaS aims at giving the client the ability to run software on the cloud infrastructure. However, in this case the consumer can deploy his own applications or create personalized ones

by using tools supported by the cloud provider. Control over the physical settings remains restricted to the cloud provider but the clients of this model are able to fully manage the deployed applications.
- *Infrastructure as a Service (IaaS)*: The consumers of this model can provision computing resources including storage, processing, and networks, which can usually be used to host any arbitrary operating system and applications. Most of the well established cloud services (such as Amazon EC2) adopt this model and charge their users according to the amount of utilized resources.

2 Authentication and Authorization

2.1 Browser Security

One of the goals of providers of Software as a Service (SaaS), as pointed out in [47], is to have a generic means of offering access to their clients, independent of their local platform. Internet browsers are the ideal choice and can be used on the client side to simply perform I/O, while the actual computation takes place in the cloud back-end. As a downside, however, known browser security issues may affect this interface for cloud authentication. We review two issues that can be exploited in this context:

- *XML-based authentication in a browser*: User authentication to the cloud via a web browser requires the issuing of cryptographically valid Extensible Markup Language (XML) tokens. As the first step of the process, the user provides his login credentials, which upon verification and validation, allow the cloud to issue the security tokens which the client will later provide to the cloud server as means of authentication. This process, however, requires the assistance of a trusted third party as browsers are not capable of issuing XML-based security tokens on their own. There are multiple services that can be used as an intermediary between users and the cloud, including Microsoft's Passport, which Jensen et al. [47] use to illustrate the security shortcomings of this interface. In that scenario, when the user wishes to authenticate to the cloud, an Hypertext Transfer Protocol (HTTP) redirect is sent to the Passport login server. After the user enters information, the server will translate those credentials into a Kerberos token which is then, via another HTTP redirect, sent to the cloud server, finally allowing the user to be authenticated. The downside of this approach lies in the mechanism via which the tokens are made secure. The Kerberos tokens sent by Microsoft Passport to the cloud server are only protected by the Same Origin Policy (SOP), which has been shown to be vulnerable [91]. Hence, if an adversary gains control of these tokens, he will be granted full access to all of his victim's services. To counteract this problem, the paper [47] discusses a few methods via which XML tokens can be made safe with the use of Transport Layer Security (TLS). This paper also suggests that a more optimal solution to these limitations would be to further enhance the capability of web browsers, adding `XML Encryption` and `XML Signature` to the browser security API. In [57], a widely known attack on XML signatures was described in which the attacker intercepts a given Simple Object Access Protocol (SOAP) message being sent to the server by some client, replacing its content with some arbitrary request on behalf of the attacked client; Amazon EC2 is vulnerable to a variation of this signature wrapping attack [38].

- *Federated Identity Management based on insecure SOP*: The legacy Same Origin Policy (SOP) is mainly used for defining access rights to web scripts. Read/write operations are only permitted to requests originated from the same source, which in a Web context is usually defined jointly by the domain name, the protocol, and the port. As previously stated, the tokens used during client authentication to the cloud are often protected by this policy. Jensen et al. [47], however, point out that Domain Name Systems (DNS) can be infected in such a way that would cause domain names to become unreliable, compromising the functioning of the security policy. For instance, it has been shown that an attacker can host a malicious website, which upon being visited, will attempt to change the DNS settings of the user's broadband network router [71]. Therefore, using SOP, exclusively for the purpose of protecting these authentication tokens does not strictly guarantee security. To mitigate this problem, [47] elaborates on better ways of integrating SOP and TLS to improve security of these protocols.

In order to ensure the safety of their clients' information, cloud providers have claimed to address the above forms of attack. However, a recent work by Somorovsky et al. [70] reports two successful variants of the signature wrapping attack on both the Amazon cloud services and Eucalyptus [30]. The first attack variant used a single eavesdropped SOAP request for the `MonitorInstances` operation with a duplicated body to perform arbitrary operations. The SOAP interface used by Amazon EC2 validated the XML signature for the original (unmodified) SOAP body, but performed the operation described in the maliciously injected body. To circumvent the timestamp verification performed before the execution of the SOAP messages, the authors similarly duplicated the timestamp element in the security header and were able to execute arbitrary operations even when the original timestamp had already expired. Upon being notified of the issue, the Amazon Web Services (AWS) security team quickly provisioned a patch that prevented the execution of messages containing duplicated timestamp elements. The second attack variant, however, showed that it was possible to issue SOAP messages with several `<soap:Body>` elements and technically achieve the same result. The modified message used in that case had three `<soap:Body>` bodies, with the second body containing the originally signed timestamp. That timestamp and the whole third body were verified by the signature verification component, while a different timestamp located in the security header was tested for expiration and the first body was interpreted to determine the operation to be performed.

The SOAP interface used by Eucalyptus is slightly different and validates the format of incoming SOAP messages against an XML schema. In such a scenario, messages containing duplicated body or timestamp elements were automatically rejected and the previously mentioned attacks were shown to be infeasible. Nevertheless, by modifying the classical signature wrapping attack [57] and by using a duplicated security header element that did not violate the XML schema, the authors were able to place the signed body and timestamp elements into a new location and replace the original ones with their own.

Besides describing these successful attack implementations, [70] also shows how advanced Cross Site Scripting (XSS) techniques can be used to steal data credentials from Amazon and Eucalyptus cloud users. In the Amazon case, Somorovsky et al. demonstrate that it is possible for an attacker to utilize the public Amazon shop discussion forums as a means for exposing user information needed to perform the previously mentioned signature wrapping attacks. The attacker simply needs to create a new discussion topic containing Java Script code to obtain cookie data via `document.cookie`. Despite the fact that Amazon utilizes padding techniques to prevent malicious code from being executed, a careful formatting of the code block was able to deceive that mechanism. Once in possession of this data, the attacker can immediately obtain access to that user's cloud services since the same log-in credentials are shared across the Amazon shop and Amazon cloud control interface.

2.2 *Binding Issues*

In order to accurately maintain and instantiate VMs (in the case of IaaS) or specific modules (in the case of PaaS), the cloud provider needs to store metadata descriptions associated with each operation. A user may refer to these metadata descriptions in order to determine the detailed functionality of a certain service upon having a particular need. Jensen et al. [46] suggest a spoofing attack that aims at reengineering this metadata.

A *metadata spoofing attack* [46] allows an adversary to modify metadata descriptions, potentially causing severe damage to the user's services. An example of this attack given in [47] suggests that an attacker may attempt to modify the server's Web Service Definition Language (WSDL) descriptions. This can, for instance, cause a call to `deleteUser` to generate a SOAP message that mimics and may be interpreted as a call to `setAdminRights` on the server side. Therefore, in that scenario, a user who intended to delete a certain account may inadvertently grant extra privileges to the adversary.

Clouds are especially susceptible to these types of attacks because they use WSDL repositories that are dynamically utilized by all users, increasing the probability of an attacker successfully spreading his malicious WSDL file. As a countermeasure, cloud providers need to cautiously check all metadata documents for authenticity, which may be somewhat troublesome as there are no standard mechanisms for doing so. Another possible approach would be to ensure the establishment of trust relationships with users prior to accepting their requests, although this may not be applicable to some scenarios [46].

3 Hardware Virtualization-Related Issues

In order to satisfy the demand for computation from multiple concurrent users, cloud service providers often need to multiplex the use of their limited available hardware. Amazon EC2 and Microsoft Azure, for instance, allow users to dynamically instantiate VMs which can be tuned to their specific needs and easily managed while running on the top of a shared physical infrastructure. Although the general use of virtualization normally implies an isolation between the workloads involved, Ristenpart et al. [63] show that new vulnerabilities can arise when an adversary is capable of determining the location of a particular target VM in the cloud infrastructure and subsequently instantiating a malicious co-resident VM. Though not particularly trivial to exploit, these vulnerabilities can vary from the extraction of cryptographic keys to DoS, depending on the adversary's intent.

Another virtualization-related security issue, raised by Christodorescu et al. [22], is that vulnerabilities arise from the fact that in highly multiplexed cloud environments, providers often have no control over what types of VMs are being deployed by their clients. Conversely, users also have a very limited knowledge of the underlying physical infra-structure. We discuss how attackers can instantiate VMs with modified OS images, how that may constitute a threat to other users, and what solutions were put forth by Christodorescu et al. [22]. Along similar lines, Wei et al. [83] explored the possible risks concerning how VM images are stored and managed in the cloud. Specifically, they showed that an attacker can build and share images that are contaminated with malware and other types of threats. A more recent work [16] gives practical examples of how these risks can be exploited on Amazon EC2 and how they should be mitigated.

We next describe the above three virtualization-related security issues in more detail.

3.1 Exploiting VM Co-residence

Ristenpart et al. [63] investigate the possible means by which a cross-VM attack – the one that originates from a given VM and targets another – may be performed. There are four distinct steps to be carried out by the adversary in the suggested attack: (1) determining where the target VM is physically located in the cloud, (2) launching VMs that will be co-resident to the targeted one, (3) verifying that the VMs involved are indeed co-resident, and (4) exploiting cross-VM information leakage. In the following we describe each of these steps.

- *Determining the target VM's location*: Using simple network probing tools, Ristenpart et al. showed that it is possible to infer considerably accurate information about the location of VMs based on their assigned IP addresses. To do so, they first sampled a large pool of IP addresses, isolated those that were responsive and associated with EC2, and used Domain Name Systems (DNS) lookups from within the Amazon EC2 network to obtain their internal IP. In parallel to that, the researchers also launched a large number of VM instances and observed that distinct IP address ranges appeared to be assigned to each different zone. Thus, they maintain that by simply having the internal IP address of a given VM, one can make fairly accurate deductions regarding its physical location. To prevent such a form of physical mapping, cloud providers need only to eliminate the static assignment of IP addresses based on availability zone, instance type, or any other feature of the VMs.

- *Instantiating co-resident VMs*: Two methods for instantiating co-resident malicious VMs are proposed by Ristenpart et al. [63]. The first method consists of launching a large number of VMs in a brute-force manner until co-residency is achieved. The second method is more refined and takes advantage of the Amazon EC2 placing algorithm's tendency to assign machines belonging to a common small set to VMs being instantiated within a short period of time from each other. Deliberately targeting VMs that had been recently launched yields a successful co-residency rate of nearly one half for certain types of instances. One way to mitigate such issues, as suggested by Ristenpart et al., is to allow users to make the decision on the placement of their VMs.
- *Verifying co-residence*: Ristenpart et al. verify or discard co-residence with a high level of correctness. To assert that any two VMs are co-resident, Ristenpart et al. check that the VMs have matching Dom0 IP addresses by tracing messages sent to the target VM to see whether the addresses of the first and last hops are the same. Co-residency is also confirmed by an analysis of the round-trip times and internal IP ranges. Preventing these techniques requires configuring the Dom0 machine to ignore `traceroute` requests, randomly assigning internal IPs, or using virtual LANs to isolate accounts.
- *Exploiting cross-VM information leakage*: Assuming the attacker was able to place his VM on the same physical machine as the target instance, [63] shows that a malicious adversary may learn information about a co-resident VM via cache-based side channels (which could possibly be used to steal cryptographic keys), as well as other physical resources that are multiplexed among the co-resident instances. This work has been recently expanded by Xu et al. [87] who were able to create similar covert channels with noticeably higher bit rates than previously reported. Despite that, the later work shows that even at higher bit rates, the harm that can be caused by such techniques is limited to the extraction of very small blocks of data (e.g., private keys).

3.2 Exploiting the Cloud's Limited Knowledge of User VMs

The use of virtualization in a cloud computing environment enforces an isolation between different workloads, protecting them from other users who may attempt to compromise their information. Another security layer deals with protecting individual workloads within themselves, which is especially important when these are exposed to the Internet. Providing such protection requires a deeper knowledge of the workload, which consists of the guest OS in the case of hardware virtualization. Information such as what operating system is running in the VM, however, is often unavailable to the cloud provider as in many cases the users are allowed to upload and run their own VMs.

The most widely available techniques for virtualization-based security are designed to work when the operating system running in the VM of interest is known in advance, allowing the virtual machine monitor (VMM) to correctly track its activity. In addition, in order to assure the integrity of the supervising tool being used, these techniques require that the monitoring be initialized via a safe boot, while the guest OS is first being put to use, which obviously becomes impractical when users are allowed to start their VMs from a guest OS snapshot, as is the case in most cloud settings. To remedy this situation where a provider cannot offer appropriate security measures due to the unknown configuration and integrity of its clients' VMs, Christodorescu et al. [22] propose a new architecture based on VM introspection. The concept of VM introspection was first introduced by Garfinkel et al. [31] and more recently formalized in [61]. Through the use of such techniques, it is possible to inspect a VM from a safer location outside the monitored machine and evaluate what is happening to it internally. This mechanism works with the aid of the hypervisor, which is modified to collect data describing its interactions with the VMs being monitored and send it to safe (specially designed) VMs that will carry out the actual analysis and monitoring process.

The solution put forth by Christodorescu et al. does not rely on any assumptions about the state of the running VMs and works even when the cloud user does not give the provider any information about the software running in their VMs. Their approach requires that the guest VMs sit on the top of a correct, trusted and unbreachable hypervisor controlled by the cloud provider, who also maintains special VMs referred to as secure VMs, which are assumed to be unbreachable and host both a guest OS identification and a rootkit detection application whose functionality is based on VMI. The OS identification application uses a whitelist of known executable kernel code obtained from several different operating systems (running on clean VMs) to match and validate those originated from the monitored VM, and while doing so it can accurately determine which guest OS is running in that specific VM. Using the same principle, the rootkit detector checks for the presence of unauthorized code in kernel space, which upon detection is compared to a database (blacklist) of known malicious code for identification. Experimental tests [22] showed that these applications, running in a secure VM and using the proposed hardware-based VM introspection, were able to both identify the OS running in a monitored VM with high confidence, and detect unauthorized code running in the guest kernel space. Another important feature of this approach is that, because it uses secure introspection and runs outside the guest OS, it will still keep the system safe from rootkits (and other malware) even in the event of an attack that disables the traditional security software running in the guest.

Another work in the realm of virtualization security [74] proposed a system called NoHype, in which the need for hypervisors is eliminated in its entirety, consequently excluding the hypervisor attack surface. In order to allow VMs to run natively on the underlying hardware while still maintaining the support for multiple concurrent VMs, NoHype places guest VMs in more direct contact with the hardware to avoid indirection. It implements slight modifications to the guest OS, allowing it to perform all system discovery during bootup. In addition, CPU and physical memory need to be pre-allocated and only virtualized I/O devices are used.

3.3 Exploiting VM Images

Despite the numerous services that can be provided by clouds and the apparent complexity of their environments, the resources offered by cloud providers can be narrowed down to primarily three: (1) a pool of VMs from which users can choose the most suitable for their specific needs; (2) a set of servers that are configured to efficiently run these VMs; and in some cases (3) a means for non-volatile storage that can be used to share data across multiple VMs or simply for backup. This section has already covered the security threats related to VMs that are currently in execution and their mitigation. Persistent data storage in the cloud and the corresponding security issues are the topic of Section 6. An area that has been paid considerably less attention is that of threats that can result in compromising VM images in the cloud; this area is discussed next.

Wei et al. [83] were the first to expose the risks involved with the publishing and large scale usage of images. The authors emphasize that the sharing of VM images is of great importance since it largely simplifies administrative tasks and costs related to installing and configuring software (i.e., instead of doing all the work, a user can simply choose a VM that is preloaded with all the software that is needed). There are, however, risks to all parties involved in this process. For example, the image *publisher* may inadvertently release sensitive information while personalizing the images he wishes to share. *Retrievers* of these pre-configured images are highly vulnerable to infections that could have been loaded into the VMs by malicious publishers and which can be used, for instance, as vectors for propagating Trojan horses. Lastly, cloud *administrators*, who are entrusted with maintaining the integrity of these images, may be liable to legal action if it is shown that malicious or illegal content was associated with the images used in their system. In light of the aforementioned threats, Wei et al. proposed an image management system called Mirage that provides a series of security features. First, it offers image publishers a framework by which they can define a set of trusted users to whom they wish to allow access rights. To further protect publishers from accidentally leaking private information, Mirage also contains an elaborate image filter that removes from each image all private (e.g., passwords) and malicious (e.g., malware and pirate software) information. Next, it also contains a provenance tracking system that keeps a detailed history of each image, including all operations performed on it after its release. In addition to these features, the system is also capable of providing a set of maintenance services, such as anti-viruses, that can be periodically run over the entire pool of images. To mitigate some of the heavy overhead induced by these new features, the authors suggest running filters and services at the repository level, where scaling can more efficiently take place by taking advantage of the similarities between different images.

The filtering systems proposed in [83] are already embedded into some of today's cloud services, but their advantages have not yet outweighed the disadvantages of implementing them. This is especially true for providers not liable for issues that may occur to shared user data as in the case with Amazon EC2. A recent work [16] presented practical examples of how these vulnerabilities can be exploited and argued that the increasing competition between cloud providers may become the driving force that will finally push for a security enhancement with regards to cloud VM images. In this work, Bugiel et al. [16] successfully developed an automated tool that was able to search for and extract highly sensitive information from public Amazon Machine Images. The cost of running the attacks was very low, and the data the authors were able to obtain contained source code repositories, administrator passwords, and other types of credentials. Bugiel et al. [16] also highlighted the discovery of several widespread vulnerabilities caused by incorrect use of Secure Shell (SSH). An alarming number of Amazon Machine Images were found to have SSH backdoors that allowed their publisher to remotely login to instances of VMs belonging to other users. Tools providing countermeasures to these threats are suggested in [83] and cloud providers can consider the trade-off between the benefits and costs of implementing these tools.

With regard to the security of VMs, we point out the work of Schiffman et al. [65], which provides an architecture that allowed for the performance of runtime integrity proofs in general purpose distributed systems. Their prototype, based on the Xen VM, uses the Clark-Wilson integrity model [24] and a main component was defined as *VM verifier*, whose job was to enforce upon running VMs the integrity requirements as defined in [24]. Though the topic of integrity protection is discussed in more depth in Section 7, we briefly provide an intuition to Schiffman's solution. As a first step, the architecture monitors each VM through a secure boot process at the end of which the system's integrity is verified by a base system controlled by the provider. The provider has full knowledge of the system's integrity properties and can leverage that information to enforce VM compliance. In order to give end users complete assurance that their VMs' integrity is preserved under that architecture, providers simply need to prove that the secure initialization process was successfully completed and that the previously mentioned integrity property enforcement was carried out through runtime.

4 Availability

Among the most attractive features of cloud computing is the fact that computational power can be easily supplied on demand. If a certain service running from the cloud suffers a sudden increase in the workload, additional hardware support can be provisioned on the fly. In spite of its appeal, this "elastic" characteristic of clouds can be exploited by attackers.

Jensen et al. [47] describe how *flooding attacks* are a real threat to clouds. This issue rises in such environments when, for instance, all requests to a certain service need to be individually checked for validity, thereby causing service overloading. Consequences of service overloading pointed out by Jensen et al. are denial of service (DoS) (which may be direct or indirect) and discrepancies related to accounting and accountability; these consequences are discussed below.

- *Direct DoS*: When a service is being flooded, the cloud operating system will likely start to supply extra computational power to it. By attacking a single cloud-based address, an adversary may eventually cause a full loss of availability to the entire service if, for instance, a large enough portion of the globally available hardware becomes dedicated to the flooded application.
- *Indirect DoS*: A possible side effect of the above attack, caused by the fact that hardware is multiplexed in the cloud, is that services that are co-resident to the one being flooded may also experience similar overloading problems. Hence, although a service may not be the direct target of a flooding attack, when its available hardware is exhausted by continuous requests sent to the attacked service, it will no longer be able to perform its tasks.
- *Accounting and accountability*: Arguably the most severe consequence to the cloud user who owns the flooded service can be caused by what is thought to be another major advantage of cloud computing. Most cloud providers charge their users according to the actual usage of their infrastructure during a pre-determined time slice. In the case of a service that is being flooded, this usage will be obviously high, which, in its turn, will most likely translate to bills that are much higher than expected.

Though it may not be an easy task to remedy flooding attacks without compromising the overall performance of the cloud to some extent, Zunnurhain et al. [95] provide a feasible alternative. The authors suggest an organization that groups different servers into fleets, each of which is designated for a distinct type of job. By structuring the available hardware in this way, the provider imposes an isolation that would prevent an overload of bogus requests from affecting the performance of the computational fleet. However, a downside to this alternative is that while a certain fleet of servers may become overloaded with a large number of valid jobs, it would not be able to offload some of its tasks to other fleets that may have several idle nodes available.

A different form of a DoS attack in cloud systems is proposed and discussed by Liu [56]. The author points out that it is possible to take advantage of the fact that cloud data centers are typically under-provisioned (i.e., the hosts only use a fraction of their resources or interface speed). In the attack proposed in [56], the adversary gains control of a few hosts in a certain subnet (i.e., set of nodes connected via a common router) and then simply transmits enough traffic to hosts located elsewhere. Despite its simplicity, this attack is likely to quickly saturate the targeted network since the uplink capacity of any given router is much smaller than the aggregate uplink of the hosts in its subnet. Since uplinks are often under-provisioned in both directions, this attack will, by symmetry, also saturate the network of the subnet that is receiving the bogus traffic. Furthermore, there are usually a sizeable number of traffic bottlenecks on cloud networks, such as routers, that are connected to a large number of hosts or other routers. In addition to this, this attack can potentially be refined to aim at specific targets. In the case of a targeted attack, the adversary needs to gain control of hosts located within the same subnet of the service that he wishes to bring down.

Due to the large size of cloud data centers, which can cause them to be significantly under-provisioned, Liu [56] asserts that this attack can be particularly harmful when targeted at this type of environment. Other components that allow for the effectiveness of the proposed attack are the fact that cloud data centers are concurrently used by many different organizations and that these users do not have control over the underlying network, making them incapable of providing counter-measures when needed. Liu also proposes a couple of prevention strategies to an attack that targets specific hosts. One involves a fast dynamic migration architecture that would leverage the already available dynamic provision capabilities of the cloud. Alternatively, a detection mechanism based on bandwidth estimation tools is proposed. Traditional tools for detecting DoS attacks usually rely on the monitoring of signals such as CPU usage and available memory, neither of which is affected by this specific attack. Therefore, [56] introduces a novel mechanism that allows a host to accurately estimate how much bandwidth it has available. With that information in hand, it becomes possible to detect an imminent attack and trigger the appropriate avoidance countermeasures.

5 Accountability

Accountability has long been considered to be a property of trustworthy computer systems. An accountable system is capable of detecting misuse and exposing the responsible entity. It should be able to generate undeniable evidence of the entity's wrongful activity. Moreover, an innocent party in an accountable system should be capable of defending itself against any false accusations. Such properties are particularly valuable in environments where clients are running their tasks or services within the infrastructure owned or maintained by third parties which is the case for public clouds and distributed systems. Several types of systems in different domains can be used to exemplify the significance of accountability:

- In a shared file system outsourced to the cloud, users would like to be assured that the cloud service provider honestly propagates their updates on the data and that other users also do not tamper with the shared data in unauthorized ways.
- In collaborative and competitive systems (e.g., on-line games [41]) delegated to the cloud, each user would like to be able to know that other users indeed follow the agreed upon protocol.

- Users relying on (storage or computing) services offered by the cloud may want to be assured that the cloud service provider faithfully carries out the tasks as directed, utilizing the promised amount of resources.

To build an accountable system, several features should be taken into consideration:

- *Identity binding*: In order to undeniably link each action to the party that performed it, a binding mechanism [89] can be used. A widely utilized approach for achieving this consists of signing each action with the private key of the action originator. This will allow other entities who retain such records to present them to an auditing authority in order to prove their innocence or accuse the action originator of wrong doing.
- *Tamper-evident logs*: To investigate occurrence of malicious behavior of a certain entity, the auditing authority is normally presented with the history of that entity's past actions in the form of a log. One prominent property of this log is tamper-evidence. After making a commitment to the current state of the log, any subsequent attempt in tampering with the previous log entries invalidates the commitment. In [41], this feature is realized by building a hash chain from the log entries, where the current commitment is computed as a hash of the previous commitment and the current log entry in the ordered log entry list. In other words, the ith commitment is computed as $C_i = H(e_i || C_{i-1})$, where here and in the rest of this work, $H(\cdot)$ denotes a collision-resistant hash function, e_i is the ith log entry, $||$ denotes concatenation, and C_0 can be set in any desired way.
- *Execution verification*: When a cloud service provider executes a task on behalf of a client and produces an outcome as the task's result, the client might be interested in verifying correctness of the result. For that purpose, Haeberlen et al. [41] suggest that the client could delegate the verification task to an auditing authority, and provide it with the log contents and records of non-deterministic events. In this context, non-deterministic events [15] are unpredictable events (e.g., hardware interrupts) that occur throughout the execution process which, together with deterministic events (e.g., messages sent or received), contribute to the final outcome of the task. With all such inputs, the auditing authority will re-execute the entire task using its trusted local resources to identify whether the outcome claimed in the log is producible. To save time for executing the complete log contents, the log can be split into several segments based on the snapshots taken during the task's execution process, so that the auditing authority can later perform spot checking of a few segments that begin and end at different snapshots. In addition to this spot-checking approach, in Section 7.2 we discuss other available mechanisms for verifying integrity of the result of an outsourced task, which can be more efficient than this solution.

There are several publications that treat accountability in both the cloud [39, 41, 54, 76] and distributed systems [40, 89, 90] environments. They target either a server which does not fulfill the tasks delegated to it in a faithful manner, or peer entities who do not follow the agreed upon protocol, as being accountable. We next briefly review selected results in this domain.

Yumerefendi and Chase [90] proposed a network storage service that allows clients to access shared objects maintained by the server while achieving strong accountability. With this solution, any malicious behavior of either the server (e.g., incorrect execution of the clients' write requests) or a client (e.g., denying execution of unauthorized operations on shared data) can be captured. To assure clients' accountability, every request contains a digital signature that can be used to uniquely identify its sender and assures the integrity of the message. The collection of such signatures from all clients corresponding to their write requests constitutes the so-called action history. It is later used during an audit request on a particular object, during which the server presents a sequence of cached, signed write requests that prove to the auditor what the correct state of the object in question should be based on the actions of its clients. In that sense, after issuing write requests on a data object, a client cannot later deny his actions.

Regarding the server's accountability, the server is required to generate a new version of a data object for every write request on that object issued by a client and to organize all data objects in the form of a Merkle hash tree. In the standard Merkle hash tree structure, a leaf node corresponds to a hash value of an individual data object, and each internal node is assigned a value that corresponds to the hash of the concatenation of its children's values. For instance, a node containing two children with their respective values v_1 and v_2 will be assigned value $H(v_1||v_2)$. After building the Merkle hash tree in this manner, the server periodically publishes the root value of the Merkle hash tree to a trusted medium from which a client can independently retrieve it to verify the correctness of a requested data object as follows. When a client triggers a read request on a specific data object, the server sends the client the object itself, together with the sibling nodes that lie on the path from the leaf node (corresponding to the retrieved data object) to the root. The client then recomputes the root value based on the received information and compares it with the one retrieved from the trusted medium. If the two values match, the client is assured that the retrieved data object was indeed included in the server's data structure. To further ensure that the current content of the data object is the result of correct execution of the clients' prior write operations on the object, the client performs an audit on the object between two consecutive commitments by the server at, say, times t_1 and t_2. By verifying the history of the write operations performed on the object, the client is assured of the valid transition of the object's content during the time interval $[t_1, t_2]$. Lastly, to prevent the server from hiding the clients' write requests on the objects it stores, the clients themselves can communicate with each other and learn whether any client's operation is missing from the action history.

Jana and Shmatikov [45] introduced an approach for verifying whether interactive web applications hosted by the cloud were executed correctly. Their design is based on two key ideas: First, it provides a mapping mechanism for transforming every HTTP request received from users through its front-end to a read or write operation on the objects stored at the back-end storage. Second, to ensure accountability when dealing with multi-user web applications, a subset of users are selected to be the "witnesses" who keep the logs of their interactions with the cloud-hosted web applications. Later, these logs are sent to a trusted verifier for examination. Since the witnesses cannot be distinguished by the cloud service provider from other users, if the cloud provider's faults are random and independent, then it will be highly probable that the cloud provider will tamper with requests of the witnesses and will correspondingly be detected.

Xiao and Xiao [86] proposed P-Accountability – a metric for quantifying the degree of accountability in a networked system. It was developed to provide a trade-off between the level of accountability that can be achieved and the cost of building an accountable system. In particular, achieving perfect accountability typically involves a large amount of overhead. Furthermore, the noisy nature of network communication due to message loss and delay makes it difficult to link every event that occurred to the responsible entity. Accounting for these factors, the authors define the P-accountability metric as $\mathrm{P} = (\sum_{e \in E} I(e))/|E|$, where E is the event set, $e \in E$, and $I(e)$ returns 1 if e could be correctly traced back to the responsible entity, and 0 otherwise. Based on application demands, such as the amount of available resources and tolerable overhead, the system can specify different P values to achieve the desired level of accountability.

Recently, Sekar and Maniatis [67] proposed a systematic approach to account for the resources consumed by the cloud when executing clients' tasks. The problem the authors address in verifiable resource accounting is two-fold: (i) verifying whether the resources consumed for an outsourced task indeed matched the expenditures the client was charged for (i.e., "Did I use what I was charged for?"), and (ii) verifying whether the resource consumption can be justified based on the agreed upon policy (i.e., "Should I have used what I was charged for?"). Here, a policy determines a "reasonable" allocation of resources for a delegated workload. For instance, the client may describe an upper bound on the resources the server can dedicate to a task (which can be specified as a function of that task). The framework that the authors offer for answering these two questions relies on the use of trusted hardware. In more detail, for a given task, the server specifies the granularity of resource tracking and upon completion of the task the server generates a consumption report. The report describes the utilized resources with the specified granularity, and attestation of the hardware on which the task is run guarantees the integrity of the report. To verify the correctness of the billing, the client provides a third-party verifier with specification of the task and the server's report receives a Yes/No answer that indicates whether the server's resource usage was appropriate for the task.

6 Remote Storage Protection

Cloud storage refers to the increasingly prevalent on-line storage services hosted at the cloud. Thanks to the enormous storage capacity, high availability, and stable performance offered by the cloud, the applications resorting to cloud storage range from general file storage such as archiving and backup to web operating systems and databases. Using Amazon Simple Storage Service (S3) as an example, cloud storage providers tend to offer plain file-system like interfaces to the end users without exposing them to the complicated management of physical servers and facilities in which they reside. Moreover, because all storage servers of a cloud storage provider are centrally controlled through dynamic provisioning, the users are also relieved from the direct oversight of component reliability and outsider security threats.

While cloud computing offers appealing advantages, it also brings new security challenges. In particular, integrity of clients' outsourced data becomes a concern, which is caused by a lack of transparency and direct control in the cloud environment. While data confidentiality can be assured through traditional mechanisms such as encryption or secret sharing, and integrity of data transmission can be assured through data authentication, integrity of data stored in the cloud is significantly more difficult to verify. For that reason, the rest of this section deals with techniques for ensuring integrity of outsourced data.

After a client moves its data to the cloud, the client relinquishes its ultimate control over the data, which is now entirely managed by the cloud service provider. Thus, it is essential for the client to be able to verify that her data is still available at the cloud in its original form and is ready for retrieval when necessary. For instance, the client might want to make sure that her data has not been corrupted (deleted or modified) or moved to an off-line unavailable storage medium, that could be caused by either an attempt of a dishonest provider to save storage costs or by outages and security breaches within the cloud services themselves.

One possibility for assuring high availability of outsourced data is through simple replication, where a client stores its data with multiple server providers. This, however, results in unnecessarily high storage overhead and cost to the client. As an alternative, the client can store her data at a single provider, but either periodically retrieve and examine the entire data, or ask the server to periodically compute and send a one-way function of the stored data which the client then compares to a previously computed value that it expects. It is clear that these two approaches are also unsatisfactory as in the first one both the server and the user suffer from a substantial communication overhead and the second approach is prone to replay attacks by the server which can conceal the fact that the server tampered with the data. To mitigate these problems, a widely utilized approach [5–7, 13, 14, 18, 25, 27, 29, 48, 66, 69, 78, 85, 92] is to employ a challenge-response mechanism consisting of the following phases:

- *Challenge*: The client creates and sends to the cloud service provider (CSP) one or more unpredictable challenges, which are used to ensure that the CSP indeed retains the client's data at the time of auditing.
- *Response*: Upon receiving the challenge, the CSP derives its response from the challenge and the stored data, and sends the computed result back to the client.
- *Response verification*: After receiving a response from the CSP, the client compares it to the expected value that it previously pre-computed from the original data and its metadata. If the two values coincide, the client is assured that the data is stored at the CSP in its intact form.

A design objective of this framework is that a cloud service provider who is highly likely to store incomplete or incorrect data will be unable to respond to the challenges correctly, and thus the client will detect data integrity violations.

While security (i.e., guaranteeing that the server's misbehavior is detected) and efficiency (i.e., minimizing the client's and server's computation and communication) should be viewed as the main design criteria, in order to build a reliable and versatile protocol, it is desirable that several other properties are taken into consideration:

- *Unbounded number of audit queries*: A robust scheme should be able to support an unbounded number of audit protocol interactions to ensure that the server's misbehavior, which might occur at any time, will be detected. If a protocol only allows for a limited number of audit queries, upon reaching the limit, the client will have to retrieve the entire data from the server and re-initiate the scheme during which the data is stored at the server again. Clearly, this imposes additional overhead on clients which can be avoided.
- *Support for dynamic data operations*: In many circumstances, the data delegated to the cloud might need to be modified through the user's update requests (unlike backup archive files that are rarely modified). In this case, a protocol that works only for static operations (i.e., upload or read) is unlikely to be applicable to dynamic operations. The update operations can be insertions or deletions of new or existing blocks at arbitrary positions in the file, or modifications of existing blocks. With cloud storage, support for dynamic data operations can be of vital importance to both remote storage and database services.
- *Public verifiability*: Oftentimes, when performing integrity verification of outsourced data, the clients themselves are unreliable and are unable to consistently perform integrity checks. They may also lack necessary expertise for this task. For these reasons, it is desirable to develop mechanisms for public verification of data integrity by a trusted auditing server. The auditing server is a reliable and independent entity that challenges the cloud service provider on behalf of the clients and assures correctness of data storage, while not learning any information contained in the stored data. For improved efficiency, the auditing server could also perform batch auditing during which it simultaneously processes auditing requests from multiple users [79].

To ensure integrity of outsourced storage and address the above mentioned properties, many existing publications [5–7, 14, 18, 19, 27, 29, 43, 48, 66, 69, 77–79, 92, 94] proposed a number of techniques. In the following, we categorize prior work based on the design objectives and highlight cryptographic details and data structures used in these publications.

- *Basic scheme.* Juels and Kaliski [48] proposed one of the first solutions to this problem by implementing the basic functionality of proofs of retrievability (POR). In their sentinel-based scheme, the client first applies an error-correcting code to the original file, encrypts each file block, and then inserts a predefined number of randomly chosen sentinels into different positions in the file.

Following a challenge-response protocol, the user challenges the cloud service provider by specifying a number of positions of sentinels, so that a distinct set of sentinels is selected for each challenge and their positions are unpredictable to the server. The user then requires the cloud service provider to return information about the blocks at the chosen positions and verifies their correctness by comparing the returned values with their locally stored copy. The scheme is considered secure if a server without the possession of the file can only succeed in producing correct responses with low probability. As can be expected, the security of the sentinel-based approach depends on the number of sentinels (defined by the security parameter) in each challenge, and the number may vary upon the user's demand on integrity assurance (e.g., 90% of user's data should be kept intact). Given a sufficient number of sentinels in a challenge, if the server tampers with a large portion of the file, it will inevitably corrupt some of the sentinels that are being checked and the scheme will correspondingly detect the integrity violation with high probability. Since the number of embedded sentinels is independent of the length of the file stored at the server, the communication and storage costs for verifying the file's integrity remain a small constant. Another notable feature of the scheme is that a small corruption of the file that may not be detected using the spot-checking mechanism can still be recovered using the error-correction code embedded into the file. In spite of this basic functionality, the scheme has several disadvantages: it does not permit an unbounded number of audit queries, lacks support for dynamic operations, and does not allow for public verifiability.

- *Schemes with support for public verifiability.* The solutions in [5, 14, 29, 69, 78, 79, 92] support both public verifiability and an unlimited number of auditing interactions. A common approach for achieving these two properties is to utilize the so-called homomorphic authenticators [5, 69, 78, 79]. A homomorphic authenticator is an unforgeable metadata derived from a file block, with the property that several homomorphic authenticators can be securely aggregated. When a client would like to verify the integrity of a number of file blocks, it is sufficient for the client to verify only the aggregated homomorphic authenticator that corresponds to a linear combination of the individual blocks. Since the only information necessary for carrying out verification of correct data storage using a challenge-response protocol is the client's public key and the challenge data, the auditing process can be delegated to a third party who is equipped with sufficient computing and storage resources.
- *Schemes with support for dynamic operations.* Publications [6, 29, 43, 77, 78, 94] provide schemes that support dynamic updates on the outsourced data. The work [6, 77] implemented this functionality by using carefully designed sentinels (unlike [48], where the sentinels are assigned random values) and outsourcing them together with the client's data. Another line of research [29, 43, 78, 94] that supports dynamic operations utilizes specialized data structures such as Merkle hash trees or skip lists to organize information about the data blocks outsourced to a cloud service provider. When a Merkle hash tree structure is used for this purpose, each leaf node corresponds to the hash of an individual data block in the

file being outsourced, and each internal node is assigned a value that corresponds to the hash of the concatenation of its children's values. The client only needs to store the root value of the tree. When the client retrieves a data block from the cloud server, the server sends the values of the sibling nodes lying on the path from the data block to the root. Having this information, the client verifies the correctness of the retrieved data block by recomputing the root value and compares it to the one that it locally stores.

7 Privacy and Integrity of Outsourced Computation

This section treats security issues that arise in the context of outsourced computation, and in particular, we address the broad topics of privacy and integrity protection of outsourced computation.

7.1 Privacy Protection of Outsourced Computation

As companies, individuals, and even governments increasingly outsource computation to the cloud, they knowingly relinquish the ultimate control over their data and the computation. If the computation involves proprietary, private, or otherwise sensitive data, cloud users might wish to protect the privacy of the data being used in the process, particularly when the infrastructure provider cannot be fully trusted. Although the concept of cloud computing is relatively new, privacy protection of outsourced computation has been studied for many years and most of the developments in this research area can be directly applied to the cloud architecture. There are currently a number of solutions that allow different types of operations (i.e., matrix multiplication, modular exponentiation, etc.) or more general functionality to be performed by the cloud provider on data that has been preprocessed by the user to ensure the safety of sensitive information. In what follows, we provide a more detailed description of some of these approaches.

All existing solutions in the literature can be grouped into the following categories: (1) schemes where the client outsources its computation to a single server, (2) solutions where the client uses multiple (normally two) servers for its computational task, but there is no interaction between the servers, and (3) protocols where a task is outsourced to multiple (two or more) servers that jointly carry out the computation. The solutions from the second category normally utilize two servers to ensure correctness of the result in the presence of faulty servers rather than in a more fundamental way. We start our description with the literature where the computation is outsourced to a single server followed by multi-server solutions.

Homomorphic encryption schemes have special properties that allow computation on ciphertexts to translate to meaningful operations on the plaintexts that they encrypt, and can be used in the context of secure outsourcing to a single server.

While an overview of homomorphic schemes is beyond the scope of this chapter, we mention that the existing techniques have limitations either in the class of functions that are able to evaluate on the encrypted data or in their performance.

In recent years, there has been an increasing interest in privately outsourcing complex (scientific) operations. Atallah et al. [4] proposed techniques for securely outsourcing several computations, with applications ranging from linear algebra to string pattern matching. As no single proposed mechanism was broad enough to be applied to a large variety of scientific problems, that work resorts to a number of different disguise techniques that are tuned to specific situations to enable secure outsourcing of most scientific computations at low cost. In [2], the protocols for outsourcing linear algebra operations are further improved to work with a single (or multiple) external servers that, unlike in previous protocols, no longer require expensive cryptographic operations. Another attractive property of this protocol is that it allows for a continuous (and safe) chain of outsourcing. For instance, by using this technique, a client can outsource a large matrix multiplication to the cloud which in its turn, can also delegate part of that computation to another remote server. In each step the client outsourcing the computation does only linear in the size of the matrix work, and can detect any attempt by the server to corrupt the answer.

As we next proceed with techniques for secure outsourcing to multiple servers, where the servers jointly carry out the computation, we note that many techniques from secure multi-party computation (SMC) can be used in this setting. Their overview, however, is beyond the scope of this chapter. We only mention that the client often can split its input between the computational servers, who can then proceed with the secure computation and return the result to the client.

We also mention works that deal with important applications. Atallah et al. [3] describe a secure outsourcing technique via which a client that does not have the necessary computational power can outsource large sequence comparisons (e.g., DNA matching) to two remote computing nodes, such that the result of the comparison and the two sequences being compared are known only by the client. This problem was revisited in [12], where a solution of improved performance was put forth. Also, the work [10] describes the means by which oblivious evaluation of a finite automaton on a private input can be securely outsourced to two or more servers. Though there are many operations that can be approached by the given technique such as searching for a regular expression in a database, the paper focuses on DNA searching and gives optimizations for that specific setting.

Finally, a couple of recent publications propose privacy-preserving architectures for MapReduce-based [26] computation. The work by Roy et al. [64] recognized the need for an automated system that could be used by clients with no knowledge of security to enforce personalized access control policies to their data. With that premise, the authors designed and implemented Airavat, which incorporated the mandatory access control features of SELinux to MapReduce, and also provided measures to avoid the leakage of information through the output of computations. Zhang et al. [93] considered a slightly different scenario, in which a given entity wishes to outsource a large computation where only a portion of the data is sensitive and needs to be processed by a private cloud, whereas the remaining information can be safely outsourced to a public cloud. Also building upon MapReduce, the authors proposed Sedic, a generic secure computing framework for hybrid clouds. Sedic allows the end user to continue to use the legacy data-processing code with the added functionality of having the data being automatically partitioned between private and public and adequately outsourced according to that labeling. Several techniques are also suggested in order to maximize the amount of computation done in the public cloud side, which increases the overall performance.

7.2 Outsourced Verifiable Computation

As mentioned earlier, a prominent advantage of the cloud paradigm is providing computing resources to resource-constrained users who are now able to outsource their computation-intensive tasks to the cloud with abundant processing, storage, and bandwidth resources. With this architecture, however, the clients no longer have direct control over the outsourced data and computation, and there is a lack of transparency in the current cloud services. The cloud provider can be incentivized to delete rarely accessed data or skip some of the computations to conserve resources (for financial or other reasons), which is especially true for volunteer-based computational clouds. Furthermore, unintentional data or computation corruption might also take place for a variety of reasons including malware, security break-ins, etc. From that perspective, it is important for the clients to be able to verify the correctness of the result of the outsourced task. Furthermore, the verification mechanism should be such that it does not require the client to perform a computation comparable in size to the outsourced task itself. Techniques that enable verification of correctness of the result for outsourced computation are treated in this section.

To formulate the problem, a client delegates the task of computing $f(x)$ for a set of inputs $X = \{x_1, x_2, \ldots, x_n\}$ to a cloud service provider. A dishonest cloud provider evaluates the function on a subset of the inputs $X' \subset X$, and for the rest of the inputs it returns the result of evaluation of $f'(\cdot)$, where $f'(\cdot)$ is cheaper to compute than $f(\cdot)$. This type of behavior is termed "lazy." Alternatively, a part of the result might be corrupt, but normally it is assumed that the cloud server will not maliciously modify the result (by performing work equivalent to or exceeding that of computing the result correctly). After receiving the results, the client would like to verify the output in an efficient manner and detect any deviation from the prescribed computation by the cloud provider.

Prior literature contains a number of techniques [2, 9, 28, 35, 37, 50, 53, 55, 73, 80, 82, 84] for efficiently verifying the result of an outsourced computing task. We categorize them into two groups based on the type of computation to which the techniques are applied.

7.2.1 Function-Independent Computation

The techniques in this category can be applied to general types of computation. The common approach of these techniques consists of assigning the extra workload to the server that is identical to computation being outsourced, and later analyzing the computation results of both the original task and additional workload to determine any deviation from the prescribed computation by the service provider. This extra workload can take the form of either insertion of so-called ringers [28, 35, 50] or duplicate task execution [36, 37, 53, 73, 82, 84]. We provide additional descriptions of some of the typical mechanisms.

Insertion of ringers was among the first proposals for addressing this problem and is now commonly used. Ringer r [35] in this context, is a value chosen by the client in the domain of the function f that it outsources to a server, and the client is assumed to be equipped with the knowledge of $f(r)$. We note that our definition of a ringer is different from the original definition in [35] to ensure that it can be applied to any functionality f that the client wishes to outsource. Furthermore, the computational server should be unable to distinguish ringers from the inputs of the original task. As a result, if the cloud provider corrupts a non-negligible portion of the computation, then with high probability, it will tamper with the computation of the ringers, which will be detected by comparing the returned results with the client's pre-computed $f(r)$. Golle and Mironov [35] applied the ringer-based technique to search for rare events for the computation that involved inverting a one-way function h (i.e., the client outsources the computation of h^{-1}).

Szajda et al. [73] used an alternative technique that consists of duplicate task execution for a number of sequential workloads that can be subdivided into independent sub-tasks. The approach assumes that multiple servers which do not collude with each other, execute a task on behalf of a client. To guarantee redundancy, it is also necessary that the number of servers is greater than the number of workloads. Under these assumptions, the proposed mechanism guarantees that the result of the outsourced computation is correct with high probability. By the law of large numbers, the correct result for each sub-task is highly likely to be the result returned by the majority of the servers; the correct computational task can then be ascertained.

Canetti et al. [17] recently proposed a general strategy for verifiable task outsourcing where a client outsources the computation to multiple servers and is guaranteed to obtain the correct result as long as at least one server is honest. The approach proposed by the authors works as follows: Given the number of instructions that compose the computation task, the client uses a binary search to determine the exact location where the intermediate states of two (or more) executions begin to disagree. Upon locating the inconsistency, the client is able to determine which execution was correct by carrying out one step of the computation himself. One requirement of this scheme is that execution of the delegated task must be fully deterministic as it will be non-trivial to distinguish an outcome that is incorrect as a result of a stochastic event from an outcome that is incorrect as a result of a malicious event. In addition, Gennaro et al. [32] and Chung et al. [23] recently proposed solutions that explicitly combine privacy with integrity verification of outsourced computation and are applicable to any computable function. Gennaro et al. [32] describe a non-interactive verifiable computation outsourcing scheme that combines Yao's garbled circuit evaluation [88] with a fully homomorphic encryption scheme (e.g., Gentry's proposal [34]). Because evaluating a garbled circuit on two different inputs cannot provide adequate security, the authors propose that the circuit is evaluated on encrypted inputs. That is, after creating a garbled Boolean circuit for the desired functionality, the client stores it at the server. When the client desires to evaluate the circuit on an input, it encrypts the labels associated with the circuit's input bits using the public key of a fully homomorphic encryption scheme and sends the encrypted labels to the server. The cloud server evaluates the function on the encrypted data using the homomorphic properties of the encryption scheme. After the server evaluates the entire function and returns encryptions of the output wire labels to the client, the client decrypts the ciphertexts, recovers the labels, and interprets the result.

Setty et al. [68] recently proposed another type of general strategy for verifiable outsourced computation based on an argument system with probabilistically checkable proofs (PCP). In an argument system, a computationally bounded prover tries to convince the verifier that the delegated computation was computed correctly, and with PCP, the verifier is able to guarantee correctness of the outsourced computation with high probability by probing only a constant number of positions within the proof. The solution builds on the work of Kilian [51, 52] in which the prover is asked to make a commitment to his computation that should be consistent with the queries that will be issued afterwards by the verifier. Setty et al. constructed a scheme that aims to work for a general computation with practical performance by incorporating several efficiency improvements to the naive implementation of [44]. The improvements include using arithmetic circuits with compact gates instead of Boolean circuits to represent the function, and amortizing the verifier's query costs through batching.

7.2.2 Function-Specific Computation

The techniques in this category exploit knowledge of the domain function of the computation to efficiently validate the results of an outsourced task. Algebraic computations [2, 9] and linear programming [80] are typical computation domains that have been studied in some depth. Also, many of the available techniques combine privacy protection of the data used throughout the outsourced computation with verification of the result of the computation.

Wang et al. [80] proposed a result verification mechanism for linear programming (LP) in cloud computing. In this approach, outsourcing the computation of linear programming means sending all parameters of the LP problem to the cloud service provider in obfuscated form to preserve privacy. To verify the returned solution, [80] utilizes the duality theorem to construct a secondary LP problem and derive necessary and sufficient conditions for an outcome to be valid. By outsourcing these two LP problems to the cloud service provider and later comparing the returned results, the client can verify the correctness of the original LP problem. In a similar vein, Benjamin et al. [9] and Atallah et al. [2] proposed solutions that deal with secure and private outsourcing of linear algebra computations (matrix multiplication).

Blanton et al. [11] recently studied the problem of verifiable outsourcing of large-scale biometric computations. The authors treat the computation of all-pairs distances between two sets of biometric data that produces a distance matrix, as well as the corresponding distribution computation that calculates the frequency of each distinct distance appearing in the distance matrix. The work focuses on three distance metrics, namely the Hamming distance, the Euclidean distance, and the set intersection cardinality. In this solution, to verify the integrity of both the all-pairs and the distance distribution computation, the client inserts fake biometric items into the computation, which are indistinguishable from the real biometric items in the datasets. The fake biometrics are carefully designed so that the range of distances between two real biometric items is separated from the range of distances between real and fake biometrics. By comparing several values returned by the server with their expected values that the client precomputes, the client achieves the desired level of assurance that the remaining values were computed honestly.

In addition to the above literature, [8] provides verifiable computation schemes for evaluation of high degree polynomials, which can also be applied to verifiable keyword searches and proofs of retrievability. Another work [59] develops a mechanism for verifying the result of outsourced set operations using authenticated data structures, with applications ranging from efficient verification of keyword search to database queries. Finally, Wang et al. [81] proposed a solution for securely outsourcing large-scale systems of linear equations. The solution is based on an iterative method and incurs only $O(n)$ local computations on the customer's side for a linear system with $n \times n$ coefficient matrix.

8 Conclusion

Cloud security has emerged as an important topic both in the research community and in practice due to the recent rapid growth in the availability and popularity of cloud computing and storage providers. It is often cited as a top reason for hesitance in adopting cloud computing services by companies and is a difficult problem that re-surfaces old security issues and brings new ones. In the attempt to bring wider attention to the topic, this work provides the first comprehensive treatment of the area by providing a literature overview of cloud security as well as secure remote storage and computation. The topics covered in this survey include:

- Authentication and authorization attacks and countermeasures
- Virtualization-related security issues
- Denial of service attacks

- Solutions that address accountability
- Schemes for integrity verification of remote storage
- Techniques for privacy and integrity protection of outsourced computation.

As the field matures, we expect to see techniques for more complex and interoperable cloud interactions such as multi-user access to remote storage and tasks' configuration automation for use with cloud providers, which will need to be treated in the security literature as well.

Acknowledgements This material is based upon work partially supported by the Air Force Office of Scientific Research (AFOSR) grant FA9550-09-1-0223.

References

1. Armbrust, M., Fox, A., Griffith, R., Joseph, A., Katz, R., Konwinski, A., Lee, G., Patterson, D., Rabkin, A., Zaharia, M.: Above the clouds: a Berkeley view of cloud computing. Technical report UCB/EECS-2009-28, University of California at Berkeley (2009)
2. Atallah, M., Frikken, K.: Securely outsourcing linear algebra computations. In: Proceedings of the 5th ACM Symposium on Information, Computer and Communications Security (ASIACCS'10), Beijing, pp. 48–59. ACM, New York, (2010). doi:10.1145/1755688.1755695
3. Atallah, M., Li, J.: Secure outsourcing of sequence comparisons. Int. J. Inf. Secur. **4**(4), 277–287 (2005)
4. Atallah, M., Pantazopoulos, K., Rice, J., Spafford, E.: Secure outsourcing of scientific computations. Adv. Comput. **54**, 216–272 (2001)
5. Ateniese, G., Burns, R., Curtmola, R., Herring, J., Kissner, L., Peterson, Z., Song, D.: Provable data possession at untrusted stores. In: Proceedings of the 14th ACM Conference on Computer and Communications Security, CCS'07, Alexandria, pp. 598–609. ACM, New York (2007). doi:10.1145/1315245.1315318
6. Ateniese, G., Di Pietro, R., Mancini, L., Tsudik, G.: Scalable and efficient provable data possession. In: Proceedings of the 4th International Conference on Security and Privacy in Communication Networks, SecureComm'08, Istanbul, pp. 9:1–9:10. ACM, New York (2008). doi:10.1145/1460877.1460889
7. Ateniese, G., Kamara, S., Katz, J.: Proofs of storage from homomorphic identification protocols. In: Proceedings of the 15th International Conference on the Theory and Application of Cryptology and Information Security: Advances in Cryptology, ASIACRYPT'09, Tokyo, pp. 319–333. Springer, Berlin/Heidelberg (2009). doi:10.1007/978-3-642-10366-7_19
8. Benabbas, S., Gennaro, R., Vahlis, Y.: Verifiable delegation of computation over large datasets. In: Proceedings of the 31st Annual Conference on Advances in Cryptology, CRYPTO'11, Santa Barbara, pp. 111–131. Springer, Berlin/Heidelberg (2011)
9. Benjamin, D., Atallah, M.: Private and cheating-free outsourcing of algebraic computations. In: Proceedings of the 6th Annual Conference on Privacy, Security and Trust, PST'08, New Brunswick, pp.240–245. IEEE Computer Society, Washington, DC (2008). doi:10.1109/PST.2008.12
10. Blanton, M., Aliasgari, M.: Secure outsourcing of DNA searching via finite automata. In: Proceedings of the Annual IFIP WG 11.3 Working Conference on Data and Applications Security, DBSec'10, Rome, pp. 49–64. Springer, Berlin/Heidelberg (2010). doi:10.1007/978-3-642-13739-6_4
11. Blanton, M., Zhang, Y., Frikken, K.: Secure and verifiable outsourcing of large-scale biometric computations. In: Proceedings of the IEEE International Conference on Information Privacy, Security, Risk and Trust, PASSAT'11, Pisa, pp. 1185–1191 (2011). doi:10.1109/PASSAT/SocialCom.2011.13

12. Blanton, M., Atallah, M., Frikken, K., Malluhi, Q.: Secure and efficient outsourcing of sequence comparisons. In: Proceedings of the 17th European Symposium on Research in Computer Security, ESORICS'12, Boston, pp. 505–522. Springer, Berlin/Heidelberg (2012). doi:10.1007/978-3-642-33167-1_29
13. Bowers, K., Juels, A., Oprea, A.: HAIL: A high-availability and integrity layer for cloud storage. In: Proceedings of the 16th ACM Conference on Computer and Communications Security, CCS'09, Chicago, pp. 187–198. ACM, New York (2009). doi:10.1145/1653662.1653686
14. Bowers, K., Juels, A., Oprea, A.: Proofs of retrievability: theory and Implementation. In: Proceedings of the ACM Workshop on Cloud Computing Security, CCSW'09, Chicago, pp. 43–54. ACM, New York (2009). doi:10.1145/1655008.1655015
15. Bressoud, T., Schneider, F.: Hypervisor-based fault tolerance. In: Proceedings of the 15th ACM Symposium on Operating Systems Principles, SOSP'95, Cooper Mountain, pp. 1–11. ACM, New York (1995). doi:10.1145/224056.224058
16. Bugiel, S., Nürnberger, S., Pöppelmann, T., Sadeghi, A.R., Schneider, T.: AmazonIA: when elasticity snaps back. In: Proceedings of the 18th ACM Conference on Computer and Communications Security, CCS'11, Chicago, pp. 389–400. ACM, New York (2011). doi:10.1145/2046707.2046753
17. Canetti, R., Riva, B., Rothblum, G.: Practical delegation of computation using multiple servers. In: Proceedings of the 18th ACM Conference on Computer and Communications Security, CCS'11, Chicago, pp. 445–454. ACM, New York (2011). doi:10.1145/2046707.2046759
18. Chang, E., Xu, J.: Remote integrity check with dishonest storage server. In: Proceedings of the 13th European Symposium on Research in Computer Security, ESORICS'08, Malaga, pp. 223–237. Springer, Berlin/Heidelberg (2008). doi:10.1007/978-3-540-88313-5_15
19. Chen, B., Curtmola, R., Ateniese, G., Burns, R.: Remote data checking for network coding-based distributed storage systems. In: Proceedings of the ACM Cloud Computing Security Workshop, CCSW'10, Chicago, pp. 31–42. ACM, New York (2010). doi:10.1145/1866835.1866842
20. Chen, Y., Paxson, V., Katz, R.: What's new about cloud computing security? Technical report UCB/EECS-2010-5, Electrical Engineering and Computer Sciences, University of California at Berkeley (2010). http://goo.gl/YbgOz5
21. Chow, R., Golle, P., Jakobsson, M., Shi, E., Staddon, J., Masuoka, R., Molina, J.: Controlling data in the cloud: outsourcing computation without outsourcing control. In: Proceedings of the ACM Cloud Computing Security Workshop, CCSW'09, Chicago, pp. 85–90. ACM, New York (2009). doi:10.1145/1655008.1655020
22. Christodorescu, M., Sailer, R., Schales, D., Sgandurra, D., Zamboni, D.: Cloud security is not (just) virtualization security. In: Proceedings of the ACM Cloud Computing Security Workshop, CCSW'09, Chicago, pp. 97–102. ACM, New York (2009). doi:10.1145/1655008.1655022
23. Chung, K.M., Kalai, Y., Vadhan, S.: Improved delegation of computation using fully homomorphic encryption. In: Proceedings of the 30th Annual Conference on Advances in Cryptology, CRYPTO'10, Santa Barbara, pp. 483–501. Springer, Berlin/Heidelberg (2010)
24. Clark, D., Wilson, D.: A comparison of commercial and military computer security policies. In: Proceedings of the IEEE Symposium on Security and Privacy, Oakland, pp. 184–194 (1987). doi:10.1109/SP.1987.10001
25. Curtmola, R., Khan, O., Burns, R., Ateniese, G.: MR. PDP: multiple-replica provable data possession. In: Proceedings of the 28th International Conference on Distributed Computing Systems, ICDCS'08, Beijing, pp. 411–420. IEEE Computer Society, Washington, DC (2008). doi:10.1109/ICDCS.2008.68
26. Dean, J., Ghemawat, S.: MapReduce: simplified data processing on large clusters. Commun. ACM **51**(1), 107–113 (2008). doi:10.1145/1327452.1327492
27. Dodis, Y., Dadhan, S., Wichs, D.: Proofs of retrievability via hardness amplification. In: Proceedings of the 6th Theory of Cryptography Conference, TCC'09, San Francisco, pp. 109–127. Springer, Berlin/Heidelberg (2009). doi:10.1007/978-3-642-00457-5_8

28. Du, W., Goodrich, M.: Searching for high-value rare events with uncheatable grid computing. In: Proceedings of the 3rd International Conference on Applied Cryptography and Network Security, ACNS'05, New York, pp. 122–137. Springer, Berlin/Heidelberg (2005). doi:10.1007/11496137_9
29. Erway, C., Kupcu, A., Papamanthou, C., Tamassia, R.: Dynamic provable data possession. In: Proceedings of the 16th ACM Conference on Computer and Communications Security, CCS'09, Chicago, pp. 213–222. ACM, New York (2009). doi:10.1145/1653662.1653688
30. eucalyptus.com, Open source private and hybrid clouds. http://goo.gl/FpdG5 (2012)
31. Garfinkel, T., Rosenblum, M.: A virtual machine introspection based architecture for intrusion detection. In: Proceedings of Network and Distributed Systems Security Symposium, NDSS'03, San Diego, pp. 191–206 (2003)
32. Gennaro, R., Gentry, C., Parno, B.: Non-interactive verifiable computing: outsourcing computation to untrusted workers. In: Proceedings of the 30th Annual Conference on Advances in Cryptology, CRYPTO'10, Santa Barbara, pp. 465–482. Springer, Berlin/Heidelberg (2010)
33. Gens, F.: blogs.idc.com, IT cloud services user survey, pt. 2: Top benefits & challenges. http://goo.gl/mtcSQ (2008)
34. Gentry, C.: Fully homomorphic encryption using ideal lattices. In: Proceedings of the 41st Annual ACM Symposium on Theory of Computing, STOC'09, Washington, DC, pp. 169–178. ACM, New York (2009). doi:10.1145/1536414.1536440
35. Golle, P., Mironov, I.: Uncheatable distributed computations. In: Proceedings of the 2001 Conference on Topics in Cryptology: The Cryptographer's Track at the RSA Conference, CT-RSA'01, San Francisco, pp. 425–440. Springer, London (2001)
36. Golle, P., Stubblebine, S.: Secure distributed computing in a commercial environment. In: Proceedings of the 5th International Conference on Financial Cryptography, FC'01, Grand Cayman, pp. 289–304. Springer, London (2001)
37. Goodrich, M.: Pipelined algorithms to detect cheating in long-term grid computations. Theor. Comput. Sci. **408**(2–3), 199–207 (2008). doi:10.1016/j.tcs.2008.08.008
38. Gruschka, N., Iacono, L.: Vulnerable cloud: SOAP message security validation revisited. In: Proceedings of the IEEE International Conference on Web Services, ICWS'09, pp. 625–631. IEEE Computer Society, Washington, DC (2009). doi:10.1109/ICWS.2009.70
39. Haeberlen, A.: A case for the accountable cloud. ACM SIGOPS Oper. Syst. Rev. **44**(2), 52–57 (2010). doi:10.1145/1773912.1773926
40. Haeberlen, A., Kouznetsov, P., Druschel, P.: PeerReview: practical accountability for distributed systems. In: Proceedings of the 21st ACM SIGOPS Symposium on Operating Systems Principles, SOSP'07, Vancouver, pp. 175–188. ACM, New York (2007). doi:10.1145/1294261.1294279
41. Haeberlen, A., Aditya, P., Rodrigues, R., Druschel, R.: Accountable virtual machines. In: Proceedings of the 9th USENIX Conference on Operating Systems Design and Implementation, OSDI'10, Stevenson, pp. 119–134. USENIX Association, Berkeley (2010)
42. Hamlen, K., Kantarcioglu, M., Khan, L., Thuraisingham, B.: Security issues for cloud computing. Int. J. Inf. Secur. Priv. (IJISP) **4**(2), 36–48 (2010)
43. Heitzmann, A., Palazzi, B., Papamanthou, C., Tamassia, R.: Efficient integrity checking of untrusted network storage. In: Proceedings of the 4th ACM International Workshop on Storage Security and Survivability, StorageSS'08, Alexandria, pp. 43–54. ACM, New York (2008). doi:10.1145/1456469.1456479
44. Ishai, Y., Kushilevitz, E., Ostrovsky, R.: Efficient arguments without short PCPs. In: Proceedings of the 22nd Annual IEEE Conference on Computational Complexity, CCC'07, San Diego, pp. 278–291. IEEE Computer Society, Washington, DC (2007). doi:10.1109/CCC.2007.10
45. Jana, S., Shmatikov, V.: EVE: verifying correct execution of cloud-hosted web applications. In: Proceedings of the 3rd USENIX Conference on Hot Topics in Cloud Computing, HotCloud'11. USENIX Association, Berkeley (2010)
46. Jensen, M., Gruschka, N., Herkenhoner, R.: A survey of attacks on web services. Comput. Sci. Res. Dev. **24**(4), 185–197 (2009)

47. Jensen, M., Schwenk, J., Gruschka, N., Iacono, L.: On technical security issues in cloud computing. In: Proceedings of the 2009 IEEE International Conference on Cloud Computing, CLOUD'09, Bangalore, pp. 109–116. IEEE Computer Society, Washington, DC (2009). doi:10.1109/CLOUD.2009.60
48. Juels, A., Kaliski, B.: PORs: proofs of retrievability for large files. In: Proceedings of the 14th ACM Conference on Computer and Communications Security, CCS'07, Alexandria, pp. 584–597. ACM, New York (2007). doi:10.1145/1315245.1315317
49. Kandukuri, B., Paturi V, R., Rakshit, A.: Cloud security issues. In: Proceedings of the IEEE International Conference on Services Computing, SCC'09, Bangalore, pp. 517–520. IEEE Computer Society, Washington, DC (2009). doi:10.1109/SCC.2009.84
50. Karame, G., Strasser, M., Capkun, S.: Secure remote execution of sequential computations. In: Proceedings of the 11th International Conference on Information and Communications Security, ICICS'09, Phoenix, pp. 181–197. Springer, Berlin/Heidelberg (2009). doi:10.1007/978-3-642-11145-7_15
51. Kilian, J.: A note on efficient zero-knowledge proofs and arguments (extended abstract). In: Proceedings of the 24th Annual ACM Symposium on Theory of Computing, STOC'92, Victoria, pp. 723–732. ACM, New York (1992). doi:10.1145/129712.129782
52. Kilian, J.: Improved efficient arguments. In: Proceedings of the 15th Annual International Cryptology Conference on Advances in Cryptology, CRYPTO'95, Santa Barbara, pp. 311–324. Springer, London (1995)
53. Kim, H., Gil, J., Hwang, C., Yu, H., Joung, S.: Agent-based autonomous result verification mechanism in desktop grid systems. In: Proceedings of the International Workshop on Agents and Peer-to-Peer Computing, AP2PC'07, Honolulu, pp. 72–84. Springer, Berlin/Heidelberg (2007). doi:10.1007/978-3-642-11368-0_6
54. Ko, R., Jagadpramana, P., Mowbray, M., Pearsoni, S., Kirchberg, M., Liang, Q., Lee, B.: TrustCloud: a framework for accountability and trust in cloud computing. In: Proceedings of the IEEE World Congress on Services, SERVICES'11, Washington DC, pp. 584–588. IEEE Computer Society, Washington, DC (2011). doi:10.1109/SERVICES.2011.91
55. Kuhn, M., Schmid, S., Watterhofer, R.: Distributed asymmetric verification in computational grids. In: Proceedings of the IEEE International Symposium on Parallel and Distributed Processing, IPDPS'08, Miami, pp. 1–10 (2008). doi:10.1109/IPDPS.2008.4536244
56. Liu, H.: A new form of DOS attack in a cloud and its avoidance mechanism. In: Proceedings of the ACM Cloud Computing Security Workshop, CCSW'10, Chicago, pp. 65–76. ACM, New York (2010). doi:10.1145/1866835.1866849
57. McIntosh, M., Austel, P.: XML signature element wrapping attacks and countermeasures. In: Proceedings of the 2005 ACM Workshop on Secure Web Services, SWS'05, Fairfax, pp. 20–27. ACM, New York (2005). doi:10.1145/1103022.1103026
58. Mell, P., Grance, T.: The NIST definition of cloud computing Natl. Inst. Stand. Technol. (NIST) Special Publication, **800**(145), 7 (2011)
59. Papamanthou, C., Tamassia, R., Triandopoulos, N.: Optimal verification of operations on dynamic sets. In: Proceedings of the 31st Annual Conference on Advances in Cryptology, CRYPTO'11, Santa Barbara, pp. 91–110. Springer, Berlin/Heidelberg (2011)
60. Pearson, S., Benameur, A.: Privacy, security and trust issues arising from cloud computing. In: Proceedings of the International Workshop on Cloud Privacy, Security, Risk, and Trust, CLOUDCOM'10, Indianapolis, pp. 693–702. IEEE Computer Society, Washington, DC (2010). doi:10.1109/CloudCom.2010.66
61. Pfoh, J., Schneider, C., Eckert, C.: A formal model for virtual machine introspection. In: Proceedings of the 1st ACM Workshop on Virtual Machine Security, VMSec'09, Chicago, pp. 1–10. ACM, New York (2009). doi:10.1145/1655148.1655150
62. Popović, K., Hocenski, Z.: Cloud computing security issues and challenges. In: Proceedings of the 33rd International Convention on Information and Communication Technology, Electronics and Microelectronics, MIPRO'10, Opatija, pp. 344–349 (2010)

63. Ristenpart, T., Tromer, E., Shacham, H., Savage, S.: Hey, you, get off of my cloud: exploring information leakage in third-party compute clouds. In: Proceedings of the 16th ACM Conference on Computer and Communications Security, CCS'09, Chicago, pp. 199–212. ACM, New York (2009). doi:10.1145/1653662.1653687
64. Roy, I., Setty, S., Kilzer, A., Shmatikov, V., Witchel, E.: Airavat: security and privacy for MapReduce. In: Proceedings of the 7th USENIX Conference on Networked Systems Design and Implementation, NSDI'10, San Jose, pp. 297–312. USENIX Association, Berkeley (2010)
65. Schiffman, J., Moyer, T., Shal, C., Jaeger, T., McDaniel, P.: Justifying integrity using a virtual machine verifier. In: Proceedings of the Computer Security Applications Conference, ACSAC'09, Honolulu, pp. 83–92. IEEE Computer Society, Washington (2009). doi:10.1109/ACSAC.2009.18
66. Sebe, F., Domingo-Ferrer, J., Martinez-Belleste, A., Deswarte, Y., Quisquater, J.J.: Efficient remote data possession checking in critical information infrastructures. IEEE Trans. Knowl. Data Eng. **20**(8), 1034–1038 (2008). doi:10.1109/TKDE.2007.190647
67. Sekar, V., Maniatis, P.: Verifiable resource accounting for cloud computing services. In: Proceedings of the 3rd ACM Cloud Computing Security Workshop, CCSW'11, Chicago, pp. 21–26. ACM, New York (2011). doi:10.1145/2046660.2046666
68. Setty, S., McPherson, R., Blumberg, A., Walfish, M.: Making argument systems for outsourced computation practical (sometimes). In: Proceedings of the Network and Distributed Systems Security Symposium, NDSS'12, San Diego (2012)
69. Shacham, H., Waters, B.: Compact proofs of retrievability. In: Proceedings of the 14th International Conference on the Theory and Application of Cryptology and Information Security: Advances in Cryptology, ASIACRYPT'08, Melbourne, pp. 90–107. Springer, Berlin/Heidelberg (2008). doi:10.1007/978-3-540-89255-7_7
70. Somorovsky, J., Heiderich, M., Jensen, M., Schwenk, J., Gruschka, N., Iacono, L.: All your clouds are belong to us: security analysis of cloud management interfaces. In: Proceedings of the ACM Cloud Computing Security Workshop, CCSW'11, Chicago, pp. 3–14. ACM, New York (2011). doi:10.1145/2046660.2046664
71. Stamm, S., Ramzan, Z., Jakobsson, M.: Drive-by pharming. In: Proceedings of the International Conference on Information and Communications Security, ICICS'07, Zhengzhou, pp. 495–506. Springer, Berlin/Heidelberg (2007)
72. symantec.com, 2011 State of cloud survey. http://goo.gl/Q5Wv9 (2011)
73. Szajda, D., Lawson, B., Owen, J.: Hardening functions for large scale distributed computations. In: Proceedings of the IEEE Symposium on Security and Privacy, SP'03, Oakland, pp. 216–225. IEEE Computer Society, Washington (2003)
74. Szefer, J., Keller, E., Lee, R., Rexford, J.: Eliminating the hypervisor attack surface for a more secure cloud. In: Proceedings of the 18th ACM Conference on Computer and Communications Security, CCS'11, Chicago, pp. 401–412. ACM, New York (2011). doi:10.1145/2046707.2046754
75. Vaquero, L., Rodero-Merino, L., Morán, D.: Locking the sky: a survey on IaaS cloud security. Computing **91**(1), 93–118 (2011). doi:10.1007/s00607-010-0140-x
76. Wang, C., Zhou, Y.: A collaborative monitoring mechanism for making a multitenant platform accountable. In: Proceedings of the 2nd USENIX Conference on Hot Topics in Cloud Computing, HotCloud'10, Shanghai. USENIX Association, Berkeley (2010)
77. Wang, C., Wang, Q., Ren, K., Lou, W.: Ensuring data storage security in cloud computing. In: Proceedings of the International Workshop on Quality of Service, Minneapolis (2009). doi:10.1109/IWQoS.2009.5201385
78. Wang, Q., Wang, C., Li, J., Ren, K., Lou, W.: Enabling public verifiability and data dynamics for storage security in cloud computing. In: Proceedings of the 14th European Symposium on Research in Computer Security, ESORICS'09, Charleston, pp. 355–370. Springer, Berlin/Heidelberg (2009)
79. Wang, C., Wang, Q., Ren, K., Lou, W.: Privacy-preserving public auditing for data storage security in cloud computing. In: Proceedings of the 29th IEEE International Conference on Computer Communications, INFOCOM'10, San Diego, pp. 525–533. IEEE, Piscataway (2010)

80. Wang, C., Ren, K., Wang, J.: Secure and practical outsourcing of linear programming in cloud computing. In: Proceedings of the IEEE International Conference on Computer Communications, INFOCOM'11, Boston, pp. 820–828 (2011). doi:10.1109/INFCOM.2011.5935305
81. Wang, C., Ren, K., Wang, J., Urs, K.: Harnessing the cloud for securely solving large-scale systems of linear equations. In: Proceedings of the 31st International Conference on Distributed Computing Systems, ICDCS'11, Saint Malo, pp. 549–558. IEEE Computer Society, Washington, DC (2011). doi:10.1109/ICDCS.2011.41
82. Watanabe, K., Fukushi, M., Horiguchi, S.: Collusion-resistant sabotage-tolerance mechanisms for volunteer computing systems. In: Proceedings of the 2009 IEEE International Conference on e-Business Engineering, ICEBE'09, Macau, pp. 213–218. IEEE Computer Society, Washington (2009). doi:10.1109/ICEBE.2009.36
83. Wei, J., Zhang, X., Ammons, G., Bala, V., Ning, P.: Managing security of virtual machine images in a cloud environment. In: Proceedings of the ACM Cloud Computing Security Workshop, CCSW'09, Chicago, pp. 91–96. ACM, New York (2009). doi:10.1145/1655008.1655021
84. Wei, W., Du, J., Yu, T., Gu, X.: SecureMR: a service integrity assurance framework for MapReduce. In: Proceedings of the Annual Computer Security Applications Conference, ACSAC'09, Genova, pp. 73–82. IEEE Computer Society, Washington (2009). doi:10.1109/ACSAC.2009.17
85. Wei, L., Zhu, H., Cao, Z., Jia, W., Vasilakos, A.: SecCloud: bringing secure storage and computation in cloud. In: Proceedings of the IEEE International Conference on Distributed Computing Systems Workshops, ICDCSW'10, Honolulu, pp. 52–61. IEEE Computer Society, Washington (2010). doi:10.1109/ICDCSW.2010.80
86. Xiao, Z., Xiao, Y.: P-accountable networked systems. In: Proceedings of the IEEE International Conference on Computer Communications Workshops (2010). doi:10.1109/INFCOMW.2010.5466640
87. Xu, Y., Bailey, M., Jahanian, F., Joshi, K., Hiltunen, M., Schlichting, R.: An exploration of L2 cache covert channels in virtualized environments. In: Proceedings of the 3rd ACM Cloud Computing Security Workshop, CCSW'11, Chicago, pp. 29–40. ACM, New York (2011). doi:10.1145/2046660.2046670
88. Yao, A.: How to generate and exchange secrets. In: Proceedings of the 27th Annual Symposium on Foundations of Computer Science, SFCS'86, Toronto, pp. 162–167. IEEE Computer Society, Washington (1986). doi:10.1109/SFCS.1986.25
89. Yumerefendi, A., Chase, J.: The role of accountability in dependable distributed systems. In: Proceedings of the 1st USENIX Conference on Hot Topics in System Dependability, HotDep'05, pp. 3–13. USENIX Association, Berkeley (2005)
90. Yumerefendi, A., Chase, J.: Strong accountability for network storage. Trans. Storage **3**(3), 11:1–11:33 (2007). doi:10.1145/1288783.1288786
91. Zalewski, M.: code.google.com, Browser security handbook. http://goo.gl/N72wl (2009)
92. Zeng, K.: Publicly verifiable remote data integrity. In: Proceedings of the 10th International Conference on Information and Communications Security, ICICS'08, Birmingham, pp. 419–434. Springer, Berlin/Heidelberg (2008). doi:10.1007/978-3-540-88625-9_28
93. Zhang, K., Zhou, X., Chen, Y., Wang, X., Ruan, Y.: Sedic: privacy-aware data intensive computing on hybrid clouds. In: Proceedings of the 18th ACM Conference on Computer and Communications Security, CCS'11, New York, pp. 515–526. ACM, New York (2011). doi:10.1145/2046707.2046767
94. Zheng, Q., Xu, S.: Fair and dynamic proofs of retrievability. In: Proceedings of the 1st ACM Conference on Data and Application Security and Privacy, CODASPY'11, San Antonio, pp. 237–248. ACM, New York (2011). doi:10.1145/1943513.1943546
95. Zunnurhain, K., Vrbsky, S.: indiana.edu, Poster from CloudCom 2010: security attacks and solutions in clouds. http://goo.gl/rX3L8 (2010)

Moving Target Defense for Cloud Infrastructures: Lessons from Botnets

Wei Peng, Feng Li, and Xukai Zou

Abstract While providing elasticity to clients through on-demand service and cost-effectiveness to service providers through efficient resource allocation, current cloud infrastructures are largely homogeneously and statically configured for ease of administration. This leaves ample opportunities for attackers to reconnoiter and penetrate the security perimeter of cloud services. This chapter (1) explores the evolution in botnet technologies from the early static architectures to the recent dynamic and resilient architectures that employ various moving target defense (MTD) techniques to circumvent crackdowns, and (2) draws lessons from botnets in identifying cloud security challenges and proposed solutions to MTD for cloud infrastructures, in which the cloud infrastructure configuration constantly evolves to confuse attackers without significantly degrading the quality of service. Proposed solutions may increase the cost for potential attackers by complicating the attack process and limiting the exposure of network vulnerability in order to make the network more resilient against novel and persistent attacks.

1 Introduction

Cloud computing has emerged as a mainstream computing and storage service model for personal, business, and government affairs out of its roots on autonomic computing [50, 59, 67], grid computing [15] and utility computing [10]. One characteristic of cloud computing is multi-tenancy. While providing elasticity to clients through on-demand allocation and cost-effectiveness to cloud service

W. Peng • F. Li (✉) • X. Zou
Indiana University-Purdue University Indianapolis, Indianapolis, IN, USA
e-mail: pengw@iupui.edu; fengli@iupui.edu; xkzou@iupui.edu

K.J. Han et al. (eds.), *High Performance Cloud Auditing and Applications*,
DOI 10.1007/978-1-4614-3296-8_2, © Springer Science+Business Media New York 2014

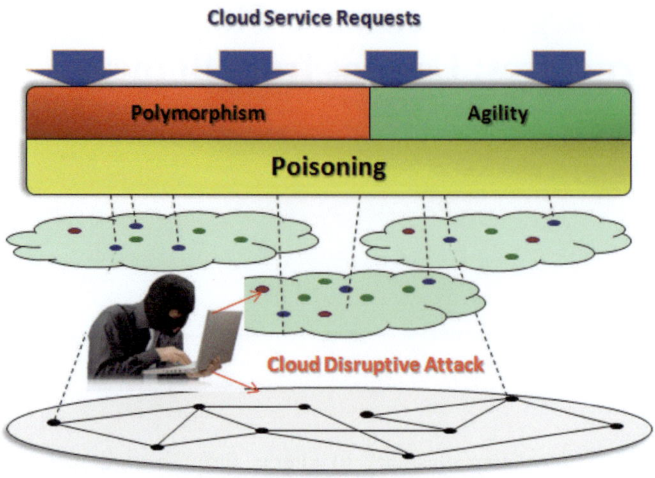

Fig. 1 Botnet moving target attack technology

providers (CSPs) through efficient resource allocation, multi-tenancy has its own security implications. For example, client virtual machine (VM) instances running on the same physical machines are susceptible to side-channel attacks [32].

The automated process of allocating resources on client demands creates many server instances with identical or very similar configurations, in which a single VM-level compromise may quickly scale up to a service-level breach due to the high homogeneity. Homogeneity and static configurations provide attack opportunities to botmasters. Botnets [24, 60, 63, 107] have plagued the Internet for over a decade. Studies on discovered botnets put the number of bots on the order of hundreds of thousands [46] to millions [34]. Though only a fraction of the whole bot population may be online at the same time [93] due to diverse geographical distribution and diurnal pattern of the bots [23], the cumulative bandwidth and computational capacity of the bots at the disposal of botmasters enable numerous nefarious activities, including email spam campaigns [86,88,114], distributed denial of service attacks [9,105], key logging [48], and identity theft [73]. Despite intensive cyber security research efforts to mitigate botnets, botnets are still active [88, 108]. Dainotti et al. [24] reported the botnet's scanning behavior, including general methods to correlate, visualize, and extrapolate botnet behavior across the global Internet. "Botnets rival the power of today's most powerful cloud computing platforms. These *dark clouds*, controlled by cybercriminals, are designed to silently infect your network." [19]

Botnets are fast-moving targets (see Fig. 1) that are difficult to detect with conventional security tools. Therefore, *moving target defense (MTD)* has become a major theme in cyber-security researches since mid-2010 [55, 56, 101].

According to the U.S. Homeland Security Cyber Security R&D Center [30]:

> Moving target defense (MTD) is the concept of controlling change across multiple system dimensions in order to: increase uncertainty and apparent complexity for attackers, reduce their window of opportunity and increase the costs of their probing and attack efforts. MTD assumes that perfect security is unattainable. Given that starting point, and the assumption that all systems are compromised, research in MTD will focus on enabling the continued safe operation in a compromised environment and to have systems that are defensible rather than perfectly secure.
>
> MTD enables us to create, analyze, evaluate, and deploy mechanisms and strategies that are diverse and that continually shift and change over time to increase complexity and cost for attackers, limit the exposure of vulnerabilities and opportunities for attack, and increase system resiliency.

In the context of cloud infrastructures, MTD is motivated by the *asymmetric* [55] costs borne by defenders and attackers. While the defenders need to secure the entire system against potential attacks, a single vulnerability is enough for the attackers to break into the system [55, 56]. This is exacerbated by the growing complexity of modern systems. MTD tries to tilt the balance towards defenders over attackers by dynamically and proactively changing configurations of the cloud infrastructures.

Research and development of MTD in cloud infrastructures are still in an early stage. How to interpret and implement MTD in the context of cloud infrastructure security is still an open challenge. The objectives of this chapter are to provide a comprehensive botnet survey and an introduction to MTD cloud infrastructures for cloud researchers, administrators and developers.

This chapter is organized as follows: Section 2 provides the evolution in botnet design from the early static architectures to the recent dynamic and resilient ones through various MTD techniques. Section 3 identifies cloud security challenges, describes botnets (dark clouds) versus clouds, and provides illustrative MTD techniques for secure clouds. Section 4 presents the proposed solutions to MTD for cloud infrastructures by drawing lessons from botnets. Conclusions are presented in Section 5.

2 Moving Target Defense: Lessons from Botnets

Research on botnet detection/mitigation evolves over time. Literatures on botnet research appeared in academia around 2003 [71]. Early signature-based techniques [64, 122], which were and still are employed in many Honeynet projects [54], fail to detect both polymorphic [119] variants of old botnets and completely new botnets. Later development of botnet detection methodology includes anomaly-based [42], DNS-based [123], mining-based [36] and heuristic-based [72] techniques, and the techniques specially designed to cope with fast IP/domain flux used by botnets [4, 129].

The following subsections provide a few of the aspects of botnets that have been investigated.

2.1 Botnet Formation and Communication

Consensus on the life cycle of a botnet [44] consists the following stages [120]:

- *Botnets Attack Vectors*: Botnets share the attack vectors with other forms of malware, including server exploits, trojan/rootkit piggyback, social engineering through spamming [90, 94], and other advanced forms of attack vectors [31]. Studies on captured bot samples indicate that modern botnets employ multiple attack vectors to maximize the chance of propagation [1]. A common trend in this stage is the increasing emphasis on *social network* vulnerabilities [12].
- *Rallying*: Methods for a newly infected bot to join the existing botnet include hard-coded IP/domain/IRC-channel names (e.g., Akbot [85]), external configuration file (e.g., Trojan.Peacomm [41, 46]), and dynamically generated rendezvous (e.g., Torpig [107]). This stage is called *rallying*. A common trend in this stage is the transition from *random* sampling, such as consecutive scanning of a whole IP block, to *targeted* probing, such as hard-coded rendezvous.
- *Command and Control (C&C)*: Existing botnet C&C channels include public Internet services such as IRC, HTTP, DNS, and various P2P protocols [132]. A common trend in this stage is the migration from traditional *centralized* IRC or web-based channels [22] to more robust *distributed* P2P or *tiered hybrid* channels.

2.2 Botnet Population Measurements

The size of a botnet is characterized by two metrics such as *footprint* and *live population* [93]. Footprint measures the cumulative number of bots over the entire lifetime of a botnet; live population measures the dynamics of online bot population reachable from the botmaster. In addition, the temporal/spatial distribution of the bots in a botnet is also of interest to both the botmaster and the defender. Measurement techniques include both *passive* detection [18] and *proactive* infiltration [1, 107]. Measurement results indicate that a significant portion of the bot population are behind Network Address Translation (NAT) firewall, possibly in home or small-office/home-office (SOHO) settings [31, 57].

2.3 Botnet Technologies

Recent developments in botnet design, especially the rallying and C&C stages, embody the principles of MTD. Detailed discussions are in Section 2.5. The other stage, initial infection of bot, is usually accomplished by a combination of drive-by download [91, 103], software vulnerabilities [131], and social networks [12, 124]. The following is a list of a few real cases.

2.3.1 Drive-by Download

Tidserv rootkit [113], used by the *TDL4* botnet [100], is bundled with rogue security software and infects low-level system printer and filesystem drivers. It also blocks system update and disables some anti-malware programs. Then, Tidserv modifies the boot record, i.e., Master Boot Record (MBR), on the hard disk so that it is loaded and executed prior to the operating system every time the system reboots. Through this technique, Tidserv circumvents the system's mandatory driver signing mechanism. *Mebroot* rootkit [43], used by the *Torpig* botnet [107], infects systems through injecting malicious HTML and JavaScript scripts that exploit vulnerabilities of Web browser plugins. If any such exploit is successful, a copy of the Mebroot rootkit is downloaded and executed on the victim's computer. Like Tidserv, Mebroot also modifies the MBR to circumvent detection by anti-malware programs. Early variants of the *Zeus* botnet [2, 11] also adopt drive-by download for initial infection by redirecting victims to a webpage that contains a malicious PDF file that exploits known vulnerabilities in the Adobe Reader software [77]. Similar vulnerabilities on Adobe Reader are also exploited by the *Gumblar* botnet [112]. *Asprox* botnet [92] launches SQL injection attacks [13] against vulnerable pages based on Microsoft Active Server Page (MSASP) to inject malicious scripts for propagating malware.

2.3.2 Software Vulnerabilities

In some variants of the *Conficker* botnet [21,89], new victim computers are infected by existing bots through specially crafted Remote Procedure Call (RPC) requests. The requests will trigger buffer overflows, which allow the existing bots to send and install the malware on the victim computers without victim users' knowledge. Other variants of Conficker launch dictionary attacks against default shared resources.

2.3.3 Social Networks

Email spamming has been the most popular malicious activity of botnets. Botnet owners make money through renting their botnets for spamming, and compromise computers or web sites for botnet expansions. For examples, the *Srizbi/Reactor* botnet [111] was behind the Ron Paul spam campaign [87]. Zeus uses Facebook phishing [27] and fakes billing emails from Verizon Wireless [51] to initiate drive-by download. The Nugache botnet [31, 109] lures victims into downloading and installing the malware (packaged into a popular video editing application) by using existing bot population to boost the visibility of the malware on popular software download aggregation sites through fake downloading [31]. A wide variety of social network-based spam campaign measurement studies and detection techniques are in [39, 86, 88].

Next, we will first discuss a few moving target defense techniques observed in real-world botnets (Section 2.5), and then sample a few botnet technologies from proactive botnet-mitigation research efforts (Section 2.6). We will notice a common trend of adopting dynamic and resilient moving-target defense mechanisms in both real-world and research botnet technologies. We conclude by relating these recent developments of moving-target defense in botnets to the rise and fall of early static IRC-based botnets (Section 2.4).

2.4 Rise and Fall of IRC-Based Botnets

According to Ferguson [37], the *Sub7* torjan and *Pretty Park* worm, which both surfaced in 1999, first introduced the concept of *malicious* bots: Victim machines connect to an IRC channel, waiting for commands issued by a remote attacker. Prior to that, bots were used on IRC channels for benevolent purposes such as automating channel administration and providing help to new users. Some notable developments in early IRC bots include Gtbot, Agobot and Spybot. In 2000, the Gtbot, which was based on the mIRC client, could initiate rudimentary DoS attacks due to the possibility of running IRC-event-triggered scripts, and having raw access to Transmission Control Protocol (TCP) and User Datagram Protocol (UDP) socket programming interface provided by the mIRC client. In 2002, the commercialization of *Sdbot* allowed development of a new botnet to be based on previous ones. In the same year, *Agobot* introduced the concept of a modular staged attack packaged as additional malicious payloads over an underlying backdoor. The initial attack is to set up the backdoor, which paves the way for later attacks that are packaged into small modules. In 2003, *Spybot* introduced key-logging, data mining, and instant messaging spam into botnet. Later that year, *Rbot* introduced SOCKS proxy and the use of compression and encryption to evade detection. Despite expansion in functionality, these early botnets all use the IRC protocol as the C&C channel. The former open nature and wide adoption of the IRC protocol proved to be vital for the success of early botnets. However, as more IRC-based botnets surfaced, system administrators became alerted to unauthorized IRC traffic and, hence, IRC ports were disabled by default and IRC traffic served as a sign for malware infection.

A real-world example is the “Operation: Bot Roast”, an international investigation against botnets led by the United States Federal Bureau of Investigation (FBI), in which five botnet authors and operators were arrested and charged, including the 18-year old New Zealand author of *Akbot* [70], an IRC-based botnet [35]. In Akbot, after a victim is infected through remote exploitation of the Windows operating system [74], the malware downloads and installs itself through an FTP server, attempts to join three fixed IRC servers on TCP port 6584 and waits for the botmaster’s instructions. Akbot was detected because of its constant IRC traffic and static IRC server/port. This is a vivid example of how static configurations in early IRC-based botnets expose the botnets and compromise their operation; this is why later botnets move on to more dynamic settings.

In late 2007, Zhuge et al. [133] reported their 1-year measurement study of IRC-based botnets. In their study, a honeypot-based, distributed, and fully-automated measurement system, consisting of 50 malware sensors deployed on 17 nodes, was used to track botnet activities on their C&C IRC channels. During the study, an average of 2,800 samples were captured every day, and 3,290 unique (in terms of DNS name, port number, and channel name) IRC-based botnets were identified from the samples. IP-address-based geographic analysis indicated that almost 38.8 % of the C&C channels were hosted in the United States, followed by China, Korea, Germany, Netherlands, Canada, Sweden, Great Britain, and other countries, and almost 15.1 % of the bots were located in Brazil, followed by China, Malaysia, Taiwan, Korea, Mexico, Russia, Argentina, India, and other countries. Only about 36.1 % of the discovered botnets used the standard IRC port 6667 to host the C&C channel; the majority other used non-standard ports for C&C. This confirmed earlier observations that IRC-based botnets started to use non-standard configuration to evade simple port-based detection. For example, 1.3 % of the discovered botnets used TCP port 135, which is commonly used by Windows for file sharing. The study found an average lifetime of 54 days for the C&C servers and, among the 3,290 botnets they monitored, 378 were still alive in mid-June, 2007. This indicates the resilience of the botnets. During the study, a total of 1,520,000 distinct bot IDs were observed for 1,904 (57.9 %) out of the 3,290 botnets. For 1,110 (33.7 % of the total) of the 1,904 botnets, 700,700 distinct IP addresses were observed, and the largest observed botnets consisted of more than 50,000 IP addresses. Though unique IP addresses do not translate to unique bot due to churn effect and dynamic IP assignments, this still shows the prevalence of botnets.

The reason that the researchers could measure the size of the 1,904 botnets is because these botnets did not disable the user listing IRC command on their C&C channel. While taking the advantage offered by the IRC infrastructure, these botnets did not customize the protocol to suit their needs and thus suffered the consequences of being easily detectable. Had they changed the default configuration and customized the underlying IRC protocol, and thus moved away from being *static* to being *dynamic*, it would have been much harder to infiltrate them.

2.5 Moving-Target Design Examples in Real-World Botnets

After the fall of IRC-based botnets (Section 2.4) due to the easy-to-detect C&C traffic on IRC channels, botnets adapted by migrating away from IRC to other HTTP-based or customized C&C channels. Therefore, botnet network topologies are hard to be identified. At first, a single C&C server, corresponding to one IP address or domain name, was used. For example, initially, Gamblar connected to a fixed domain gumblar.cn, and the taking down of this domain in May 2009 apparently shut down Gamblar [20]. Gamblar reappeared later, and used multiple domain names for rallying, making it harder to detect and stop. *Fast flux* [49] was devised by botnet creators in response to the crackdown of individual C&C servers. With fast flux, a bot queries a known domain name, which is associated with a DNS record with short time to live (TTL) value and thus would be translated into a series of IP addresses in a quick and round-robin fashion. This reduces the impact of a single C&C server being taken down on the operation of the whole botnet. Some traditional system defense mechanisms, like IP-based access control, become ineffective against fast flux. Storm [28, 38, 40] and Warezov/Stration [110] were among the first botnets that adopted fast flux, with Wibimo [45] being a more recent example.

Although fast flux solves the single-point-of-failure problem (for the botnet) with regard to C&C server's IP address, the unique domain name still leaves a single point of failure for authorities to take down the botnet. New generations of botnets solve this problem with a technique commonly known as *domain flux*. With domain flux, each bot independently and periodically computes a list of domain names with a customized *domain generation algorithm* (DGA). The bot proceeds to contact the hosts with these domain names one by one until a host responds and validates itself as a C&C server under the botnet protocol. Due to dynamic assignment and decentralized management of the rendezvous, even if one of the C&C domain names is blocked by ISPs or taken down by authorities, it is likely that another C&C domain name is still valid, and thus could be used by the bots to locate the C&C server. DGAs were previously used as the primary network evasion technique in many highly publicized and well studied botnets [26], including Conficker [21, 89], Murofet [16], and Torpig [53, 107]. Recent variants of Zeus use DGAs as backup strategies for rallying should their primary rallying mechanisms, such as peer-to-peer channels or hard-coded IP addresses, fail [25]. In a case study on a previously unreported botnet sample conducted by Antonakakis et al. [3], the botnet uses the date as the seed to its pseudo-random-number generator and generates approximately 1,000 domains, employing purely alphanumeric characters and a particular top-level domain (TLD) for a given day. Another example is Torpig. Torpig uses a two-tier DGA. Each bot first generates a weekly domain name, which only depends on the current week of the year, and appends a few common top-level domains (such as ".com" and ".net") to the weekly domain name. The bot tries to contact each of these domain names until an attempt succeeds. If all fail, the bot generates another daily domain name, appends the TLDs, and contacts the resulting domain names in turn. If all fail, the bot resorts to a number of hard-coded domain names in its configuration file. The DGA used in Torpig is completely deterministic. This, along with the weak obfuscation over the C&C communication channel, allowed researchers to take over the botnet for 10 days between 25 January 2009 and 4 February 2009 [107].

The network topology used by botnets to organize the bots and C&C servers also evolve, from the centralized architectures (early variants of Zeus, Torpig, and Grum [79]) to more dynamic and robust P2P (used in TDL-4 [100], later variants of Zeus [61], Nugache [31, 109], and Storm [28, 38, 40]) or hybrid (Waledac [81, 102] and Sality [34]) ones. Interestingly, a partial reversion to hard-coded IP lists for initial rallying in some recent P2P botnets (e.g., recent variants of Zeus [61]) was observed. This is, perhaps, due to botmasters' aversion to inherent latency in coordination and control introduced by the P2P architecture. This indicates the need to strike a balance in real applications between the security gain and the performance hit, both introduced by the adoption of moving-target defense. In the rest of this section, we will use Nugache and Waledac to illustrate the operations of P2P and hybrid botnets, respectively.

Nugache has evolved over time. The original Nugache, first documented in late April 2006 [80], was largely dismissed by the security community as trivial to detect due to a few distinctive *invariants* such as connections to certain TCP ports. After the malware was updated to use random high-numbered port for communication, Nugache stayed largely undiscovered until the arrest of its author in September 2007 [78]. Unlike the *inside* study used on Torpig [107] for infiltration and takeover, which exploited weak obfuscation in the communication protocol and deterministic domain-name probing in the domain-name generation algorithm used by Torpig, the study on Nugache was more difficult, due to its strongly encrypted communication channel. Traditional honeypots [47], which worked well in detecting centralized botnet and collecting information (e.g., captured packet, intrusion detection signatures, and passive operating system fingerprints), fell short in detecting P2P botnets. In their study on Nugache [31], Dittrich and Dietrich specifically customized the honeynets to deal with the peculiarities of Nugache. Unable to infiltrate the C&C structure and examine the encrypted message exchanged by peering bots, the analyses on Nugache were largely based on external traffic analysis which included P2P connections, probing associated with remote vulnerability exploitation, and DDoS attacks through the P2P network [31]. Traffic analysis on the Nugache botnet trapped in the honeypots indicates that Nugache does not have a centralized C&C structure. Each bot makes infrequent inbound/outbound connections at the rate of dozens per day. The P2P structure with strong encryption allows the botmaster to control a significant number of hosts with only a small percentage of them actively probing and forwarding commands at the same time. Also, each bot only maintains a few peers to minimize exposure, in case that the bot is captured by the defender. IRC logs recorded for an early IRC-based variant of Nugache, which contain systematic probing in the non-routable IP address blocks reserved for intranet by RFC 1918 [95], indicate that the small-office/home-office (SOHO) networks were targeted by Torpig.

Waledac emerged in late 2008 after the infamous Storm botnet [28], which gained its notoriety through its large infection base (various sources put the number of bots in Storm from 250,000 up to a few million [38]), through its deliberate counter-attack against investigation (Storm was known to launch DDoS attacks against security vendors and researchers who participated in its investigation [40]), and through its enormous cumulative computing resources and bandwidth (some claimed that the cumulative computing capability was greater than some supercomputers [115]). An early variant of Waledac was delivered through the same backdoor used for carrying Storm; thus, Waledac is considered a descendant of Storm. In Waledac, the bots are divided into two layers, *spammers* and *repeaters*, based on whether a bot is behind network address translations (NATs): Those behind NATs, which do not have a public-accessible IP address, are the spammers and the other publicly accessible bots serve as repeaters. Within the botnet, each spammer communicates exclusively with an upper-tier repeater; repeaters, besides serving a few spammers, communicate among themselves. This structure is similar to the tiered architecture used in the Skype VoIP system [6]. P2P communication is restricted to the repeaters, which are responsible for collecting data from spammers and distribute commands from the botmaster. Unlike Storm, Waledac does not communicate through the decentralized *Overnet P2P networks using Kademlia* [69], but exclusively through encrypted HTTP with fast flux. The rallying for Waledac is much like Storm. Each newly infected bot finds a neighboring repeater through probing a hard-coded list of IP addresses. If the probing fails, the bot will download an IP-address list through a hard-coded URL, which is fortified with fast flux to reduce the chance of being taken down. Repeaters are solely responsible for coordinating communication between the spammers and upper-tier nodes, which are under close control of the botmaster. Early research speculated that the upper tier was also tiered [102], which was later verified when the researchers Nunnery et al. [81], collaborating with two of the affected hosting providers in the Netherlands, were able to obtain the file-system images and network traces of the servers serving as upper-tier botnet nodes. Sinclair et al. [102] defined that "TSL is the name of the Windows registry entry that the Waledac binary uses to store a list of servers for this tier. As such, we named the list of these servers as the TSL layer." Note that the meaning of this acronym is unclear. The upper tier consists of two additional layers: several TSL servers and a single Upper Tier Server (UTS). The UTS server is the ultimate C&C server directly controlled by the botmaster. The TSL servers are responsible for coordinating communication between the UTS server and the repeaters, and take their name from an entry in the repeaters' local configuration listing their corresponding TSL servers. The TSL servers are set up with pre-packaged customized software stack (which includes the operating system) and are hosted on third-party hosting services. They insulate the UTS server, which is the C&C center, from lower-tier bots. In case a TSL server is taken down, the botmaster can set up a new one and relegate the repeaters corresponding to the compromised TSL server to the new server through the repeater-layer P2P channel. Waledac is considerably more resilient against crackdown than earlier centralized versions, due to its dynamically tiered hybrid C&C structure.

2.6 Towards More Resilient Botnets

The ongoing battle between the botmasters and the security professionals prompts some researchers to take a more proactive approach. The motto for this approach is "forewarned is forearmed": Rather than conducting *postmortem* analysis when the distress has been caused, the best way to defend against an unknown botnet is to explore advanced botnet design and mitigation before they have been seen in the real world. In this section, we sample two such proposals.

Wang et al. [121] propose a hybrid botnet design. Their design is motivated by the following challenges faced by botmasters [121]:

- How to reduce the chance of being detected by the defenders via communication traffic analysis?
- How to minimize the exposure of the whole network to the defender, if some bots are captured?
- How to maintain a robust network for the rest, if a substantial number of bots are taken down?
- How to monitor the botnet given the constraints implied by the above challenges?

The key ideas are [31, 121]:

- The differentiation between two types of bots, *slaves* and *servants*, based on whether the IP address is publicly accessible. Only the servants, which have publicly accessible IP addresses, will appear in the peer-list of a bot. This is similar to the spammer-repeater distinction in Waledac.
- Infection is through a worm-like channel, in which the infector and the infectee can directly communicate. The infector shares with its infectee its peer-list; if the infector is a servant itself, the infectee adds the infector to its peer-list. This eliminates the bootstrap phase, which is often the Achilles' heel of a botnet due to the staticity in this phase.
- The number of peers in a peer-list does not exceed a system parameter. Thus, each bot only knows a small portion of the whole botnet population. This reduces the chances of exposing the whole botnet in case one bot is compromised.
- A botmaster could monitor the entire botnet by issuing a special report command, which instructs the bots to report to a compromised machine called the *sensor* host. The sensor host changes every time to avoid being compromised by the defender.
- For each botnet, the service port for incoming connections is randomly chosen, and every connection is encrypted by a locally negotiated symmetric key to prevent sniffing. The increased dynamicity reduces the impact of infiltration and poisoning attacks.

Dittrich and Dietrich [31] report that most of these ideas, perhaps with the exception of the monitoring mechanism using sensor hosts, have already been used by Nugache. Wang et al.'s design was partially inspired by the deficiencies in an earlier variant of the Nugache botnet with a IRC C&C channel.

Inspired by the works of Wang et al. and others, Liu et al. [66] proposed a recoverable hybrid botnet design. Their design is motivated by two challenges: (1) to recover from the event of C&C being taken down and (2) to reduce the impact of P2P routing table poisoning. The key ideas are [66]:

- The C&C structure consists of two independent but coordinated mechanisms: *decentralized hybrid P2P-based C&C (HPCC)* and *centralized domain-flux-based C&C*, i.e., URL Flux-based C&C (UFCC).
- The HPCC uses the servant/slave distinction proposed by Wang et al. [121].
- The UFCC is hosted on a few robust Web 2.0 services, which are used to publish command and links to malicious payloads.

- A propagation-based reputation system is used in peer-list exchange to avoid Sybil attacks [28] and P2P routing table poisoning against the botnet.

Unlike real-world botnets, these hypothetical designs lack convincing verification on their effectiveness, which, paradoxically, is available only if they have been implemented, released, and tested in the real world. This is a dilemma faced by researchers taking the proactive approach to botnet mitigation. This echoes the call for action from Aviv and Haeberlen [5] for a PlanetLab-like botnet-research testbed, where researchers can test and verify their ideas without disrupting the whole Internet. Nevertheless, these designs, along with the real-world botnets before them, clearly indicate moving-target defense as the unifying theme in future research on botnet design and mitigation.

3 Moving Target Defense: Towards Secure Clouds

Elasticity has attracted many organizations to migrate their computing and storage services to clouds, which provide on-demand provision of computing resources such as processor time, memory, and mass storage. The ever increasing information assets at stake have raised security concerns over clouds. In this section, we will discuss cloud infrastructure security challenges based on lessons from Botnets for cloud security.

3.1 Cloud Infrastructure Security Challenges

The transfer of management responsibility from customers to CSPs, while providing elasticity to clients through on-demand allocation and cost-effectiveness to service providers through efficient resource allocation, introduces numerous security challenges that permeate the whole cloud infrastructure, from the bottom hardware layer, through the medium VM layer, up to the operating system layer that supports customer applications. We briefly discuss a few such challenges below.

Virtual local area networks (VLANs) are often used for traffic isolation and for providing layer-2 QoS in the data center Ethernets. Rouiller [97] summarizes some cloud infrastructure security challenges in layer-2 VLANs as follows.

- *MAC Flooding Attack*: Attackers send numerous fake Media Access Control (MAC) address queries to the switches. This will saturate the MAC table associated with each VLAN port, after which the Ethernet switch will essentially degenerate to a hub by broadcasting every MAC message, and thus the frames could be sniffed.
- *Layer-2 Routing Manipulation Attack*: Some versions of the Spanning Tree Protocol (STP) are used by Ethernet switches to implement loop-free layer-2 routing. It is possible to corrupt the configuration data such that the Bridge

Protocol Data Unit (BPDU) messages used in the STP cause a switch to be elected as the root of the spanning tree and thereby have the traffic directed as desired by the attacker.

Besides the layer-2 Ethernet switches used for internal interconnection, layer-3 [84] routers are used to connect cloud data centers to the external world. These routers become more important in infrastructures with distributed data centers, to support a seamless cloud computing environment. Cloud infrastructure challenges in layer-3 routing include the following:

- *QoS Misconfigurations*: Improperly configured IP-flow and QoS related settings, such as Multi-Protocol Label Switching (MPLS) signaling [83, 98, 128] and differentiated services [62].
- *Layer-3 Routing Poisoning Attack*: Forged or tampered routing messages, which would lead to catastrophic consequences if the messages are accepted without authentication.

The complexity associated with the rich features provided by inter-domain routing protocol such as Border Gateway Protocol (BGP), coupled with the reluctance of some Internet service providers (ISPs) to share their configuration, leads to plenty of chances of misconfigurations [7], manifested as misconfigured address-prefix advertisements, alternative routes, or packet filtering rules. For example, in *BGP prefix hijacking* [83, 98, 128], an autonomous system address space is incorrectly announced without the owner's permission, perhaps due to misconfigurations or deliberate attacks. This will negatively affect the availability of cloud-based resources. Studies indicate that BGP prefix hijacking occurs several 100 times per month due to misconfigurations and less than a 100 times per month due to deliberate attacks. A real-world instance of BGP prefix hijacking is that in 2008, Pakistan Telecom attempted to block YouTube in the country, due to suspected blasphemous video being hosted there, by announcing an incorrect route to the service; YouTube became unavailable world-wide for 2 h as a result [106].

The increased external DNS querying due to outsourcing services to clouds makes DNS-based attacks particularly challenging in cloud environments. For example, vulnerabilities in many deployed DNS servers were discovered that allowed attackers to direct legitimate DNS queries from users to malicious domains under attackers' control by poisoning the DNS caches [99]. The vulnerabilities were rooted in the lack of both strong authentication (that prevents attackers from manipulating queries) and sufficient protocol randomness (that prevents attackers from being able to fake query responses with correct IDs) in deployed DNS software implementations.

In clouds, the abstraction of infrastructures and services also means that customers usually do not have the ability to precisely control the visibility and lifetime of some underlying resources, such as IP or physical address caches. There are lags between changing resource IP address and updating the new address in DNS caches, and similarly, between changing physical (MAC) address and the clearing of old entries in Address Resolution Protocol (ARP) caches. This means that some resources that have migrated or are being removed might still be accessible in the caches. According to Mather et al. [68], earlier reports on IP-address aging problems were likely the impetus behind the announcement of Elastic IP services from Amazon Web Services (AWS) in 2008 [14].

Currently, virtualization is implemented by one of several models, including OS-level virtualization (e.g., Solaris containers and Linux/BSD jails), paravirtualization (a combination of hardware/software virtualization), and hardware virtualization (Xen, VMware, and Microsoft Hyper-V). Fortifying virtualization is critical to securing cloud infrastructures. In a real-world incident in 2009, attackers erased over 10^6 websites hosted by UK based web hosting service provider VAserv, by exploiting a zero-day vulnerability in the hypervisor manager used by the company [117].

Client VM instances running on the same physical machines are susceptible to side channel attacks. Using the Amazon Elastic Compute Cloud (Amazon EC2) service, Ristenpart et al. [96] demonstrate that it is possible to map the internal cloud infrastructure, to identify where a particular target VM is likely to reside, and then to instantiate new VMs until one is placed co-resident with the target. Such placement can be used to mount cross-VM side-channel attacks to extract information from a target VM on the same machine.

Similar to centralized C&C center in the botnets, cloud infrastructures maintain configuration information in centralized databases. When new servers are allocated, such databases are automatically duplicated for ease of management and allocation efficiency. The automated process of allocating resources on client demands creates many server instances with identical or very similar configurations, in which a single VM-level compromise may quickly scale up to a service-level breach due to the high homogeneity.

3.2 Similarities and Dissimilarities: Botnets and Clouds

Security is a hidden feature, where "no news is good news", and is often shadowed by more visible features such as usability and performance. The explosive growth of cloud services has outpaced the security measures used to protect their infrastructures. This situation bears resemblance to the early golden age for the IRC botnets, in which botnet authors pushed the boundary of what an IRC botnet could do without concern for the vulnerabilities in the underlying IRC infrastructure. For example, almost a third of all the botnets studied by Zhuge et al. [133] used the default IRC port 6667, and over one half of the botnets allowed unauthorized query for channel membership provided by the IRC protocol; this directly led to the exposure of whole botnets. When the IRC C&C channel became the single point of failure of a botnet, botnets began to adopt MTD techniques like fast IP/domain fluxes and more resilient hybrid or purely distributed C&C architectures. Conceivably, something similar will happen for clouds.

In the rest of this subsection, we will identify a few similarities and differences between clouds and botnets, and suggest a few lessons that can be learned from botnets for securing cloud infrastructures. In both clouds and botnets, we have adversaries, albeit of different characters. In clouds, adversaries are attackers who disrupt and compromise cloud services for fame or for gain. In botnets, adversaries are usually security professionals with the intention of protecting the Internet community at large from the botnet. Intentions and nature aside, similar techniques could be used by malicious attackers for evil as well as by security professionals for good.

Both clouds and botnets need to maintain a known entry. In a botnet, this entry is used in rally for newly infected bots to join the botnet. This could an IRC channel in an IRC-based botnet, the IP address in a web-service-based botnet, the domain name in a fast-IP-flux-based botnet, the DGA in a fast-domain-flux-based botnet, the repeater/servant/super-node in a hierarchical hybrid botnet [130], and the hard-coded rendezvous or peer-exchanged peer-list in a P2P-based botnet. In a cloud infrastructure, the entry is usually the web-service-based management gateway. Fast-flux-like techniques could be adapted to cloud services to increase the uncertainty for malicious attackers without compromising usability for regular users.

The large number of computational elements (i.e., bots in a botnet, and service instances in a cloud infrastructure) brings forth another similarity. The similarity in configuration between the elements allows an attacker to quickly scale up the attack once a vulnerability is identified and exploited. In modern botnets, polymorphic programming techniques, such as memory randomization, code obfuscation, and encryption, which are traditionally associated with advanced computer viruses, are all used to frustrate security analysts' attempts to understand the malware. On the network level, traditional fixed C&C communication ports are being replaced by dynamically assigned ones. Also, the hard-coded rendezvous points in traditional centralized botnets are supplemented by localized peer-list in recent P2P botnets. Similar measures should be taken in cloud infrastructures for the distributed, static, duplicated, and largely identical configurations to thwart scalable attacks. As long as the agreed resources are guaranteed, the service provider is free to deploy MTD mechanisms, which thwart incremental reconnaissance and vulnerability exploitation by making user-invisible changes to the infrastructure.

Communication protocols and channels are usually the Achilles' heel in both clouds and botnets. In botnets, many security analyses begin with profiling them in terms of communication dynamics. The C&C protocol and structure could be easily identified if such communication is not properly protected, like in early IRC-based botnets and some early MTD botnets such as Torpig. A strong encryption goes a long way in protecting the integrity of the botnet, as demonstrated in the case of Nugache, in which the botnet stays under the radar for a long time, and security researchers have to resort to speculation in its analysis. A lesson for securing cloud infrastructures is that it pays off to encrypt communication, both externally and internally, at every possible level.

A difference between cloud infrastructures and botnets is the ownership. In botnets, with the possible exception of the C&C center, most hosts, including all the bots and most C&C proxies, usually do not belong to the botmasters. Thus, the botmasters adopt techniques to hide the botnet malware from the owners of these hosts. In typical clouds, the numerous computational, storage, and communication elements in the infrastructures all belong to the cloud service providers. The security mechanisms on these elements can be more rich in functionality and aggressive in resource appropriation.

This difference in ownership means that cloud infrastructures could afford better security mechanisms at the cost of increased resource consumption. This also means that the best security practices for botnet designs are not necessarily that for cloud infrastructures. Though we have witnessed, in botnets, a steady transition from centralized architectures to distributed architectures for resilience against crackdowns, the lack of real-time control over the whole infrastructure, inherent in the distributed architectures, is often not acceptable for clouds. Thus, when mapping MTD designs from botnets to cloud infrastructures, we shall carefully consider the impact of these differences on the mapping.

In the next section, we are going to present a few botnet-inspired MTD techniques in detail to illustrate the application of the above lessons to securing cloud infrastructures.

3.3 Illustrative MTD Techniques for Secure Cloud

In this section, we present four complementary techniques to adapt lessons from botnets for securing cloud infrastructures. Specifically, these techniques are:

- *Heterogeneous VM Replication* to deliberately introduce diversity to the multiple replications of a client VM instance, to reduce the chances of the same vulnerability being exploited across the replications.
- *Proactive VM Deployment Evolution* to thwart reconnaissance and penetration from attackers by constantly monitoring active client instances for security exposure, and migrating high-risk active VM instances to heterogeneous but compatible low-risk replications to neutralize attackers' reconnaissance and penetration attempts.
- *Agile, or Security-Context-Aware, Opportunistic Migration* to minimize the chances of the migration process being exploited at a lower data/control plane level through randomization.
- *Dynamic Authentication* to seamlessly protect the other techniques from unauthorized access, tampering, theft and poisoning.

4 Secure Cloud Infrastructures

Homogeneity and static configurations have been used for simple cloud administration and management tasks, but attackers have ample opportunities to reconnoiter and penetrate the security perimeter of cloud services. Figures 1 and 2 indicate that security design objectives of cloud and *dark clouds (botnets)* are similar because both clouds face constant threats. Bot creators sometimes use polymorphism to increase computational complexities to identify and remove bots. Polymorphic malicious codes or files are functionally identical but differ from one another in file size, content, or other respects [75]. There are two general types of malicious polymorphism: "(1) Server-side polymorphism, in which a server is configured to serve a slightly different version of a file every time it is accessed, possibly by

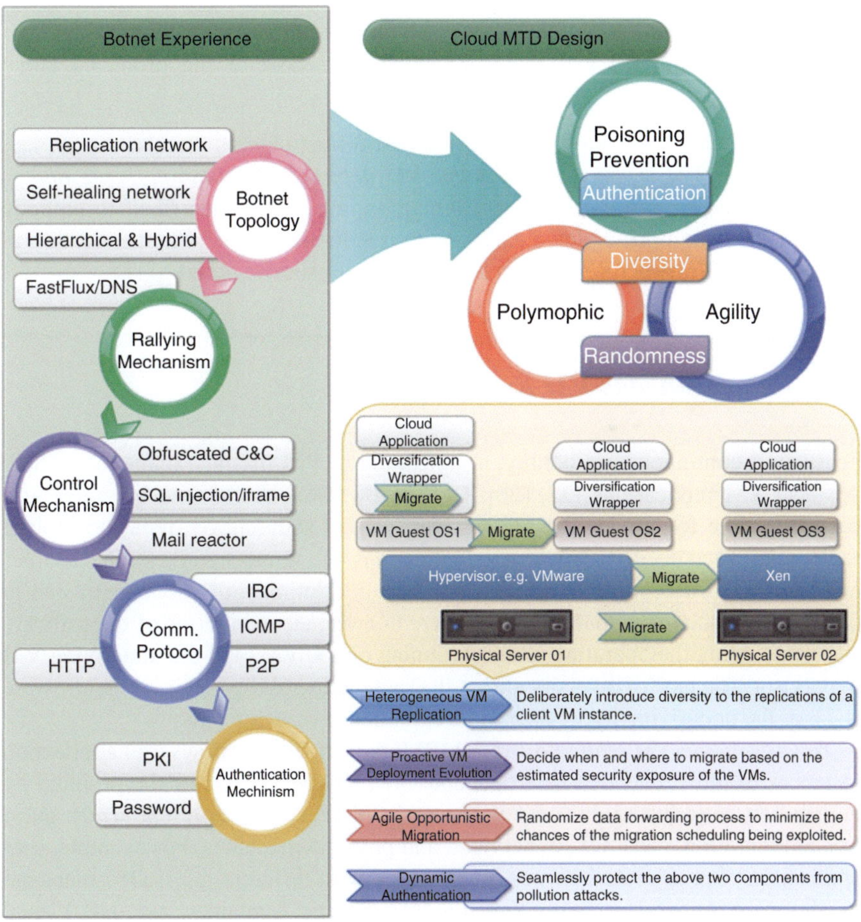

Fig. 2 Cloud moving target defense (MTD)

changing the file name of a component to a new random value, or by encrypting or compressing it in a slightly different way. (2) Malware polymorphism, in which the malware itself is designed to change slightly every time it replicates [75]." Cloud defense mechanisms have not been very effective because botnet technologies have been evolving rapidly using novel moving-target attack methods such as polymorphism and agility to protect botnet operations (see Fig. 1).

Our MTD research aims to develop moving-target defense technologies for secure cloud infrastructures as follows:

- *Polymorphism*: Explore botnet polymorphism techniques, i.e., server-side polymorphism and malware polymorphism [8, 52, 116]. Develop the novel cloud defense polymorphism techniques to protect cloud infrastructures from attackers.
- *Agility*: Investigate botnet agility behaviors. Develop the rapid provisioning technologies of cloud resources to provide high resource availability to cloud customers.
- *Poisoning Prevention*: Probe botnet poisoning mechanisms. Develop the tamper-evident technologies that make unauthorized access to the protected cloud resources easily detected.

In this section, our on-going secure cloud infrastructures research areas, i.e., *heterogeneous VM replication*, *proactive VM deployment evolution*, *Agile opportunistic migration* and *dynamic authentication*, are briefly described to embody three MTD design principles, i.e., diversity, randomization, and authentication in secure cloud infrastructures. Figure 2 depicts the illustrative cloud MTD.

4.1 Heterogeneous VM Replication

To support continuous availability, a logical client VM instance maps to multiple physical VM replications [33, 126]. Only a few replications need be active at a particular time to support the service; the rest are standing by, ready to replace faulty ones when needed. Only active replications will interact with external sources and hence are susceptible to attacks. For ease of administration, the status quo in cloud infrastructures is that replications are homogeneous in terms of the software stack (including the operating system) running on the VM, and the underlying VM hypervisor. The lack of diversity allows the same vulnerability in either the software stack or the underlying hypervisor to be exploited across replications.

Heterogeneous VM replication (HVMR) [65, 126], which allows user application instances to move between heterogeneous VM images, alleviates the problem by deliberately introducing diversity in the replications at four different levels: guest OS configuration, guest OS, hypervisor, and physical machines. For example, user applications can migrate between VM images with different guest OSs managed by the same hypervisor (e.g., Xen, QEMU, Solaris Container, and HP Virtual Partitions), or same guest OSs managed by different hypervisors. HVMRs, when coupled with *proactive infrastructure deployment evolution* and *agile opportunistic migration* to be discussed later, thwart attacks by changing the profile of the targeted client instance to the disadvantage of attackers.

A major challenge is to support transparent migrations between the heterogeneous replications. Ideally, users should not notice any differences in the heterogeneous physical instances. In reality, it is challenging due to the fact that subtle differences in implementations may result in an application behaving differently. Yet another challenge is to decide on the amount of diversity sufficient to thwart attackers without disturbing clients.

4.2 Proactive VM Deployment Evolution Strategies

Inspired by the dynamic C&C structures in emerging botnets, to mitigate reconnaissance and penetration, *proactive VM deployment evolution* (PVMDE) constantly monitors individual client instances for security exposure, and proactively migrates online instances to a heterogeneous but compatible replication.

4.2.1 Globally Coordinated Evolution Strategy

Similar to centralized C&C structure in a botnet, a central scheduler in the cloud collects information from all active replications. Due to the redundancy of physical replications in supporting a logical client instance, the active replications, called the *backbone*, at a particular time are a subset of all the partitions. All the other replications are standing by, and are beyond the reach of attackers.

The reduced security exposure of the stand-by replications comes at the cost of the active ones. Thus, it is necessary to rotate the backbones to amortize the risks. A straightforward solution is to construct multiple disjointed backbones and let them work alternatively. This can be implemented by, for example, Connected Dominating Set (CDS) [76, 125, 127]. With centralized coordination, backbones can be enumerated and alternately activated to amortize security exposure. However, global coordination requires a control channel to collect status information from replications and send activation/deactivation commands to them. As in botnets with a centralized C&C channel, the control channel is susceptible to attacks.

4.2.2 Locally Coordinated and Attack-Surface-Based Probabilistic Evolution Strategy

The heterogeneous VM replications created for logical client instances have diverse attack surfaces. The *locally coordinated, attack-surface-based, probabilistic evolution strategy* makes use of this in scheduling migrations. Intuitively, a replication with a small attack surface can be in service longer before being replaced by (i.e., migrating to) other replications.

In this strategy, an active replication will be in service until its accumulated risks, characterized by security exposure, exceed a predefined value. Then, it will make a probabilistic migration decision. In making the migration decision, each potential replacement will be chosen by a probability based on its attack surface: Potential replacements with smaller attack surfaces are more likely to be chosen over those with larger attack surfaces. Quantitatively, for two potential replacement replications *a* and *b*, if *a* has an attack surface half the size of *b*, *a* will be twice as likely to be chosen as the replacement than *b*.

4.2.3 Locally Coordinated and Switching-Based Probabilistic Evolution Strategy

In this strategy, an active replication j (corresponding to a logical client instance i) makes a probabilistic migration decision based on a *switching* probability P_s.

$$P_s = \begin{cases} 1 - \frac{\overline{E_{i,R[j]}}}{E_{i,j}} & \text{if } \overline{E_{i,R[j]}} \leq E_{i,j}, \\ 0 & \text{otherwise} \end{cases},$$

in which $E_{i,j}$ is replication i's security exposure when the active application is j, and $\overline{E_{i,R[j]}}$ is the average security exposure of all the replacements $R[j]$ of replication j.

4.3 *Agile Opportunistic Migration*

The aforementioned locally coordinated evolution strategies introduce diversity and randomness in the *logical* migration process. However, since mutually replaceable replications may be managed by different hypervisors or reside in different machines or even geographically distant data centers, the *physical* migration process over the underlying data communication network needs to be protected from disruptive attacks that are prevalent in static networks. For example, an attacker may launch man-in-the-middle attacks on the migration process by logically positioning himself on the migration path using a number of techniques such as ARP spoofing, DNS poisoning, and route hijacking [82, 118]. Another possibility is jamming attacks.

Agile opportunistic migration (AOM) provides the desired protection by randomizing the data forwarding process over the underlying network during migration. By opportunistic, we mean that, instead of setting up a deterministic route from the source replication to the target one (i.e., source routing), the data forwarding decision is made en route. By agile, we mean that the data forwarding process is security-context and replication-status aware. More specifically, based on the instantaneous disruption level on the link, and replication status (whether the replication is active or standing by), priorities will be assigned to candidate next-hop forwarders; those replications with high priorities are favored as the next-hop data forwarders in a probabilistic forwarding process.

4.3.1 Locally Coordinated Opportunistic Data Forwarding

When a link (i, j) is subject to disruption, the probability $p_{i,j}$ that a migration data packet will successfully pass through that link is a number between 0 and 1. Suppose the candidate next-hop forwarders of i are $(i+1, i+2, \ldots, i+k)$. With a probability $p_{i,i+1}$, $i+1$ will receive and hence forward the packet. Otherwise, with a probability $p_{i,i+2} \cdot (1 - p_{i,i+1})$, $i+2$ will receive and hence forward the packet, and so on.

Since more than one candidate next-hop forwarders may receive the data packet, the opportunistic data forwarding process is subject to the following local coordination.

- A node assigns a priority to each candidate next-hop forwarder based on agility.
- The node appends the IDs of the candidate next-hop forwarders to the packet header, and sends the packet out.
- Upon receiving the packet, the candidate next-hop forwarder sets a timer, based on its priority.
- Upon timeout, a candidate next-hop forwarder will forward the packet *if and only if* no other nodes have done so.

A candidate next-hop forwarder that indeed forwards the packet upon timeout will notify all other candidate next-hop forwarders of the forwarding. By the coordination rule, this will reduce the chances of the same packet being transmitted over the network, and hence decreases the chances of the packet being intercepted by attackers.

4.3.2 Adaptive Next-Hop Forwarder Selection

A possible next-hop forwarder selection strategy is to enlist every candidate. A more sophisticated, and potentially more efficient and secure, strategy is to consider:

- Candidates' active/stand-by status at the time of forwarding.
- Current link disruption conditions.
- Progressiveness towards the intended destination.

Only those candidates that will be active at the time of forwarding shall be selected. Whether a candidate will be active at the time of forwarding depends on the chosen PVMDE strategy.

After estimating the candidates' active/stand-by status, a challenge is to choose the number of candidates to forward the data packet, which, ideally, should adapt to the current security context: If the links to the candidates are severely jammed, more candidates should be enlisted to increase the likelihood of successfully forwarding the data packet; otherwise, fewer candidates shall be enlisted to reduce redundancy and local coordination overhead.

"Progressiveness towards the intended destination" means the chosen next-hop forwarders should be topologically closer to the destination. If not so, a data packet may be stuck in a loop due to opportunistic routing. Existing measures of an intermediate forwarder's distance to the destination include hop count and the expected transmission count metric (ETX) [29].

4.3.3 Lottery-Like Priority Assignment

Priorities that are assigned to the candidate next-hop forwarders determine the preference of the candidates. To make the forwarding process less susceptible to disruption by increasing attackers' uncertainty, we can assign priorities by *drawing lotteries*. The winning probability of a candidate i is commensurate with its potential contribution/utility u_i to the migration data forwarding process.

Suppose the set of candidate next-hop forwarders is C. The winning probability of a candidate i is $u_i / \sum_{c \in C} u_c$. After the first candidate is chosen, and if more candidates are needed, the lottery will determine other candidates.

4.4 Dynamic Authentication

Besides disruption, an even more serious attack against replication migration (the *data/control plane*) is unauthorized access (theft) and tampering (poisoning). Theft violates confidentiality and poisoning violates integrity. For example, a confidential document can be stolen if an attacker can read the migrating VM's file system image, or a backdoor can be planted if the attacker can modify the VM's memory image. In addition, if an attacker can inject or tamper with control signals, the whole VM migration [104] process can be hijacked. VM migration is also susceptible to theft at the data plane (e.g., passive snooping) and poisoning at the control plane (e.g., incoming/outgoing migration control attack, false resource advertising) [82].

A comparable case comes from botnet design. Although new botnets, especially those with peer-to-peer (P2P) C&C structures, are more dynamic than their predecessors to circumvent a single point of failure at the central C&C host, many of them are susceptible to poisoning, due to inherent design decisions of the underlying P2P protocols on which they are based. Examples: (1) Some P2P botnets allow bots to access or modify the peer index without any authentication; (2) in some botnets, keying materials used to encrypt data and authenticate peers are stored in plaintext on the bots, and the whole mechanism is compromised if the keying materials are obtained by security professionals from a captured bot sample.

Unlike botnets, in clouds, infrastructures are under the control of administrators. Cloud administrators have some unique advantages over botmasters in defending against theft and poisoning. For example, cloud administrators can implement traditional secret-key-based encryption and authentication mechanisms [17] to protect against outside theft and poisoning threats. However, an insider attack [58] requires more secure key management; mechanisms for doing this are currently being developed by cloud security researchers.

5 Conclusion

Homogeneity and static configurations in cloud infrastructures leave attackers ample opportunities to reconnoiter and penetrate the security perimeter of current cloud services. This chapter has discussed cloud infrastructure security through learning from botnets. Lessons were drawn by studying the evolution of botnets from early static designs to the latest MTD designs that circumvent ever intensifying crackdowns. Challenges to cloud infrastructure security were identified. Illustrative MTD techniques were presented to inspire further research on improving availability, resiliency and data integrity of clouds. Then, secure cloud infrastructures were briefly discussed. Heterogeneous VM replication, proactive VM deployment, agile opportunistic migration, and dynamic authentication technologies will be further developed to shift secure cloud infrastructures, which reduce attackers' understanding of the systems and their ability to launch attacks, while maintaining satisfactory cloud service performance.

Acknowledgements This material is based upon work partially supported by the Northrop Grumman Cybersecurity Research Consortium grant, the Air Force Office of Scientific Research (AFOSR) and the Air Force Research Laboratory (AFRL) Visiting Faculty Research Program (VFRP) extension grant LRIR 11RI01COR.

References

1. Abu Rajab, M., Zarfoss, J., Monrose, F., Terzis, A.: A multifaceted approach to understanding the botnet phenomenon. In: Proceedings of the 6th ACM SIGCOMM Conference on Internet Measurement, IMC'06, New York, pp. 41–52. ACM, New York (2006). doi: 10.1145/1177080.1177086
2. abuse.ch, ZeuS gets more sophisticated using P2P techniques. http://goo.gl/ugThA (2011)
3. Antonakakis, M., Demar, J., Elisan, C., Jerrim, J.: damballa.com, DGAs and cyber-criminals: a case study. http://goo.gl/yDG2C (2012)
4. Antonakakis, M., Perdisci, R., Nadji, Y., Vasiloglou, N., Abu-Nimeh, S., Lee, W., Dagon, D.: From throw-away traffic to bots: detecting the rise of dga-based malware. In: Proceedings of the 21st USENIX Conference on Security Symposium, Security'12, Bellevue, pp. 24–24. USENIX Association, Berkeley (2012)
5. Aviv, A.J., Haeberlen, A.: Challenges in experimenting with botnet detection systems. In: Proceedings of the 4th Conference on Cyber Security Experimentation and Test, CSET'11, San Francisco, pp. 6–6. USENIX Association, Berkeley (2011)

6. Baset, S.A., Schulzrinne, H.: An analysis of the skype peer-to-peer internet telephony protocol. In: Proceedings the 25th IEEE International Conference on Computer Communications, INFOCOM'06, Barcelona, pp. 134–146. IEEE, Washington, DC (2006).doi:10.1109/INFOCOM.2006.312
7. Bauer, L., Garriss, S., Reiter, M.K.: Detecting and resolving policy misconfigurations in access-control systems. ACM Trans. Inf. Syst. Secur. **14**(1), 2:1–2:28 (2011). doi:10.1145/1952982.1952984
8. Bayoglu, B., Sogukpinar, I.: Polymorphic worm detection using token-pair signatures. In: Proceedings of the 4th International Workshop on Security, Privacy and Trust in Pervasive and Ubiquitous Computing, SecPerU'08, Sorrento, pp. 7–12. ACM, New York (2008). doi:10.1145/1387329.1387331
9. Beitollahi, H., Deconinck, G.: Review: analyzing well-known countermeasures against distributed denial of service attacks. Comput. Commun. **35**(11), 1312–1332 (2012). doi:10.1016/j.comcom.2012.04.008
10. Bhattacharya, J., Vashistha, S.: Utility computing-based framework for e-governance. In: Proceedings of the 2nd International Conference on Theory and Practice of Electronic Governance, ICEGOV'08, Cairo, pp. 303–309. ACM, New York (2008). doi:10.1145/1509096.1509160
11. Binsalleeh, H., Ormerod, T., Boukhtouta, A., Sinha, P., Youssef, A., Debbabi, M., Wang, L.: On the analysis of the Zeus botnet crimeware toolkit. In: Proceedings of 8th Annual International Conference on Privacy Security and Trust, PST'10, Ottawa (2010). doi:10.1109/PST.2010.5593240
12. Boshmaf, Y., Muslukhov, I., Beznosov, K., Ripeanu, M.: The socialbot network: when bots socialize for fame and money. In: Proceedings of the 27th Annual Computer Security Applications Conference, ACSAC'11, Orlando, pp. 93–102. ACM, New York (2011). doi:10.1145/2076732.2076746
13. Boyd, S., Keromytis, A.: SQLrand: preventing SQL injection attacks. In: Proceedings of the 2nd Applied Cryptography and Network Security, ACNS'04, Yellow Mountain, pp. 292–302 (2004)
14. businesswire.com, Amazon Web Services launches "Elastic IPs" – static IPs for dynamic cloud computing
15. Caracas, A., Altmann, J.: A pricing information service for grid computing. In: Proceedings of the 8th ACM/IFIP/USENIX International Middleware Conference: 5th International Workshop on Middleware for Grid Computing, MGC'07, Newport Beach, pp. 4:1–4:6. ACM, New York (2007). doi:10.1145/1376849.1376853
16. Cepe, J.: trendmicro.com, The plot thickens for ZeuS-LICAT. http://goo.gl/roa3j (2010)
17. Cheng, Y., Agrawal, D.: An improved key distribution mechanism for large-scale hierarchical wireless sensor networks. Ad Hoc Netw. **5**(1), 35–48 (2007)
18. Choi, H., Lee, H., Lee, H., Kim, H.: Botnet detection by monitoring group activities in DNS traffic. In: Proceedings of the 7th IEEE International Conference on Computer and Information Technology, CIT'07, Fukushima, University of Aizu, pp. 715–720. IEEE Computer Society, Washington, DC (2007)
19. Comazzetto, A.: sophos.com, Botnets: the dark side of cloud computing. http://goo.gl/AOaoB
20. computerweekly.com, Reports of Gumblar's death greatly exaggerated. http://goo.gl/n41HQ (2009)
21. confickerworkinggroup.org, Conficker Working Group: lessons learned. http://goo.gl/bfsPZ (2011)
22. Cova, M., Kruegel, C., Vigna, G.: Detection and analysis of drive-by-download attacks and malicious javascript code. In: Proceedings of the 19th International Conference on World Wide Web, WWW'10, Raleigh, pp. 281–290. ACM, New York (2010). doi:10.1145/1772690.1772720
23. Dagon, D., Zou, C., Lee, W.: Modeling botnet propagation using time zones. In: Proceedings of the 13th Network and Distributed System Security, NDSS'06, San Diego. USENIX Association, Berkeley (2006)

24. Dainotti, A., King, A., Claffy, K., Papale, F., Pescapè, A.: Analysis of a "/0" stealth scan from a botnet. In: Proceedings of the 2012 ACM Conference on Internet Measurement Conference, IMC'12, Boston, pp. 1–14. ACM, New York (2012). doi:10.1145/2398776.2398778
25. damballa.com, ZeuS gets more sophisticated using P2P techniques. http://goo.gl/MseB7 (2011)
26. damballa.com, DGAs in the hands of cyber-criminals. http://goo.gl/MseB7 (2012)
27. Danchev, D.: zdnet.com, Facebook phishing campaign serving Zeus crimeware. http://goo.gl/dn4cb (2010)
28. Davis, C., Fernandez, J., Neville, S., McHugh, J.: Sybil attacks as a mitigation strategy against the storm botnet. In: Proceedings of the 3rd International Conference on Malicious and Unwanted Software, MALWARE'08, Fairfax. IEEE Computer Society, Washington, DC (2008). doi:10.1109/MALWARE.2008.4690855
29. De Couto, D.S.J., Aguayo, D., Bicket, J., Morris, R.: A high-throughput path metric for multi-hop wireless routing. Wirel. Netw. **11**(4), 419–434 (2005). doi:10.1007/s11276-005-1766-z
30. dhs.gov, U.S. Homeland Security Cyber Security R&D Center: Moving Target Defense (MTD) program. http://goo.gl/XuIUx (2012)
31. Dittrich, D., Dietrich, S.: P2P as botnet command and control: a deeper insight. In: Proceedings of the 3rd International Conference On Malicious and Unwanted Software, MALWARE'08, Fairfax, pp. 46–63. IEEE, Piscataway (2008)
32. Domnitser, L., Jaleel, A., Loew, J., Abu-Ghazaleh, N., Ponomarev, D.: Non-monopolizable caches: low-complexity mitigation of cache side channel attacks. ACM Trans. Archit. Code Optim. **8**(4), 35:1–35:21 (2012). doi:10.1145/2086696.2086714
33. Dong, Y., Chen, Y., Pan, Z., Dai, J., Jiang, Y.: ReNIC: architectural extension to SR-IOV I/O virtualization for efficient replication. ACM Trans. Archit. Code Optim. **8**(4), 40:1–40:22 (2012). doi:10.1145/2086696.2086719
34. Falliere, N.: symantec.com, Sality: story of a peer-to-peer viral network. http://goo.gl/kCfm5 (2011)
35. fbi.gov, Operation: bot roast. http://goo.gl/FnHZK (2007)
36. Feily, M., Shahrestani, A., Ramadass, S.: A survey of botnet and botnet detection. In: Proceedings of the 3rd International Conference on Emerging Security Information, Systems and Technologies, SECURWARE'09, Athens, pp. 268–273. IEEE Computer Society, Washington, DC (2009). doi:10.1109/SECURWARE.2009.48
37. Ferguson, R.: trendmicro.eu, The history of the botnet—Part I. http://goo.gl/nfDHl (2010)
38. Francia, R.: blorge.com, Storm worm network shrinks to about one-tenth of its former size. http://goo.gl/Jw8j7 (2007)
39. Gao, H., Hu, J., Wilson, C., Li, Z., Chen, Y., Zhao, B.Y.: Detecting and characterizing social spam campaigns. In: Proceedings of the 10th ACM SIGCOMM Conference on Internet Measurement, IMC'10, Melbourne, pp. 35–47. ACM, New York (2010). doi:10.1145/1879141.1879147
40. Gaudin, S.: informationweek.com, Storm worm botnet attacks anti-spam firms. http://goo.gl/0PtVa (2007)
41. Grizzard, J.B., Sharma, V., Nunnery, C., Kang, B.B., Dagon, D.: Peer-to-peer botnets: overview and case study. In: Proceedings of the 1st Workshop on Hot Topics in Understanding Botnets, HotBots'07, Cambridge, pp. 1–1. USENIX Association, Berkeley (2007)
42. Gu, G., Perdisci, R., Zhang, J., Lee, W.: BotMiner: clustering analysis of network traffic for protocol- and structure-independent botnet detection. In: Proceedings of the 17th Conference on Security Symposium, SS'08, San Jose, pp. 139–154. USENIX Association, Berkeley (2008)
43. Gutmann, P.: The commercial malware industry. In: Proceedings of the 2007 DEFCON Conference, DEFCON'07, Las Vegas (2007)
44. Hachem, N., Mustapha, Y.B., Granadillo, G.G., Debar, H.: Botnets: lifecycle and taxonomy. In: Proceedings of the 2011 Conference on Network and Information Systems Security, SAR-SSI'11, La Rochelle, pp. 1–8. IEEE Computer Society, Washington, DC (2011). doi:10.1109/SAR-SSI.2011.5931395

45. Higgins, K.J.: darkreading.com, New fast-flux botnet unmasked. http://goo.gl/5CpCu (2011)
46. Holz, T., Steiner, M., Dahl, F., Biersack, E., Freiling, F.: Measurements and mitigation of peer-to-peer-based botnets: a case study on storm worm. In: Proceedings of the 1st Usenix Workshop on Large-Scale Exploits and Emergent Threats, LEET'08, San Francisco, pp. 9:1–9:9. USENIX Association, Berkeley (2008)
47. honeynet.org, Honeywall. http://goo.gl/TU4vi
48. Howard, A., Hu, Y.: An approach for detecting malicious keyloggers. In: Proceedings of the 2012 Information Security Curriculum Development Conference, InfoSecCD'12, Kennesaw, pp. 53–56. ACM, New York (2012). doi:10.1145/2390317.2390326
49. Huang, S.Y., Mao, C.H., Lee, H.M.: Fast-flux service network detection based on spatial snapshot mechanism for delay-free detection. In: Proceedings of the 5th ACM Symposium on Information, Computer and Communications Security, ASIACCS'10, Beijing, pp. 101–111. ACM, New York (2010). doi:10.1145/1755688.1755702
50. Huebscher, M.C., McCann, J.A.: A survey of autonomic computing: degrees, models, and applications. ACM Comput. Surv. **40**(3), 7:1–7:28 (2008). doi:10.1145/1380584.1380585
51. hyphenet.com, Fake Verizon Wireless bill notification emails lead to malware. http://goo.gl/PrkaX (2012)
52. Jabrooth, A.U., Parvathavarthini, B.: Polymorphic worms detection using extended PolyTree. In: Proceedings of the 2nd International Conference on Computational Science, Engineering and Information Technology, CCSEIT'12, Coimbatore, pp. 532–538. ACM, New York (2012). doi:10.1145/2393216.2393305
53. Jackson, D.: secureworks.com, Untorpig. http://goo.gl/RCfvl (2008)
54. Jain, P., Sardana, A.: Defending against internet worms using honeyfarm. In: Proceedings of the CUBE International Information Technology Conference, CUBE'12, Pune, pp. 795–800. ACM, New York (2012). doi:10.1145/2381716.2381867
55. Jajodia, S., Ghosh, A.K., Swarup, V., Wang, C., Wang, X.S. (eds.): Moving Target Defense: Creating Asymmetric Uncertainty for Cyber Threats. Advances in Information Security, vol. 54. Springer, New York (2011). doi:10.1007/978-1-4614-0977-9
56. Jajodia, S., Ghosh, A.K., Subrahmanian, V.S., Swarup, V., Wang, C., Wang, X.S. (eds.): Moving Target Defense II: Application of Game Theory and Adversarial Modeling. Advances in Information Security, vol. 100. Springer, New York (2012)
57. Kang, B.B., Chan-Tin, E., Lee, C.P., Tyra, J., Kang, H.J., Nunnery, C., Wadler, Z., Sinclair, G., Hopper, N., Dagon, D., Kim, Y.: Towards complete node enumeration in a peer-to-peer botnet. In: Proceedings of the 4th International Symposium on Information, Computer, and Communications Security, ASIACCS'09, Sydney, pp. 23–34. ACM, New York (2009). doi:10.1145/1533057.1533064
58. Katz, J., Shin, J.S.: Modeling insider attacks on group key-exchange protocols. In: Proceedings of the 12th ACM Conference on Computer and Communications Security, CCS'05, Alexandria, pp. 180–189. ACM, New York (2005). doi:10.1145/1102120.1102146
59. Kephart, J.O.: Autonomic computing: the first decade. In: Proceedings of the 8th ACM International Conference on Autonomic Computing, ICAC'11, Huddersfield, pp. 1–2. ACM, New York (2011). doi:10.1145/1998582.1998584
60. Lee, S., Kim, J.: Fluxing botnet command and control channels with URL shortening services. Comput. Commun. **36**(3), 320–332 (2013). doi:10.1016/j.comcom.2012.10.003
61. Lemos, R.: eweek.com, 'Gameover' financial botnet compromises nearly 700,000 victims. http://goo.gl/izm6t (2012)
62. Li, Z., Mohapatra, P.: QoS-aware multicasting in DiffServ domains. Comput. Commun. Rev. **34**(5), 47–57 (2004). doi:10.1145/1039111.1039112
63. Li, Z., Goyal, A., Chen, Y., Paxson, V.: Automating analysis of large-scale botnet probing events. In: Proceedings of the 4th International Symposium on Information, Computer, and Communications Security, ASIACCS'09, Sydney, pp. 11–22. ACM, New York (2009). doi:10.1145/1533057.1533063

64. Liang, Z., Sekar, R.: Fast and automated generation of attack signatures: a basis for building self-protecting servers. In: Proceedings of the 12th ACM Conference on Computer and Communications Security, CCS'05, Alexandria, pp. 213–222. ACM, New York (2005). doi:10.1145/1102120.1102150
65. Liu, P., Yang, Z., Song, X., Zhou, Y., Chen, H., Zang, B.: Heterogeneous live migration of virtual machines. In: Proceedings of the International Workshop on Virtualization Technology (IWVT), Beijing (2008)
66. Liu, C., Lu, W., Zhang, Z., Liao, P., Cui, X.: A recoverable hybrid C&C botnet. In: Proceedings of the 6th International Conference on Malicious and Unwanted Software, MALWARE'11, Fajardo, pp. 110–118. IEEE Computer Society, Washington, DC (2011). doi:10.1109/MALWARE.2011.6112334
67. Maggio, M., Hoffmann, H., Santambrogio, M.D., Agarwal, A., Leva, A.: Decision making in autonomic computing systems: comparison of approaches and techniques. In: Proceedings of the 8th ACM International Conference on Autonomic Computing, ICAC'11, Karlsruhe, pp. 201–204. ACM, New York (2011). doi:10.1145/1998582.1998629
68. Mather, T., Kumaraswamy, S., Latif, S.: Cloud Security and Privacy: An Enterprise Perspective on Risks and Compliance. O'Reilly Media, Sebastopol (2009)
69. Maymounkov, P., Mazières, D.: Kademlia: a peer-to-peer information system/ based on the xor metric. In: Proceedings of the 1st International Workshop on Peer-to-Peer Systems, Cambridge, pp. 53–65 (2002)
70. mcafee.com, W32/Akbot. http://goo.gl/cbrRC (2006)
71. McCarty, B.: Botnets: big and bigger. IEEE Secur. Privacy **1**(4), 87–90 (2003). doi:10.1109/MSECP.2003.1219079
72. Mendonça, L., Santos, H.: Botnets: a heuristic-based detection framework. In: Proceedings of the Fifth International Conference on Security of Information and Networks, SIN'12, Jaipur, pp. 33–40. ACM, New York (2012). doi:10.1145/2388576.2388580
73. Mercuri, R.T.: Scoping identity theft. Commun. ACM **49**(5), 17–21 (2006). doi:10.1145/1125944.1125961
74. microsoft.com, Microsoft Security Bulletin MS04-011. http://goo.gl/DP4QB (2004)
75. microsoft.com, How Does Botnets Work? http://goo.gl/UYGQ1 (2009)
76. Misra, R., Mandal, C.: Rotation of cds via connected domatic partition in Ad Hoc sensor networks. IEEE Trans. Mob. Comput. **8**(4), 488–499 (2009). doi:10.1109/TMC.2008.128
77. Moscaritolo, A.: scmagazine.com, Zeus spreading through drive-by download. http://goo.gl/KJ4y8 (2009)
78. Mrozek, T.: justice.gov, Wyoming man charged with infecting thousands of computers with 'trojan' that he used to commit fraud. http://goo.gl/G6wtW (2008)
79. Mushtaq, A.: fireeye.com, Killing the beast – part 5. http://goo.gl/mtDH7 (2012)
80. Nazario, J.: arbornetworks.com, Nugache: TCP port 8 bot. http://goo.gl/FqF6D (2006)
81. Nunnery, C., Sinclair, G., Kang, B.B.: Tumbling down the rabbit hole: exploring the idiosyncrasies of botmaster systems in a multi-tier botnet infrastructure. In: Proceedings of the 3rd USENIX Conference on Large-Scale Exploits and Emergent Threats: Botnets, Spyware, Worms, and More, LEET'10, San Jose, pp. 1–1. USENIX Association, Berkeley (2010)
82. Oberheide, J., Cooke, E., Jahanian, F.: Empirical exploitation of live virtual machine migration. In: Proceedings of the 2008 Blackhat Conference, BLACKHAT'08, Las Vegas (2008)
83. Palmieri, F., Fiore, U.: Enhanced security strategies for MPLS signaling. J. Netw. **2**(5), 1–13 (2007). doi:10.4304/jnw.2.5.1-13
84. Pang, W.L., Chieng, D., Ahmad, N.N.: A practical layer 3 admission control and adaptive scheduling (l3-acas) for cots wlans. Wirel. Pers. Commun. **63**(3), 655–674 (2012). doi:10.1007/s11277-010-0157-7
85. Park, Y., Reeves, D.S.: Identification of bot commands by run-time execution monitoring. In: Proceedings of the 2009 Annual Computer Security Applications Conference, ACSAC'09, Honolulu, pp. 321–330. IEEE Computer Society, Washington, DC (2009). doi:10.1109/ACSAC.2009.37

86. Pathak, A., Qian, F., Hu, Y.C., Mao, Z.M., Ranjan, S.: Botnet spam campaigns can be long lasting: evidence, implications, and analysis. In: Proceedings of the 11th International Joint Conference on Measurement and Modeling of Computer Systems, SIGMETRICS'09, Seattle, pp.13–24. ACM, New York (2009). doi:10.1145/1555349.1555352
87. Paul, R.: arstechnica.com, Researchers track Ron Paul spam back to Reactor botnet. http://goo.gl/Qgk5Q (2007)
88. Pitsillidis, A., Kanich, C., Voelker, G.M., Levchenko, K., Savage, S.: Taster's choice: a comparative analysis of spam feeds. In: Proceedings of the 2012 ACM Conference on Internet Measurement Conference, IMC'12, Boston, pp. 427–440. ACM, New York (2012). doi: 10.1145/2398776.2398821
89. Porras, P., Saïdi, H., Yegneswaran, V.: A foray into Conficker's logic and rendezvous points. In: Proceedings of the 2nd USENIX Conference on Large-Scale Exploits and Emergent Threats: Botnets, Spyware, Worms, and More, LEET'09, Boston, pp. 7–7. USENIX Association, Berkeley (2009)
90. Provos, N., Holz, T.: Virtual Honeypots: From Botnet Tracking to Intrusion Detection. Addison-Wesley, Boston (2007)
91. Provos, N., Mavrommatis, P., Rajab, M.A., Monrose, F.: All your iFRAMEs point to us. In: Proceedings of the 17th Conference on Security Symposium, SS'08, San Jose, pp. 1–15. USENIX Association, Berkeley (2008)
92. Provos, N., Rajab, M.A., Mavrommatis, P.: Cybercrime 2.0: when the cloud turns dark. Queue **7**(2), 46–47 (2009). doi:10.1145/1515964.1517412
93. Rajab, M.A., Zarfoss, J., Monrose, F., Terzis, A.: My botnet is bigger than yours (maybe, better than yours): why size estimates remain challenging. In: Proceedings of the 1st Workshop on Hot Topics in Understanding Botnets, HotBots'07, Cambridge, pp. 5–5. USENIX Association, Berkeley (2007)
94. Ramachandran, A., Feamster, N.: Understanding the network-level behavior of spammers. ACM SIGCOMM Comput. Commun. Rev. **36**(4), 291–302 (2006). doi:10.1145/1151659.1159947
95. Rekhter, Y., Karrenberg, D., Groot, G., Moskowitz, B.: ietf.org, RFC 1918: address allocation for private internets. http://goo.gl/qTuQN (1996)
96. Ristenpart, T., Tromer, E., Shacham, H., Savage, S.: Hey, you, get off of my cloud: exploring information leakage in third-party compute clouds. In: Proceedings of the 16th ACM Conference on Computer and Communications Security, CCS'09, Chicago, pp. 199–212. ACM, New York (2009). doi:10.1145/1653662.1653687
97. Rouiller, S.: askapache.com, Virtual LAN security: weaknesses and countermeasures. http://goo.gl/wrCZf (2006)
98. Sanchez, F., Duan, Z.: Region-based BGP announcement filtering for improved BGP security. In: Proceedings of the 5th ACM Symposium on Information, Computer and Communications Security, ASIACCS'10, Beijing, pp. 89–100. ACM, New York (2010). doi:10.1145/1755688.1755701
99. Schneider, D.: Fresh phish. IEEE Spectr. **45**(10), 34–38 (2008). doi:10.1109/MSPEC.2008.4635052
100. securelist.com,TDL4: top bot. http://goo.gl/23BaA (2011)
101. Sheldon, F.T., Vishik, C.: Moving toward trustworthy systems: R&d essentials. Computer **43**(9), 31–40 (2010). doi:10.1109/MC.2010.261
102. Sinclair, G., Nunnery, C., Kang, B.: The Waledac protocol: the how and why. In: Proceedings of the 4th International Conference on Malicious and Unwanted Software, MALWARE'09, Montreal, pp. 69–77. IEEE Computer Society, Washington, DC (2009). doi:10.1109/MALWARE.2009.5403015
103. Song, C., Zhuge, J., Han, X., Ye, Z.: Preventing drive-by download via inter-module communication monitoring. In: Proceedings of the 5th ACM Symposium on Information, Computer and Communications Security, ASIACCS'10, Beijing, pp. 124–134. ACM, New York (2010). doi:10.1145/1755688.1755705

104. Srinivasan, K., Yuuw, S., Adelmeyer, T.J.: Dynamic VM migration: assessing its risks & rewards using a benchmark. ACM SIGSOFT Softw. Eng. Notes **36**(5), 317–322 (2011). doi: 10.1145/1958746.1958791
105. Srivatsa, M., Iyengar, A., Yin, J., Liu, L.: Mitigating application-level denial of service attacks on web servers: a client-transparent approach. ACM Trans. Web **2**(3), 15:1–15:49 (2008). doi: 10.1145/1377488.1377489
106. Stone, B.: nytimes.com, Pakistan cuts access to YouTube worldwide. http://goo.gl/qG0Hn (2008)
107. Stone-Gross, B., Cova, M., Cavallaro, L., Gilbert, B., Szydlowski, M., Kemmerer, R., Kruegel, C., Vigna, G.: Your botnet is my botnet: analysis of a botnet takeover. In: Proceedings of the 16th ACM Conference on Computer and Communications Security, CCS'09, Chicago, pp. 635–647. ACM, New York (2009). doi:10.1145/1653662.1653738
108. Stone-Gross, B., Holz, T., Stringhini, G., Vigna, G.: The underground economy of spam: a botmaster's perspective of coordinating large-scale spam campaigns. In: Proceedings of the 4th USENIX Conference on Large-Scale Exploits and Emergent Threats, LEET'11, Boston, pp. 4–4. USENIX Association, Berkeley (2011)
109. Stover, S., Dittrich, D., Hernandez, J., Dietrich, S.: Analysis of the storm and nugache trojans: P2p is here. Login Issue **32**(6), 18–27 (2007)
110. symantec.com, W32.Stration. http://goo.gl/RZl3e (2007)
111. symantec.com, Trojan.Srizbi. http://goo.gl/nOExB (2007)
112. symantec.com, Gumblar. http://goo.gl/GV3m0 (2009)
113. symantec.com, Backdoor.Tidserv. http://goo.gl/Z4B1Z (2012)
114. Thonnard, O., Dacier, M.: A strategic analysis of spam botnets operations. In: Proceedings of the 8th Annual Collaboration, Electronic Messaging, Anti-Abuse and Spam Conference, CEAS'11, Perth, pp. 162–171. ACM, New York (2011). doi:10.1145/2030376.2030395
115. Tung, L.: zdnet.co.uk, Storm worm: more powerful than Blue Gene. http://goo.gl/4zNr9 (2007)
116. Van Gundy, M., Balzarotti, D., Vigna, G.: Catch me, if you can: evading network signatures with web-based polymorphic worms. In: Proceedings of the 1st USENIX Workshop on Offensive Technologies, WOOT'07, Boston, pp. 7:1–7:9. USENIX Association, Berkeley (2007)
117. Vijayan, J.: computerworld.com, U.K. Web hoster, customers scramble after attack deletes 100,000 sites. http://goo.gl/fMfye (2009)
118. Wählisch, M., Maennel, O., Schmidt, T.C.: Towards detecting bgp route hijacking using the rpki. In: Proceedings of the 2012 ACM SIGCOMM Conference on Applications, Technologies, Architectures, and Protocols for Computer Communication, SIGCOMM'12, Helsinki, pp. 103–104. ACM, New York (2012). doi:10.1145/2342356.2342381
119. Wang, L., Li, Z., Chen, Y., Fu, Z., Li, X.: Thwarting zero-day polymorphic worms with network-level length-based signature generation. IEEE/ACM Trans. Netw. **18**(1), 53–66 (2010). doi:10.1109/TNET.2009.2020431
120. Wang, P., Aslam, B., Zou, C.C.: Peer-to-peer botnets, Chap. 18. In: Stavroulakis, P., Stamp, M. (eds.) Handbook of Information and Communication Security, pp. 335–350. Springer, Heidelberg (2010)
121. Wang, P., Sparks, S., Zou, C.C.: An advanced hybrid peer-to-peer botnet. IEEE Trans. Dependable Secure Comput. **7**(2), 113–127 (2010). doi:10.1109/TDSC.2008.35
122. Xie, Y., Yu, F., Achan, K., Panigrahy, R., Hulten, G., Osipkov, I.: Spamming botnets: signatures and characteristics. SIGCOMM Comput. Commun. Rev. **38**(4), 171–182 (2008). doi:10.1145/1402946.1402979
123. Yadav, S., Reddy, A.K.K., Reddy, A.N., Ranjan, S.: Detecting algorithmically generated malicious domain names. In: Proceedings of the 10th ACM SIGCOMM Conference on Internet Measurement, IMC'10, Melbourne, pp. 48–61. ACM, New York (2010). doi:10.1145/1879141.1879148
124. Yan, G., Chen, G., Eidenbenz, S., Li, N.: Malware propagation in online social networks: nature, dynamics, and defense implications. In: Proceedings of the 6th ACM Sympo-

sium on Information, Computer and Communications Security, ASIACCS'11, Hong Kong, pp. 196–206. ACM, New York (2011). doi:10.1145/1966913.1966939
125. Yang, S., Wu, J.: Efficient broadcasting using network coding and directional antennas in MANETs. IEEE Trans. Parallel Distrib. Syst. **21**(2), 148–161 (2010). doi:10.1109/TPDS.2009.44
126. Ye, K., Jiang, X., Ma, R., Yan, F.: Vc-migration: live migration of virtual clusters in the cloud. In: Proceedings of the 2012 ACM/IEEE 13th International Conference on Grid Computing, GRID'12, Beijing, pp. 209–218. IEEE Computer Society, Washington, DC (2012). doi:10.1109/Grid.2012.27
127. Yu, J., Wang, N., Wang, G., Yu, D.: Review: connected dominating sets in wireless ad hoc and sensor networks – a comprehensive survey. Comput. Commun. **36**(2), 121–134 (2013). doi:10.1016/j.comcom.2012.10.005
128. Zhang, Z., Zhang, Y., Hu, Y.C., Mao, Z.M.: Practical defenses against BGP prefix hijacking. In: Proceedings of the 2007 ACM CoNEXT Conference, CoNEXT'07, New York, pp. 3:1–3:12. ACM, New York (2007). doi:10.1145/1364654.1364658
129. Zhang, L., Yu, S., Wu, D., Watters, P.: A survey on latest botnet attack and defense. In: Proceedings of the 10th IEEE International Conference on Trust, Security and Privacy in Computing and Communications, TRUSTCOM'11, Changsha, pp. 53–60. IEEE Computer Society, Washington, DC (2011). doi:10.1109/TrustCom.2011.11
130. Zhang, Z., Lu, B., Liao, P., Liu, C., Cui, X.: A hierarchical hybrid structure for botnet control and command. In: Proceedings of the 2011 IEEE International Conference on Computer Science and Automation Engineering, CSAE'11, Shanghai, pp. 483–489. IEEE Computer Society Press, Washington, DC (2011). doi:10.1109/CSAE.2011.5953266
131. Zhang, R., Huang, S., Qi, Z., Guan, H.: Static program analysis assisted dynamic taint tracking for software vulnerability discovery. Comput. Math. Appl. **63**(2), 469–480 (2012). doi:10.1016/j.camwa.2011.08.001
132. Zhu, Z., Lu, G., Chen, Y., Fu, Z.J., Roberts, P., Han, K.: Botnet research survey. In: Proceedings of the 32nd Annual IEEE International Computer Software and Applications Conference, COMPSAC'08, Turku, pp. 967–972. IEEE Computer Society, Washington, DC (2008). doi:10.1109/COMPSAC.2008.205
133. Zhuge, J., Holz, T., Han, X., Guo, J., Zou, W.: Characterizing the IRC-based botnet phenomenon. Technical report, Universität Mannheim/Institut für Informatik (2007)

Secure Mobile Cloud Computing and Security Issues

Qijun Gu and Mina Guirguis

Abstract The proliferation of mobile devices, coupled by the increase in their capabilities, have enabled the establishment of a rich mobile computing platform that can be utilized in conjunction with cloud services. In this chapter, we overview the latest mobile computing models and architectures focusing on their security properties. In particular, we study a wide range of threats against the availability, privacy and integrity of mobile cloud computing architectures in which the mobile devices and the cloud jointly perform computation. We then present defense mechanisms that ensure the security of mobile cloud computing architectures and their applications. Throughout the chapter, we identify potential threats as well as possible opportunities for defenses.

1 Introduction

Cloud computing is emerging as a revolutionary technology that is transforming the way we sense, compute and communicate into a new era. Meanwhile, mobile devices (e.g., smartphones) are becoming more advanced and pervasive, providing an information processing platform for mobile users. Due to recent technological advances in hardware, new capabilities in mobile devices have enabled the support of a wide range of applications. Despite such advances, however, mobile devices are still limited when compared to traditional computers and cannot effectively execute compute-intensive applications. As a result, mobile cloud computing emerged as a promising solution that conjugates the advantages of mobile devices with traditional cloud computing. In a mobile cloud computing environment, mobile devices offload intensive computational jobs, that they cannot perform locally, to the cloud.

Q. Gu (✉) • M. Guirguis
Texas State University-San Marcos, San Marcos, TX, USA
e-mail: qijun@txstate.edu; msg@txstate.edu

K.J. Han et al. (eds.), *High Performance Cloud Auditing and Applications*,
DOI 10.1007/978-1-4614-3296-8_3, © Springer Science+Business Media New York 2014

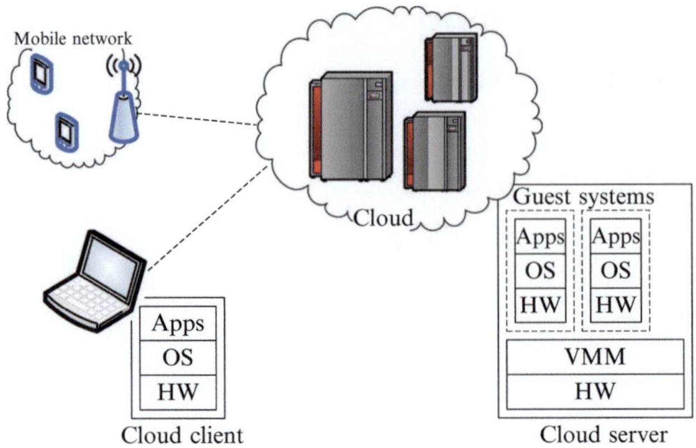

Fig. 1 Mobile computing cloud architecture

The coordination is conducted in a seamless manner so that users can experience and utilize a richer computational platform beyond the capabilities of the mobile devices.

There are many scenarios in which computation can be performed effectively on a mobile cloud computing platform. Consider the following two scenarios. The first one is when intensive computation is required. An example would be running virus scanning and application validation tools, which require significant portions of CPU time and memory on the mobile devices. A solution is to offload the execution of such tools to the cloud and let the devices perform the imminent jobs [25]. In the cloud, multiple virtualized phones can be built to scan applications for viruses and monitor running ones for validation. Another example is image stitching [62], whereby mobile users can take pictures of a scene and upload them to the cloud. Then, the cloud can realign the pictures and stitch them to produce a full-scale, detailed, and up-to-date image of the scene. The second scenario is when no feasible access to traditional cloud services is available. Consider the example outlined in [56], where a child is lost in a parade. Tourists can upload recently taken pictures to computers in a nearby police station. In the police station, image recognition software can run on the pictures to detect the location in which the child was last seen. In this example, tourists do not need to access any online cloud service to help finding the lost child.

A typical cloud system with mobile devices is illustrated in Fig. 1. In this system, mobile devices function as clients in a similar way to other regular computers. They offload computing jobs to cloud servers. In the cloud, the servers build virtualized guest systems that run mobile applications on virtual hardware and virtual operating systems. The guest systems are managed by the virtual machine manager (VMM) that bridges the host system with the guest systems. The VMM virtualizes, partitions, and allocates computing resources of the host system to the guest systems to provide computing services. In this chapter, we will discuss other proposed architectures for mobile cloud computing that support the needs of new applications.

Mobile cloud computing brings in a set of new challenges, especially when it comes to the security of the applicants, the availability of services and the privacy of the users. Due to the inherent complexity in mobile cloud computing architectures, attackers can target and exploit a much wider range of resources/protocols when compared to traditional client-server architectures. For example, on the mobile device, attackers can exploit vulnerabilities in the *mobile applications (Apps)*, the mobile operating systems and access to the mobile network. On the cloud, attackers can exploit vulnerabilities in cloud management, cloud virtualization and cloud access protocols. Recent attacks have shown that personal information can be disclosed, computing tasks can be maliciously altered and cloud services can be disabled for mobile users. Thus, it is critical to examine the security issues involving mobile cloud computing. In this chapter, our main goal is to identify threats against mobile cloud computing architectures and inspect the applicability of recent security solutions to address such threats.

This chapter is organized as follows. We overview a wide spectrum of mobile cloud computing architectures in Section 2. At one end of the spectrum, we have architectures that rely completely on mobile devices performing the computation, while at the other end of the spectrum, we have architectures that offload computation to the cloud. In Section 3, we examine security threats in mobile cloud computing along three security properties: availability, privacy, and integrity. We will examine how and what threats can compromise the security properties in mobile cloud computing architectures. In Section 4, we summarize defense mechanisms that ensure the security of mobile computing architectures against the outlined threats. We will first look into the applicable defenses to various components in the cloud and then summarize the security solutions designed specially for mobile cloud computing. Finally, we will conclude this chapter in Section 5.

2 Overview of Mobile Computing in the Cloud

In this section, we first review three newly proposed and representative cloud architectures for mobile computing. These architectures are designed to utilize the new features of mobile devices and satisfy new needs of mobile applications. Then, we discuss the performance and security issues of mobile computing in the cloud.

2.1 Mobile Client-Server Architectures

Mobile devices, such as smartphones, have enabled rich user experiences with Internet access, Global Positioning System (GPS), sensors and various applications. They naturally became an entry point for an exploding number of mobile users to cloud services to perform computing-intensive applications that are beyond the capabilities of mobile phones. A straightforward idea to support mobile users with cloud computing is to apply the classical client-server model for mobile applications. Within this model, mobile phones function as thin clients in the sense that they are similar to an ordinary client computer that only provides a user interface (UI) to browse a server, and cloud servers run all applications. The mobile phone requests the execution of computational intensive applications from the cloud in a manner similar to requesting resources from web servers. However, this traditional client-server model does not take any advantage of the smartphones and overlooks many of the new and unique features they possess.

One main feature that draws smartphones apart from traditional mobile phones and desktop computers is their context awareness with regards to mobile computing. The context awareness is enabled by rich sensors on the smartphones. In particular, the context includes three components: spatial context, activity context, and group context [10]. The spatial context is harvested by the phone's GPS sensor and location-based service (LBS). It provides current location and past mobility traces to mobile applications. The activity context represents the user's activities, for example, whether he is driving, shopping, or talking. The activity information enables customized mobile applications to provide needed functionality to the user. The group context represents the surrounding mobile devices and users. It is the collection of their spatial context and activity context that enable mobile applications to interact with other nearby mobile devices. Thus, the context provides extra information to service providers so that they can perform computing according to the needs of mobile users given their context.

Another feature that makes mobile cloud computing different from traditional client server computing model is that mobile users dictate the computation on the cloud servers. Even though mobile applications can be executed in cloud servers, it is the mobile devices that decide how to execute mobile applications. When a mobile device runs an application, it identifies the computing-intensive portion of the application, and offloads that portion to the cloud server for processing. The cloud servers are typically considered as secondary processors and storage that provide extra computing resources for performing mobile applications. The coordination of such offloaded computing is mainly controlled by the mobile devices.

To meet the new requirements of mobile computing and exploit the new features of mobile devices, two types of new client-server architectures were proposed for augmenting mobile applications in the cloud. One architecture is based on a whole image clone [12], which is illustrated in Fig. 2a. A clone of a mobile phone is created in the cloud. The state of the phone and the clone is synchronized periodically or on-demand. During execution of an application, if the phone detects a block of computation that needs to be executed in the cloud, the application process on the phone enters a sleep state. The process is then continued in the cloud. When the cloud completes the execution, it updates the state of the phone. Then, the phone resumes the application process.

Rather than cloning the complete image of a phone, another architecture modularizes mobile applications and inserts a middleware layer between the applications and the operating systems of the phone [26]. The architecture is illustrated in Fig. 2b. A mobile application is partitioned into multiple modules according to

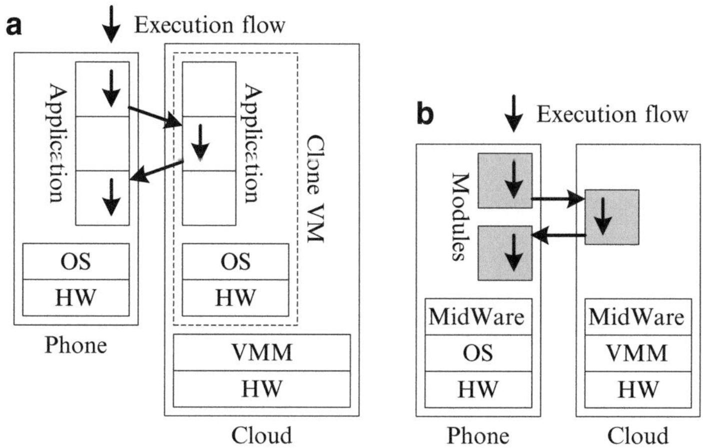

Fig. 2 Mobile cloud client-server architectures. (**a**) Whole image clone. (**b**) Code partitioning

their functionalities. Data and functionality dependencies are added as edges among modules. The modules are then labeled as movable or non-movable based on their resource requests, such as computation needs, storage needs, and amounts of interactions. Because the movable modules can migrate between the mobile devices and the cloud, the mobile applications are called *elastic applications*. Static and dynamic partition algorithms were developed to optimize the performance of offloading [13, 14, 26].

The movable modules are offloaded to the cloud when the mobile devices cannot execute them with their resources. The middleware, namely *elastic manager*, handles the migration of data and code associated with the offloaded modules. The elastic manager can be built on top of a web service [10], and correspondingly the application modules become weblets. The mobile devices request the migration and execution of the weblets through extended web requests. The elasticity manager can also be built as Java virtual machine (JVM) based modular framework [26], and the application modules are objects that can be loaded and executed as a Java class in this framework.

2.2 Cloudlet Architectures

Although the client-server cloud architecture can support the needs of mobile computing, it is affected by the inherent limitation of the access links of mobile devices. The fundamental limitation is the latency in the network. In the cloud, mobile devices interact with cloud servers over wireless wide area network (WWAN) and the Internet. Both networks incur non-negligible round-trip delay and jitter for interactive mobile applications. Such delay and jitter largely result in negative impact on the user's perception and cognitive engagement. Another latency

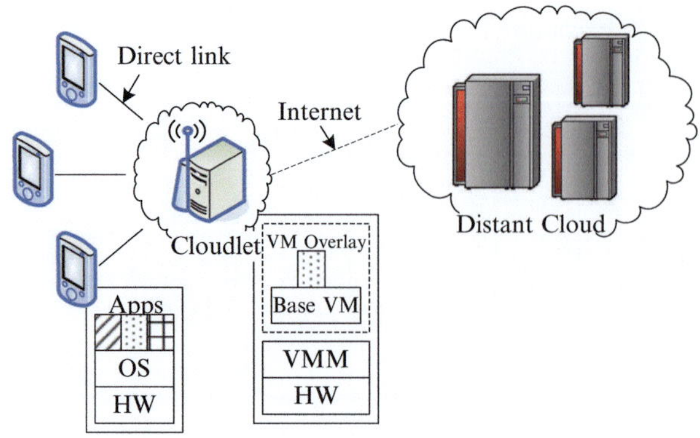

Fig. 3 Cloudlet architecture

is incurred in bulk transfers of data and code over bandwidth-limited WWAN. Even though wireless broadband technologies have been improving greatly, the bandwidth of WWAN is still significantly smaller than the bandwidth of wireless local area network (WLAN). The transfer time of bulk data is still far from satisfactory in WWAN.

To overcome the latency limitation with powerful but distant cloud servers, a new cloudlet architecture is proposed in [57], in which a resource-rich cloudlet is deployed close to the mobile users. The architecture is illustrated in Fig. 3.

A cloudlet is composed of powerful computers or a cluster that has sufficient computing capability and power. It provides cloud service to the mobile devices over a one-hop high-bandwidth wireless access such as WiFi. The advantages of using cloudlets arise from the real-time interactive response due to short and bounded-latency between the mobile devices and the cloudlets. A cloudlet can be integrated with a wireless access point (AP) for ease of management.

Utilizing cloudlets requires new mechanisms rather than simply treating cloudlets as traditional cloud servers so that the short latency of the cloudlets can be fully exploited. The authors in [57] proposed a transient cloudlet solution, namely dynamic Virtual Machines (VM) synthesis. When a mobile device needs a cloudlet to execute an application, the mobile device produces a *VM overlay* for the application. The VM overlay contains the soft state of the execution of the application. Then, the device delivers the VM overlay to the cloudlet. The cloudlet is pre-installed with a base VM that has the minimum system core components but can support running most applications. The cloudlet merges the overlay with the base and then executes the application from the state stored in the VM overlay. Once the job is done, the cloudlet returns the results to the mobile device and discards the VM overlay.

To verify the feasibility and performance of the cloudlet architecture, the proof-of-concept prototype Kimberley was implemented in VirtualBox. The prototype created VM overlays for a collection of Linux applications, such as AbiWord, GIMP and PathFind. The size of the compressed VM overlays ranges from 100 to 200 MBs in comparison with the full Linux VM image size of several gigabytes. Furthermore, the experiments with the prototype showed that the processing time of a VM overlay mainly includes the time of applying VM overlay, decompressing VM overlay, transferring VM overlay, and compressing private data. The reduction of the total time will lead to a more practical cloudlet architecture.

2.3 Ad Hoc Mobile Cloud Architectures

Although the first two cloud architectures for mobile computing are promising in managing and executing applications for mobile devices, they need an infrastructure to host the cloud service and to provide access to the mobile devices. It is not unusual that the access to such infrastructure may be unavailable to mobile users under various circumstances. For example, during a natural disaster (e.g., hurricane) that destroys part of the infrastructure, mobile devices may hardly access any services other than detecting themselves over Bluetooth and WiFi in an ad hoc mode. It may also be too expensive to access cloud service through the infrastructure, especially in areas where only high cost wireless data links (e.g., cellular link and satellite link) are available. Considering that the volume of bulk data and code to be offloaded is much higher than the amount of voice data, the cost of performing cloud computing via these high cost wireless links may largely negate the advantages obtained from the cloud services.

A new ad hoc mobile cloud solution was proposed in [31] to address the problems raised when no access to the cloud services is available. The solution is specially suitable to the needs of mobile users who are in close locations and share common activities. For example, when a tourist is visiting museums, he may find interests in the description of exhibits. He can take pictures of the text and run text-recognition software for recognition so that he can store the text on his mobile phone. However, processing the whole text requires more computing resources than his phone capabilities. He then looks for help from nearby tourists. Because other tourists may be interested in the description as well, they form an ad hoc network using the low cost WiFi communication and perform the text-recognition. The recognized text is then stored on their phones.

The adhoc mobile cloud architecture is illustrated in Fig. 4 to support this kind of location-bounded group activities [31]. The architecture consists of four main components for managing resources, applications, contexts, and offloading to the cloud. The *resource manager* profiles the needs of the applications and monitors the available resource. These profiles along with current resource information are checked upon the execution of an application to decide if the support from the cloud is needed. The *application manager* handles loading and executing applications. If the cloud is needed for launching an application, the manager adds extra information to the application for offloading. The *context manager* monitors the location and the number of mobile devices in the vicinity. This context information provides mobility traces of devices and builds social relations among them. The *offloading manager*

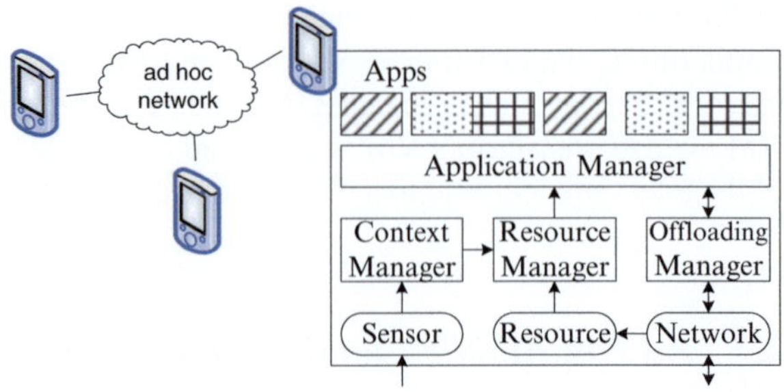

Fig. 4 Ad hoc mobile cloud architecture

partitions, sends and manages jobs from the mobile devices to other devices. It also processes and manages jobs received from other devices.

A concern of this ad hoc mobile cloud architecture is that the overall performance of the offloaded applications will be worse than running them on a single device because of the offloading overhead. To address this concern, a proof-of-concept implementation was built in JamVM and tested with Hadoop. The results showed that the execution with offloading was only 1% slower than regular execution. The execution time was further split to two parts: 44% for offloading preparation and waiting and 56% for processing and synthesizing.

2.4 *Performance and Security Issues*

In the following part, we summarize the performance and security issues in recent research in mobile cloud computing and discuss their interrelation.

2.4.1 Performance

The performance of mobile computing in the cloud is often considered along two dimensions: (1) the resources needed to support mobile computing in the cloud and (2) the overhead incurred for processing mobile applications, in both the mobile devices and the cloud.

Because mobile applications are offloaded to the cloud, the cloud needs to allocate resources to accommodate the offloaded portions of the mobile applications. The resources in the cloud are usually measured as (1) the size of memory where code and data are stored, (2) the size of the offloaded portions of the mobile application that are transferred, (3) the bandwidth needed for transferring offloaded applications, (4) the number of CPU cores or time slices for computing, among others.

The process of offloading and execution of mobile applications in the cloud is carried out through several procedures, such as partitioning, migration and execution. The overhead is often measured as processing times, which include

(1) the time for profiling and detecting the resources needed for running the applications in the mobile devices, (2) the time spent in partitioning and generating the offloaded portions of applications, (3) the time spent transferring the offloaded portions and the results between the mobile devices and the cloud, (4) the time spent merging and/or installing the offloaded portions into the guest systems in the cloud, (5) the time spent executing the offloaded portions in the cloud, (6) the interactive latency and jitter between the mobile devices and the cloud, and so on. Another unique measurement of the overhead in mobile computing is the power consumption on the mobile devices. It is as important as other performance factors in determining the feasibility of mobile computing in the cloud.

2.4.2 Security

The security of mobile cloud computing [2, 3, 7, 19, 45] covers a vast range of aspects and affects all components in the cloud. The security issues include the confidentiality and privacy of users' personal data, the integrity of the computation, the availability of the cloud services, the risk analysis of cloud configurations, the auditing of computation and information management, the accounting of the allocations of the resources, the secure management of the guest systems, the authentication of the users and devices, and key management with running applications, just to name a few.

Among all these issues, we will focus on the integrity, confidentiality and availability issues that are closely related to computing. The integrity issue arises since the mobile devices are less trustable and the mobile applications are offloaded to the cloud. The computational results from untrusted devices and offloaded applications need to be authenticated and verified. The confidentiality issue is due to sharing the code and data in cloud settings. Critical information, such as personal data, in mobile devices may be accessible from the cloud as the information is also backed up in the cloud. The availability issue is inherent to mobile computing due to the limited resources (e.g., cellular access links). In the cloud, attackers can also find ways to exhaust resources that should be used for guest systems.

2.4.3 Impact of Security on Performance

There is always a tradeoff between security and performance in computer systems and services. In mobile cloud computing, security measures typically require extra protection procedures to encrypt offloaded mobile applications and data, to authenticate mobile devices and users, and to maintain trustable relations between mobile devices and cloud through exchanging and updating credentials. Hence, security requests more resources for keeping credentials and performing security functions, and thus adds additional overhead in terms of processing time and power consumption.

3 Threats Against Mobile Cloud Computing

Mobile cloud computing architectures are vulnerable to a different and wider range of attacks than traditional computing systems [4]. This is primarily due to the following reasons:

1. *Limited resources*: Despite major technological advances, mobile devices are still constrained by computational, networking, storage and power capabilities in comparisons to regular computers. This prevents these devices from running the necessary security applications and leaves them vulnerable to attacks. For example, monitoring tools such as packet sniffing tools and intrusion detection systems may consume extensive resources that would overwhelm the whole system and shorten its lifetime. Thus an inherent tradeoff between security and performance is present.
2. *Multiple locations for storage and computation*: Data from mobile devices are often uploaded and stored in the cloud. Also, extensive computation is offloaded and executed in the cloud. This makes the data and code vulnerable to attacks at multiple locations (e.g., the mobile device, the cloudlet and the cloud). Moreover, the links utilized in transferring the data and code could be targeted by attackers. In essence the attacker can target the weakest resource in the chain to access confidential information or compromise the integrity of the computation. The involvement of third-parties complicates these issues further.
3. *Mobility as a Service*: In mobile cloud computing, users have the capability to move their computation between different public clouds in addition to some cloud providers that offer mobility as a service for its users [1]. Malicious users can infer cloud specifics (e.g., cloud topology, the bottleneck links and the load on the machines) and launch sophisticated attacks that target specific resources (e.g., bottleneck links) [41]. Thus mobility in cloud architectures brings new sets of threats to the table.

In this section, we expose threats against mobile cloud computing architectures focusing on attacks that target confidentiality, integrity and availability.

3.1 Mobile Devices

When it comes to privacy, mobile devices carry valuable personal information (e.g., photos, contact lists and communication history) as well as critical application data (e.g., health records and credit card information). When a mobile device gets in the wrong hands (even for a short duration), attackers can perform forensics analysis and extract such information. There are many legitimate tools for forensics examiners that are publicly available and can be used for malicious intent [39, 40]. Another example is the Joint Test Access Group (JTAG) attacks [54] in which attackers can access testing and debugging functionalities on the device to extract confidential information.

Perhaps one of the major incidents concerning user's privacy was that with Carrier IQ in which it was reported that Carrier IQ software captured every keystroke, location and other data on a wide range of smart phones. Collected information was transmitted and made available to Carrier IQs customers [21].

In recent years, malware has proliferated that threatens the privacy of the users through exploiting the operating system. In [48], it was demonstrated that through a drive-by download, an attacker can potentially steal the short message service (SMS) database from an iPhone 3GS. On the Android platform, GingerMaster is malware that is packaged into legitimate applications. The malware runs in the background collecting information and sending it to a remote server [33]. RootMaster is another type of malware that would gain root privilege on an Android device, connect to a command and control server and even install additional malware [34].

In general, if an attacker gains root permissions on a mobile device through malware, one could impact the computational integrity of the operating systems as well as of other applications (e.g., provide incorrect sensing information or results). To the best of our knowledge, we are not aware of specific threats that achieved this. However, it suffices to mention that the above malware that can control a mobile device can be orchestrated to compromise other applications and impact the integrity of the computation performed.

When it comes to availability, wireless communication on mobile devices is inherently vulnerable to interference and jamming attacks [47, 68]. Attackers can acquire hardware jammers to prevent the device from receiving the intended signals. Since many mobile devices rely on a wireless interface for communication, such interface may be subjected to a wide range of attacks on the Media Access Control (MAC) layer [37, 55, 63, 67] as well as traditional DoS attacks in which attackers send a large amount of traffic that would saturate the wireless link, preventing the transmission of other information on that interface. Moreover, if it is known that a mobile device is implementing a particular packet sniffing application or an intrusion detection system, an attacker may subject the device to a malicious traffic stream that triggers specific rules to be applied which, in turn, overwhelms the resources (e.g., rules that try to match strings). Another form of attack that targets the availability is through malware that executes junk instructions just to drain the power of the mobile device.

3.2 Cloudlets

When a mobile device uploads data/code on the cloud/cloudlet [64] for storage/processing, it becomes vulnerable to privacy issues and computational integrity attacks. For example, many architectures rely on creating an image clone of the mobile device on the cloud [12]. Execution can then occur on the device or in the virtualized cloud environment. According to recent report [52], mobile and cloud attacks will be dramatically increased because users and attackers shift to mobile and cloud in 2013. Georgia Tech researchers [38] also predicted that cloud botnets and mobile attacks will be the biggest cloud issues in 2013.

The integrity of the computation is directly affected by the technologies adopted in the cloud. Virtualization has been a key player in the development of cloud computing architectures. Users upload their images as virtual machines to be executed in the cloud. To save resources, multiple virtual machines may share the same server. This causes a security mismatch in the sense that users would like to keep their data and code private while the cloud needs to ensure the security and safety of the code being executed on their physical infrastructure [11]. Dynamic Data Kernel Rootkit attacks may target the VMM and control the execution of the host operating system, thus affecting all other virtual machines running on top [20, 61]. Moreover, it has been shown that certain types of malware can, in fact, detect that they are being executed in a virtual environment and would change their behavior accordingly [50].

Beyond the traditional availability attacks that can occur on access links to the cloud, new forms of threats started to surface. In [41], the authors describe an attack scheme in which attackers can carry out DoS attacks through targeting specific links within the cloud architecture and saturating them. The attack scheme starts by inferring the network topology within the cloud and then choosing specific hosts that share common bottleneck links and sends large amounts of traffic between them. The authors demonstrated this attack scheme on one cloud infrastructure. Moreover, the ability of the users to move computation between clouds/cloudlets seems to open backdoors for attacks to be mounted.

3.3 Ad Hoc Mobile Clouds

In this section, we study threats against code distribution mechanisms that enable multiple parties to execute code segments for specific users. These parties could be other mobile users (ad hoc), cloudlet components, or a hybrid.

Many applications require the help of nearby devices in executing code segments. Such applications arise in scenarios in which executing the whole application on a single device is not feasible due to various constraints such as: (1) limited battery power constraints, since a single device may not have enough battery life to execute the whole application to completion; (2) limited infrastructure access, since a single device may not have access to the infrastructure (e.g., dead zones, remote areas) or there is a high cost incurred (e.g., exceeding the cellular data allowance); (3) different working data sets [56], since nearby mobile devices may operate on local data stored on them (e.g., photos captured within a specific time frame) or acquire data from the environment around them (e.g., collecting sensor information).

Due to the reliance on a distributed manner in executing code segments, the existence of malicious nodes should be considered. Malicious nodes can also collude to reverse-engineer code segments in order to reveal the functionality and capability of the code. If the malicious nodes can share the code and the data they are operating on, the privacy of the requesting user is violated. Although in [42], the security threats of colluding applications have been studied, to the best of our knowledge, we have not yet seen colluding threats that target code partitioning. As for the integrity of the executing segments, when users ask nearby devices to execute code segments on their behalf, they run the risk of users not responding or responding with incorrect results. The requesting device cannot easily verify the correctness of the results, unless by adding a layer of redundancy (multiple devices are asked to execute the same code segments and the results are compared). Unfortunately, this adds a lot of overhead and the process of managing the segments may be itself subjected to a new set of clever attacks. Jamming attacks or Denial of service (DoS) attacks on the links between the mobile nodes can impact the code distribution modules. In particular, a subset of the mobile devices may not receive their assigned code segments. This prevents the requesting users from receiving the results and may trigger additional work to redistribute the segments that have not been executed, leading to longer latencies. Another threat that targets the availability of the systems is for malicious users to distribute code segments that causes mobile devices to drain their power (e.g., code that requires a lot of CPU computation and communication).

4 Mobile Cloud Computing Security Measures

In this section, we overview recently proposed security solutions for mobile computing in the cloud. The summary will be focused on the security aspects closely related to computing. Security works, such as vulnerability assessment [36], communication security [4], interface and session security [59], verifiable accounting [58], and so on, are not included in this section.

4.1 Mobile Devices

As discussed in [4], mobile devices are threatened by various attacks targeting their hardware, operating systems, applications software, communication links, back-end systems, and users. Although all these threats concern the security of mobile devices, this section will discuss computing-related security measures that ensure confidentiality and integrity of mobile computing in the cloud. In particular, the security of the device storage and operating system have direct impacts on mobile computing in the cloud and are the trust base of newly proposed security measures.

4.1.1 Mobile Device Storage Security

The storage of a mobile device is usually made of an on-board memory and a plugged-in memory. The on-board memory stores the code and data of the running software. The plugged-in memory includes non-volatile flash storage and subscriber identity module (SIM) card storage. The stored security-critical data includes, but is not limited to, personal information in the flash, device credential in the SIM card, and runtime keys in the on-board memory. In cloud computing, security-critical code and data could be mirrored in the cloud as well. Because mobile devices can be lost or stolen, they become a weak point in the complete system of mobile cloud computing. Attackers can obtain access to the internals of stolen devices without the need to hack into the cloud. Hence, securing the device storage is a must for protecting data confidentiality and personal privacy.

Attackers can obtain access to the code and data stored in on-board memory via JTAG functionalities [32] or through directly tapping into the circuit board [16]. The security measures in protecting on-board memory include disabling JTAG functions and implementing secure storage and secure access to user I/O facilities following the recommendation of Open Mobile Terminal Platform (OMTP) Advanced Trusted Environment [44].

Attackers can obtain access to the code and data stored in plugged-in memory via forensic analysis [65]. The protection measures rely on the encryption of either the data or the memory. The security of encryption keys needs Trusted Platform Module (TPM), which is either a stand-alone chip in the device or a part of the functionality of the SIM card.

4.1.2 Operating System Security

One major advancement in mobile phone technology is the phone's mobile operating system (MOS) that fully utilizes the advancement of the phone's hardware and enables a large variety of mobile applications. Currently, the MOS functionalities are quite close to regular computer operating system (OS). Securing MOS is the fundamental measure to ensure the confidentiality and the integrity of running mobile applications in the cloud.

The security of MOSs can be achieved with kernel hardening, security control over applications, and secure updating and installation. Because MOSs are derived from regular OSs, their kernels are hardened with similar security techniques as regular OSs, including address space layout randomization and non-executable data memory. Mandatory Access Control (MAC) can also be implemented in MOSs to enhance the overall security.

Learned from hard lessons in regular computer, MOSs should be designed with security control over applications based on the principles of lease privileges and separation of duty. For example, each application should run with only the needed privileges for completing a job. Although many applications request more privileges than needed during installation in the Android, users can install additional tool-kits to disable unwanted privileges for each application. In Android, each application runs in a process isolated from other processes through virtualization of the phone so that a compromising process will not endanger other processes or the operating system.

Most modern MOSs support updating the system remotely to update or patch drivers, the OS modules and the installed applications. Remote updating in MOSs can be protected in the same fashion as with regular OSs [5]. Installing applications, however, is a more complicated process. In Android, a list of privileges come with each application to be installed. Accepting the list is the only way to install the application. However, the privileges in the list usually give the application access to personal data without any additional mechanisms to actually inspect how the application uses the data. A category of security measures have been proposed to disable unwanted privileges [17] or assign only needed permissions at run time [46]. Another category of security measures is to profile and track the information flow of applications and discover the critical points in time when applications access sensitive information or are about to expose such information [18, 28].

4.2 *Cloud Servers*

As discussed in Section 3, the cloud can be threatened by many different attacking techniques that target different components of the cloud. Various security solutions have been proposed in the past to address these threats in the cloud. In this section, we discuss the security measures that are closely related to the security of cloud computing and that are applicable to mobile computing in the cloud as well. Because the traditional cloud is built upon the client-server model, the security measures in this section focus on the cloud side. In particular, we discuss the defense mechanisms to ensure confidentiality, integrity and availability of computing in the cloud servers.

As illustrated in Fig. 1, the system components in a cloud server include the guest systems, the VMM and the host system. Accordingly, we classify the security measures into three categories. The first category is on the security of computation in a guest system. The second category focuses on how a VMM manages the guest systems. The third category is on the security of user-controlled operations that may affect the security of the cloud.

4.2.1 Guest Systems

As more and more services are transferred to the cloud, each user's personal information is transferred to the cloud as well. When users execute their applications in guest systems with their personal information, the privacy and integrity of the computation inside the cloud are concerns that must be addressed to convince the users to use the cloud services.

A straightforward approach is access control that enforces security policies so that only authorized entities and processes can access data and applications. However, if the data and the applications are stored in the clear, administrators of the cloud can access them or even change the security policies [45]. Therefore, cryptographic solutions are desired so that data and applications are encrypted in the cloud. One promising cryptographic solution is the fully homomorphic encryption [23,24,43], where data is not only encrypted in storage but also in computation. With fully homomorphic encryption, computation is performed directly over encrypted data, and the encrypted results are returned to the users. Then, the users decrypt the results to obtain the actual results as if the computation was performed over clear data. Hence, the encryption ensures that no one in the cloud could peak at the users' personal data. The hurdle of fully homomorphic encryption solutions in practice is the computational inefficiency in that it is much slower than classical encryptions and regular data operations. Another issue is that the encryption only ensures confidentiality and privacy but not the integrity of computation. It is hard for the cloud to prove that the result was not manipulated during the computation based on a fully homomorphic encryption.

In another work [15], the authors propose to use some well-established cryptographic solutions to achieve the desired security of computation but on the client side. Personal data is encrypted as cryptographic commitments, which not only hide the data but also can be used later to verify the data in the cloud. Nevertheless, the computation is not performed in the cloud but rather on the client. When the cloud needs to process user's data, it sends the commitments to the user and requests the user to perform some computation. Then, the user sends the results together with non-interactive zero-knowledge proofs back. The cloud verifies the results with the commitments and the zero-knowledge proofs to validate the integrity of the computation. This solution is an interim solution to the needs of real-world scenarios before the fully homomorphic encryption can well satisfy the security needs.

4.2.2 Virtual Machine Managers (VMM)

The VMM is the component that manages multiple guest systems in a cloud server. The security of VMM usually refers to two aspects. One is the security among multiple guest systems, i.e., a compromised guest system shall not endanger other guest systems. The other is the security of the guest systems themselves, i.e., a guest system shall perform its own computation as expected.

To achieve the first security goal, isolation is the major mechanism, by which VMM places boundaries between computational resources of different guest systems. The boundary could be physical, i.e., each guest system is hosted on different machines that are physically locked in different locations. In practice, isolation is virtualized so that cloud providers can flexibly allocate fine-grained and requested computational resources to the guest systems. The problem of virtualized isolation is that a guest system may interfere with another guest system. Although isolation is in place, multiple guest systems in the same computer are still sharing computational resources, such as CPU cores, caches, memory, and I/O channels. Among the resources, cache is the most complicated factor to isolate, because it is inside the CPU core and often not directly controlled by QoS-provision VMM. A cache-aware isolation solution was proposed in [51] to address the sharing issue of the last level cache (LLC). The solution places the cores that share the same LLC into one core group and then assigns each core group to at most one guest system. Thereby, a guest system will not compete for caches with other guest systems. An obvious drawback of this solution is that the granularity of the computational resource is at the level of the core group. If a resource of a core group is under utilized with a guest system, this resource cannot be assigned to other guest systems.

The second security goal requires integrity checks on the running guest systems. Because applications in the guest systems may access the Internet and thus experience attacks as do regular computer systems, it is important to ensure that the guest systems are not compromised during execution. Since the cloud is built using virtualization technologies, it is fairly convenient to implement the integrity check as a part of the functionality of the VMM. Various integrity check solutions have been proposed. One type of integrity check solution is virtual machine introspection [11, 22], which is an intrusion detection module in the VMM that monitors the state of the processors, memory and disks in the guest systems to detect intrusive actions. A second type of integrity check solution is to check the execution of the guest system's code. Before executing the code, it is stored in a protected memory and fetched from that memory [53], or the control flow of the code is verified first [49]. These solutions need the behavior of the guest systems, which must be stored in the VMM so that it can compare the running guest system with the expected behavior.

4.2.3 User-Controlled Security Measures

Mobile users are involved in cloud computing in two kinds of operations: VM image dissemination, and end-user configuration. Because users create, upload and retrieve VM images, their dissemination operations may disclose their own personal information. They may also retrieve malicious images from the cloud. Note that the role of the cloud provider is just to provide images to clients with the authorization of the owners of the images. Hence, securely managing images is an important task in the cloud. Mirage [66] is a VM image management system, and consists of four components. The first component is an access control framework that regulates the sharing of VM images under the discretion of the image's owner. The second component consists of image filters that are applied to remove or hide sensitive information when sharing images. The third component is a provenance tracking mechanism that tracks the derivation history of an image when a user revises the image and generates a new one. The last component is a set of repository maintenance services that detect and fix vulnerabilities in images.

In the cloud, users are allowed to group machines and enact rules to control communication among groups. In this manner, users can set up groups with different security levels and provide customized services in the cloud. However, such end-user configuration may put users' information at risk, considering the complexity of the cloud infrastructure and services. Therefore, it is necessary to assess the security of the end-user configuration to protect users' information. In [6], a configuration audit approach was proposed for multi-tier cloud infrastructure. The approach models the security policy rules as an information flow graph, in which each vertex represents a repository of information and each edge represents the information flow. Using the graph, the approach conducts reachability analysis according to the security policy rules. A violation is detected when an information flow path exists in the graph but is not specified by the policy rules.

4.3 Mobile Clouds

Security measures in traditional cloud settings can be applied to address a significant portion of the security needs for mobile cloud computing, because the design and management of cloud servers are usually not affected by the type of the clients. Nevertheless, as discussed in Section 2, new architectures and computing models have been proposed for mobile computing in the cloud, which raise new and unique security needs of mobile computing. In particular, new security measures were developed in the following aspects of mobile computing: (1) secure code partitioning and offloading, (2) mobile user authentication, and (3) secure mobile data management. The security of these three aspects not only provide the integrity and confidentiality to mobile computing in the cloud, but also take advantage of the features present in mobile devices.

4.3.1 Secure Code Partitioning and Offloading

Code partitioning and offloading were common functionalities in all of the three mobile cloud computing architectures outlined above. Hence, securing them is an indispensable aspect of securing elastic applications in mobile cloud computing. A secure elastic framework [27, 69] was proposed to address this issue. The framework consists of four components: secure installation, module authentication, secure migration, and permission authorization. Although the framework is designed to weblet-based elastic applications, the principal ideas of its security components can be applied to other modular elastic applications.

- The *secure installation component* ensures that genuine applications are installed on the mobile devices and that applications have signed hashes. During installation, the hashes of applications are checked to ensure the integrity of applications. After successful verification, the applications are registered with elastic services.
- The *module authentication component* enables one module of an application to authenticate another module that belongs to the same application. Because modules of the same application may be migrated and executed at different

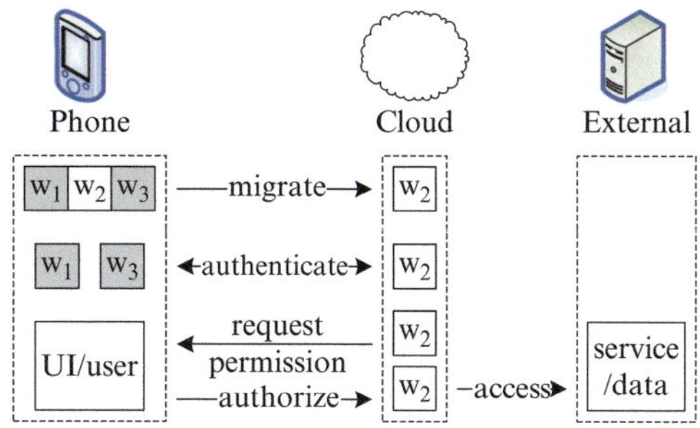

Fig. 5 Secure execution of an elastic application

locations, dependent modules need to be authenticated in the sequence of execution to ensure that the genuine modules of the same applications are executed regardless of the locations of execution. Upon executing the modules, the manager of elastic applications generates a pair of session key and session secret associated with each module. The pair of session key and secret are then used for authentication and secure communication among modules.

- The *secure migration component* ensures the security of the migration process. When a module needs to be migrated, the module saves its state and enters the migration state. During migration, the pair of key and secret associated with the module are delivered with the module to the destination for authentication. After verification, the migrated module resumes its execution from the saved state.
- The *permission authorization component* assigns different permissions to the modules of an application. Because the functionality and the location of modules are different within the same application, they should be authorized with different permissions based on their tasks. One proposed approach is that modules in the cloud, if requesting access to sensitive data, should forward such requests to the mobile device to obtain credentials from the device, and then use the credentials to authenticate and obtain authorization. Another approach is that the modules in the cloud initiate an authentication request to access sensitive data, but ask the mobile device to complete the authentication process and forward the resulting authenticated session to the modules in the cloud.

An example of executing an elastic application with the four components is illustrated in Fig. 5, where the application has three weblets w_1, w_2 and w_3. Upon execution, w_2 is migrated to the cloud for execution by the migration component. During execution, w_2 authenticates with w_1 and w_3 using the authentication component. At the time when w_2 needs to access an external server, w_2 requests authorization through the authorization component. The user may be prompted to authorize the access through the phone's user interface (UI). If granted, w_2 proceeds to access the external server.

The secure elastic framework emphasizes on the management, migration, and the execution of the offloaded modules of an application. The main security goal is the integrity of the elastic application when its modules are executed at different locations. However, the framework is weak when it comes to ensuring the security of

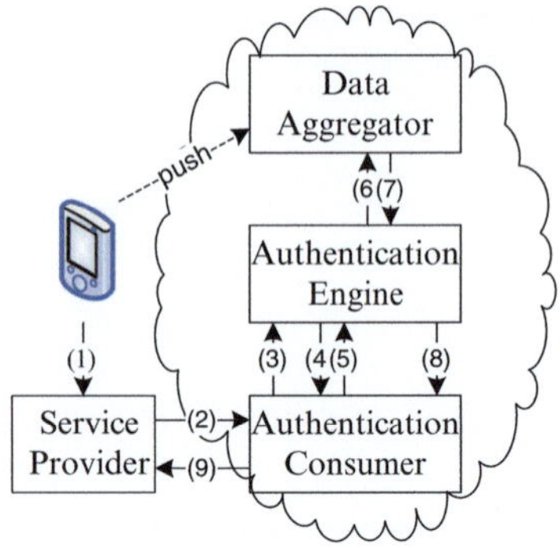

Fig. 6 Mobile client authentication

the code partitions themselves. The current focus of code partition is on optimizing the overhead of migration and execution with partitioned code modules. Security is not considered in the model of code partitioning so far. In [8], the confidentiality and integrity of the partitioned programs are studied by using security labels annotated in programs. Privacy of sensitive data in cloud computing is also studied in [35, 70]. However, research on similar security issues for elastic mobile applications in the cloud is lacking.

4.3.2 Mobile User Authentication

Mobile cloud computing requires secure authentication methods to prevent data theft. Typical mobile authentication approaches rely on identity information of ownership factors (e.g., mobile devices and security tokens), knowledge factors (e.g., password, pass phrase, or personal identification number (PIN), and challenge questions and answers), and inherence factors (e.g., fingerprint and other biometric identifiers). Some unique and critical authentication information could be stored on mobile devices. However, the identity information is insecure because mobile devices could be stolen, and/or attackers could steal identity information. To solve this problem, Song et al. [9] proposed a new authentication principle in the TrustCube infrastructure that is an end-to-end infrastructure to control the access of authenticated clients by using the history of authenticated actions and behaviors of mobile clients (users). Their approach is built on the TrustCube platform [60], which provides client authentication services in clouds with integration of various authentication methods. The authentication framework for mobile clients [9], as illustrated in Fig. 1 in [9], consists of three components in the cloud: *Data Aggregator*, *Authentication Engine*, and *Authentication Consumer*. The *Data Aggregator* collects past actions of the mobile devices. The *Authentication Engine* obtains data from the *Data Aggregator* and makes authentication decisions. The *Authentication Consumer* provides policies to the *Authentication Engine*.

We illustrate the mobile authentication procedure in Fig. 6. The client device regularly pushes the mobility traces to the *Data Aggregator* to reduce risk of

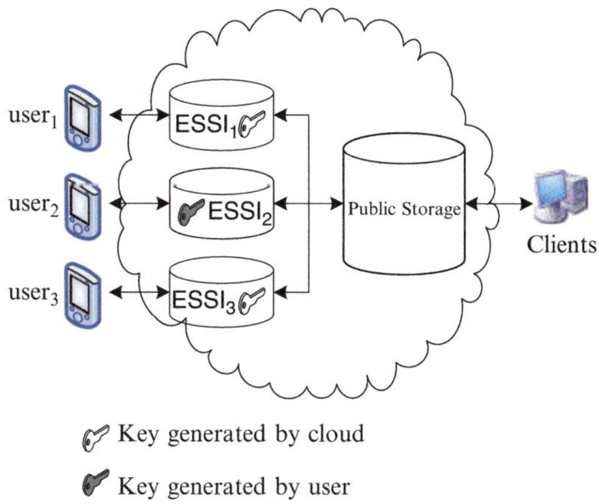

Fig. 7 Secure data processing and management in mobile cloud computing

identity theft. When the client intends to access a cloud service (*step (1)*), the request of authentication will be forwarded to the *Authentication Consumer* (*step (2)*), which redirects the request to the *Authentication Engine* (*step (3)*). The *Authentication Engine* obtains the policy for this access from the *Authentication Consumer* (*steps (4) and (5)*), and queries the *Data Aggregator* about the client device (*step (6)*). The *Data Aggregator* furnishes past authenticated actions and behaviors of a mobile client to the *Authentication Engine* (*step (7)*). The *Authentication Engine* exchanges a secret with the *Authentication Consumer* during authentication in order to later verify the authentication result. The *Authentication Engine* determines the authentication result and furnishes it to the *Authentication Consumer* (*step (8)*). Finally, the *Authentication Consumer* notifies the service provider concerning the access permission result such as *access acceptance* or *access denial* (*step (9)*).

4.3.3 Secure Mobile Cloud Computing

In mobile cloud computing, the information collected by mobile devices is processed, stored, and provided to other users as a part of cloud services. The images of mobile devices could be stored in the cloud to augment the functionalities of the devices as well. Confidentiality and privacy of this information is a concern which is gaining much attention. A secure data processing framework for mobile cloud computing [30] is proposed to address this concern.

We illustrate a secure mobile data processing and management in Fig. 7. The cloud is divided into a public service domain and a trusted domain. The public domain stores information generated from the mobile devices that can be provided to the public. The trusted domain is made of extended semi-shadow images (ESSIs) [29, 30], which are clones of the mobile devices in a secure storage. To ensure the security of a user's data, the cloud generates keys for each user to encrypt data in ESSIs. When the public service needs the data of a user, the data will be obtained from the ESSI of the user. The data will be preprocessed to remove identity information before being used in the public domain. However, security concerns on the encrypted data in ESSIs still remain since the keys are generated and controlled by the cloud providers. Therefore, the secure data processing framework [30] applies a multi-tenant data management scheme that partitions data into two security levels: critical data and normal data. The normal data is stored and secured by the data encryption keys as aforementioned. The critical data is encrypted with keys generated by the owners. Hence, access to critical data needs the authorization of the owners. The data processing in ESSI is built on the security capability model. Three capabilities are defined: cloud root, user root, and auditing root. The user root encrypts and decrypts data in ESSI.

5 Conclusion

In this chapter, we reviewed the latest research and development of secure mobile computing in the cloud. We first discussed three representative cloud architectures designed to support new mobile computing models in the cloud. We showed that new features can be realized to extend the computing capabilities of mobile devices when the advantages of the mobile devices and cloud computing are integrated into one system. We then studied a wide range of threats against the availability, privacy and integrity of mobile cloud computing architectures. We showed that attackers can target and exploit a much wider range of resources/protocols in a mobile cloud computing environment when compared to traditional client-server architectures. Finally, we summarized newly proposed defense mechanisms that ensure the security of mobile cloud computing architectures and their applications. We showed that securing mobile computing in the cloud needs not only the applicable defenses in traditional cloud, but also the security solutions designed specially for mobile cloud architectures.

Acknowledgements This material is based upon work partially supported by the One-Time Research Support Program at Texas State University-San Marcos, the National Science Foundation (NSF) grant CNS-1149397, the Air Force Office of Scientific Research (AFOSR)/the Air Force Research Laboratory (AFRL) Visiting Faculty Research Program (VFRP) extension grant LRIR 11RI01COR.

References

1. Baliga, A., Chen, X., Coskun, B., de los Reyes, G., Lee, S., Mathur, S., Van der Merwe, J.E.: VPMN: virtual private mobile network towards mobility-as-a-service. In: Proceedings of the 2nd International Workshop on Mobile Cloud Computing and Services, MCS'11, Washington, DC, pp. 7–12. ACM, New York (2011). doi:10.1145/1999732.1999735
2. Barrera, D., Kayacik, H.G., van Oorschot, P.C., Somayaji, A.: A methodology for empirical analysis of permission-based security models and its application to Android. In: Proceedings

of the 17th ACM Conference on Computer and Communications Security, CCS'10, Chicago, pp. 73–84. ACM, New York (2010). doi:10.1145/1866307.1866317. http://doi.acm.org/10.1145/1866307.1866317

3. Barrera, D., Clark, J., McCarney, D., van Oorschot, P.C.: Understanding and improving app installation security mechanisms through empirical analysis of Android. In: Proceedings of the 2nd ACM Workshop on Security and Privacy in Smartphones and Mobile Devices, SPSM'12, Raleigh, pp. 81–92. ACM, New York (2012). doi:10.1145/2381934.2381949
4. Becher, M., Freiling, F.C., Hoffmann, J., Holz, T., Uellenbeck, S., Wolf, C.: Mobile security catching up? Revealing the nuts and bolts of the security of mobile devices. In: Proceedings of the 2011 IEEE Symposium on Security and Privacy, SP'11, Oakland, pp. 96–111. IEEE Computer Society, Washington, DC (2011). doi:10.1109/SP.2011.29
5. Bellissimo, A., Burgess, J., Fu, K.: Secure software updates: disappointments and new challenges. In: Proceedings of the 1st USENIX Workshop on Hot Topics in Security, HOTSEC'06, Vancouver, pp. 37–43. USENIX Association, Berkeley (2006)
6. Bleikertz, S., Schunter, M., Probst, C.W., Pendarakis, D., Eriksson, K.: Security audits of multi-tier virtual infrastructures in public infrastructure clouds. In: Proceedings of the 2010 ACM Workshop on Cloud Computing Security Workshop, CCSW'10, Chicago, pp. 93–102. ACM, New York (2010). doi:10.1145/1866835.1866853
7. Chaudhuri, A.: Language-based security on Android. In: Proceedings of the ACM SIGPLAN 4th Workshop on Programming Languages and Analysis for Security, PLAS'09, Dublin, pp. 1–7. ACM, New York (2009). doi:10.1145/1554339.1554341
8. Chong, S., Liu, J., Myers, A.C., Qi, X., Vikram, K., Zheng, L., Zheng, X.: Secure web applications via automatic partitioning. SIGOPS Oper. Syst. Rev. **41**(6), 31–44 (2007). doi:10.1145/1323293.1294265
9. Chow, R., Jakobsson, M., Masuoka, R., Molina, J., Niu, Y., Shi, E., Song, Z.: Authentication in the clouds: a framework and its application to mobile users. In: Proceedings of the 2010 ACM Workshop on Cloud Computing Security Workshop, CCSW'10, Chicago, pp. 1–6. ACM, New York (2010). doi:10.1145/1866835.1866837
10. Christensen, J.H.: Using RESTful web-services and cloud computing to create next generation mobile applications. In: Proceedings of the 24th ACM SIGPLAN Conference Companion on Object Oriented Programming Systems Languages and Applications, OOPSLA'09, Orlando, pp. 627–634. ACM, New York (2009). doi:10.1145/1639950.1639958
11. Christodorescu, M., Sailer, R., Schales, D.L., Sgandurra, D., Zamboni, D.: Cloud security is not (just) virtualization security: a short paper. In: Proceedings of the 2009 ACM Workshop on Cloud Computing Security, CCSW'09, Chicago, pp. 97–102. ACM, New York (2009). doi:10.1145/1655008.1655022
12. Chun, B.G., Maniatis, P.: Augmented smartphone applications through clone cloud execution. In: Proceedings of the 12th Conference on Hot Topics in Operating Systems, HotOS'09, Monte Verita, pp. 1–5. USENIX Association, Berkeley (2009)
13. Chun, B.G., Maniatis, P.: Dynamically partitioning applications between weak devices and clouds. In: Proceedings of the 1st ACM Workshop on Mobile Cloud Computing & Services: Social Networks and Beyond, MCS'10, San Francisco, pp. 7:1–7:5. ACM, New York (2010). doi:10.1145/1810931.1810938
14. Cuervo, E., Balasubramanian, A., Cho, D.k., Wolman, A., Saroiu, S., Chandra, R., Bahl, P.: MAUI: making smartphones last longer with code offload. In: Proceedings of the 8th International Conference on Mobile Systems, Applications, and Services, MobiSys'10, San Francisco, pp. 49–62. ACM, New York (2010). doi:10.1145/1814433.1814441
15. Danezis, G., Livshits, B.: Towards ensuring client-side computational integrity. In: Proceedings of the 3rd ACM Workshop on Cloud Computing Security Workshop, CCSW'11, Chicago, pp. 125–130. ACM, New York (2011). doi:10.1145/2046660.2046683
16. Drimer, S., Murdoch, S.J., Anderson, R.: Thinking inside the box: system-level failures of tamper proofing. In: Proceedings of the 2008 IEEE Symposium on Security and Privacy, SP'08, Oakland, pp. 281–295. IEEE Computer Society, Washington, DC (2008). doi:10.1109/SP.2008.16

17. Enck, W., Ongtang, M., McDaniel, P.: On lightweight mobile phone application certification. In: Proceedings of the 16th ACM Conference on Computer and Communications Security, CCS'09, Chicago, pp. 235–245. ACM, New York (2009). doi:10.1145/1653662.1653691
18. Enck, W., Gilbert, P., Chun, B.G., Cox, L.P., Jung, J., McDaniel, P., Sheth, A.N.: TaintDroid: an information-flow tracking system for realtime privacy monitoring on smartphones. In: Proceedings of the 9th USENIX Conference on Operating Systems Design and Implementation, OSDI'10, Vancouver, pp. 1–6. USENIX Association, Berkeley (2010)
19. Fahl, S., Harbach, M., Muders, T., Baumgärtner, L., Freisleben, B., Smith, M.: Why eve and mallory love Android: an analysis of Android SSL (in)security. In: Proceedings of the 19th ACM Conference on Computer and Communications Security, CCS'12, Raleigh, pp. 50–61. ACM, New York (2012). doi:10.1145/2382196.2382205
20. Florio, E.: symantec.com, when malware meets rootkits. http://goo.gl/WdznF
21. forbes.com: Phone rootkit maker carrier IQ may have violated wiretap law in millions of cases. http://goo.gl/P3NJg
22. Garfinkel, T., Rosenblum, M.: A virtual machine introspection based architecture for intrusion detection. In: Proceedings of the 10th Network and Distributed Systems Security Symposium, NDSS'03, San Diego, pp. 191–206 (2003)
23. Gentry, C.: Fully homomorphic encryption using ideal lattices. In: Proceedings of the 41st Annual ACM Symposium on Theory of Computing, STOC'09, Bethesda, pp. 169–178. ACM, New York (2009). doi:10.1145/1536414.1536440
24. Gentry, C.: Computing arbitrary functions of encrypted data. Commun. ACM **53**(3), 97–105 (2010). doi:10.1145/1666420.1666444
25. Gilbert, P., Chun, B.G., Cox, L.P., Jung, J.: Vision: automated security validation of mobile apps at app markets. In: Proceedings of the 2nd International Workshop on Mobile Cloud Computing and Services, MCS'11, Bethesda, pp. 21–26. ACM, New York (2011). doi:10.1145/1999732.1999740
26. Giurgiu, I., Riva, O., Juric, D., Krivulev, I., Alonso, G.: Calling the cloud: enabling mobile phones as interfaces to cloud applications. In: Proceedings of the 10th ACM/IFIP/USENIX International Conference on Middleware, *Middleware'09*, Urbana, vol. 5, pp. 5:1–5:20. Springer, New York (2009)
27. He, S., Guo, L., Guo, Y.: Elastic application container. In: Proceedings of the 12th IEEE/ACM International Conference on Grid Computing, GRID'11, Lyon, pp. 216–217. IEEE Computer Society, Washington, DC (2011). doi:10.1109/Grid.2011.35
28. Hornyack, P., Han, S., Jung, J., Schechter, S., Wetherall, D.: These aren't the droids you're looking for: retrofitting Android to protect data from imperious applications. In: Proceedings of the 18th ACM Conference on Computer and Communications Security, CCS'11, Chicago, pp. 639–652. ACM, New York (2011). doi:10.1145/2046707.2046780
29. Huang, D., Zhang, X., Kang, M., Luo, J.: MobiCloud: building secure cloud framework for mobile computing and communication. In: Proceedings of 5th IEEE International Symposium on Service Oriented System Engineering, SOSE'10, Nanjing, pp. 27–34. IEEE Computer Society, Washington, DC (2010). doi:10.1109/SOSE.2010.20
30. Huang, D., Zhou, Z., Xu, L., Xing, T., Zhong, Y.: Secure data processing framework for mobile cloud computing. In: Proceedings of the IEEE Conference on Computer Communications Workshop, Shanghai, pp. 614–618 (2011). doi:10.1109/INFCOMW.2011.5928886
31. Huerta-Canepa, G., Lee, D.: A virtual cloud computing provider for mobile devices. In: Proceedings of the 1st ACM Workshop on Mobile Cloud Computing & Services: Social Networks and Beyond, MCS'10, San Francisco, pp. 6:1–6:5. ACM, New York (2010). doi:10.1145/1810931.1810937
32. Jack, B.: blackhat.com, exploiting embedded systems. http://goo.gl/oz7Vs (2006)
33. Jiang, X.: ncsu.edu, GingerMaster: first Android malware utilizing a root exploit on Android 2.3 (Gingerbread). http://goo.gl/uvTFT
34. Jiang, X.: ncsu.edu, security alert: new RootSmart Android malware utilizes the GingerBreak root exploit. http://goo.gl/ZTxpg

35. Ko, S.Y., Jeon, K., Morales, R.: The HybrEx model for confidentiality and privacy in cloud computing. In: Proceedings of the 3rd USENIX Conference on Hot Topics in Cloud Computing, HotCloud'11, Portland, pp. 1–5. USENIX Association, Berkeley (2011)
36. Kupsch, J.A., Miller, B.P., Heymann, E., César, E.: First principles vulnerability assessment. In: Proceedings of the 2010 ACM Workshop on Cloud Computing Security, CCSW'10, Chicago, pp. 87–92. ACM, New York (2010). doi:10.1145/1866835.1866852
37. Law, Y.W., Palaniswami, M., Hoesel, L.V., Doumen, J., Hartel, P., Havinga, P.: Energy-efficient link-layer jamming attacks against wireless sensor network MAC protocols. Trans. Sens. Netw. **5**(1), 6:1–6:38 (2009). doi:10.1145/1464420.1464426
38. Lee, W., Rotoloni, B.: Emerging cyber threats report 2013. Technical report, Georgia Institute of Technology (2012)
39. Lessard, J., Kessler, G.: Android forensics: simplifying cell phone examinations. Small Scale Digit. Device Forensics J. **4**(1), 1–12 (2010)
40. linuxsleuthing.blogspot.com, Linux Sleuthing: iPhone forensics tools. http://goo.gl/Wc31M
41. Liu, H.: A new form of DoS attack in a cloud and its avoidance mechanism. In: Proceedings of the 2010 ACM Workshop on Cloud Computing Security, CCSW'10, Chicago, pp. 65–76. ACM, New York (2010). doi:10.1145/1866835.1866849
42. Marforio, C., Francillon, A., Capkun, S.: osti.gov, application collusion attack on the permission-based security model and its implications for modern smartphone systems. http://goo.gl/0Csm2
43. Micciancio, D.: A first glimpse of cryptography's holy grail. Commun. ACM **53**(3), 96–96 (2010). doi:10.1145/1666420.1666445
44. omtp.org: OMTP advanced trusted environment. http://goo.gl/Nzf6p (2009)
45. Ongtang, M., Butler, K., McDaniel, P.: Porscha: policy oriented secure content handling in Android. In: Proceedings of the 26th Annual Computer Security Applications Conference, ACSAC'10, Austin, pp. 221–230. ACM, New York (2010). doi:10.1145/1920261.1920295
46. Ongtang, M., McLaughlin, S., Enck, W., McDaniel, P.: Semantically rich application-centric security in Android. Secur. Commun. Netw. **5**(6), 658–673 (2012). doi:10.1002/sec.360
47. Pelechrinis, K., Iliofotou, M., Krishnamurthy, V.: Denial of service attacks in wireless networks: the case of jammers. IEEE Commun. Surv. Tutor. **13**(2) (2011). doi:10.1109/SURV.2011.041110.00022
48. Portnoy, A.: tippingpoint.com, Pwn2POwn 2010. http://goo.gl/XLJN
49. Quynh, N.A., Takefuji, Y.: Towards a tamper-resistant Kernel rootkit detector. In: Proceedings of the 2007 ACM Symposium on Applied Computing, SAC'07, Seoul, pp. 276–283. ACM, New York (2007). doi:10.1145/1244002.1244070
50. Raffetseder, T., Kruegel, C., Kirda, E.: Detecting system emulators. In: Proceedings of the Information Security, Valparaíso, pp. 1–18 (2007)
51. Raj, H., Nathuji, R., Singh, A., England, P.: Resource management for isolation enhanced cloud services. In: Proceedings of the 2009 ACM Workshop on Cloud Computing Security, CCSW'09, Chicago, pp. 77–84. ACM, New York (2009). doi:10.1145/1655008.1655019
52. Ramsey, R.: tmcnet.com, as users shift to mobile and cloud, so will attackers: cybercrime in 2013. http://goo.gl/MLeuk (2012)
53. Riley, R., Jiang, X., Xu, D.: Guest-transparent prevention of Kernel rootkits with VMM-based memory shadowing. In: Proceedings of the 11th International Symposium on Recent Advances in Intrusion Detection, RAID'08, Cambridge, pp. 1–20. Springer, Berlin/Heidelberg (2008). doi:10.1007/978-3-540-87403-4_1
54. Rosenfeld, K., Karri, R.: Attacks and defenses for JTAG. IEEE Des. Test **27**(1), 36–47 (2010). doi:10.1109/MDT.2010.9
55. Sang, L., Arora, A.: Capabilities of low-power wireless jammers. In: Proceedings of INFOCOM, Rio de Janeiro (2009). doi:10.1109/INFCOM.2009.5062185
56. Satyanarayanan, M.: Mobile computing: the next decade. SIGMOBILE Mobile Comput. Commun. Rev. **15**(2), 2–10 (2011). doi:10.1145/2016598.2016600
57. Satyanarayanan, M., Bahl, P., Caceres, R., Davies, N.: The case for VM-based cloudlets in mobile computing. IEEE Pervasive Comput. **8**(4), 14–23 (2009). doi:10.1109/MPRV.2009.82

58. Sekar, V., Maniatis, P.: Verifiable resource accounting for cloud computing services. In: Proceedings of the 3rd ACM Workshop on Cloud Computing Security, CCSW'11, Chicago, pp. 21–26. ACM, New York (2011). doi:10.1145/2046660.2046666
59. Somorovsky, J., Heiderich, M., Jensen, M., Schwenk, J., Gruschka, N., Lo Iacono, L.: All your clouds are belong to us: security analysis of cloud management interfaces. In: Proceedings of the 3rd ACM Workshop on Cloud Computing Security, CCSW'11, Chicago, pp. 3–14. ACM, New York (2011). doi:10.1145/2046660.2046664
60. Song, Z., Molina, J., Lee, S., Lee, H., Kotani, S., Masuoka, R.: Trustcube: an infrastructure that builds trust in client. In: Proceedings of the 1st International Conference Future of Trust in Computing, Berlin, pp. 68–79. Vieweg+Teubner (2009). doi:10.1007/978-3-8348-9324-6_8
61. symantec.com: W32.Fanbot.A@mm. http://goo.gl/NkX5h
62. Szeliski, R.: Image alignment and stitching: a tutorial. Found. Trends Comput. Graph. Vis. **2**(1), 1–104 (2006). doi:10.1561/0600000009
63. Thuente, D.J., Acharya, M.: Intelligent jamming in wireless networks with applications to 802.11b and other networks. In: Proceedings of the 2006 IEEE Conference on Military Communications, MILCOM'06, Washington, DC, pp. 1075–1081. IEEE Press, Piscataway (2006)
64. Verbelen, T., Simoens, P., De Turck, F., Dhoedt, B.: Cloudlets: bringing the cloud to the mobile user. In: Proceedings of the 3rd ACM Workshop on Mobile Cloud Computing and Services, MCS'12, Low Wood Bay, pp. 29–36. ACM, New York (2012). doi:10.1145/2307849.2307858
65. Walls, R.J., Learned-Miller, E., Levine, B.N.: Forensic triage for mobile phones with DEC0DE. In: Proceedings of the 20th USENIX Conference on Security, SEC'11, San Francisco, pp. 1–14. USENIX Association, Berkeley (2011)
66. Wei, J., Zhang, X., Ammons, G., Bala, V., Ning, P.: Managing security of virtual machine images in a cloud environment. In: Proceedings of the 2009 ACM Workshop on Cloud Computing Security, CCSW'09, Chicago, pp. 91–96. ACM, New York (2009). doi:10.1145/1655008.1655021
67. Wilhelm, M., Martinovic, I., Schmitt, J.B., Lenders, V.: Short paper: reactive jamming in wireless networks: how realistic is the threat? In: Proceedings of the 4th ACM Conference on Wireless Network Security, WiSec'11, Hamburg, pp. 47–52. ACM, New York (2011). doi:10.1145/1998412.1998422
68. Xu, W., Ma, K., Trappe, W., Zhang, Y.: Jamming sensor networks: attack and defense strategies. Netw. Mag. Glob. Internetwkg. **20**(3), 41–47 (2006). doi:10.1109/MNET.2006.1637931
69. Zhang, X., Schiffman, J., Gibbs, S., Kunjithapatham, A., Jeong, S.: Securing elastic applications on mobile devices for cloud computing. In: Proceedings of the 2009 ACM Workshop on Cloud Computing Security, CCSW'09, Chicago, pp. 127–134. ACM, New York (2009). doi:10.1145/1655008.1655026
70. Zhang, K., Zhou, X., Chen, Y., Wang, X., Ruan, Y.: Sedic: privacy-aware data intensive computing on hybrid clouds. In: Proceedings of the 18th ACM Conference on Computer and Communications Security, CCS'11, Chicago, pp. 515–526. ACM, New York (2011). doi:10.1145/2046707.2046767

Information Fusion in a Cloud-Enabled Environment

Erik Blasch, Yu Chen, Genshe Chen, Dan Shen, and Ralph Kohler

Abstract Recent advances in cloud computing pose interesting capabilities for information fusion which have similar requirements of big data computations. With a cloud enabled environment, information fusion systems could be conducted over vast amounts of entities across multiple databases. In order to properly implement information fusion in a cloud, information management, system design, and real-time execution must be considered. In this chapter, three aspects of current developments integrating *low/high-level information fusion (LLIF/HLIF)* and cloud computing are discussed: (1) agent-based service architectures, (2) ontologies, and (3) metrics (timeliness, confidence, and security). We introduce the *Cloud-Enabled Bayes Network (CEBN)* for *wide area motion imagery target tracking and identification*. The Google Fusion Tables service is also selected as a case study to illustrate commercial cloud-based information fusion applications.

"Approved for Public Release; Distribution Unlimited: 88ABW-2013-1114, 08-Mar-2013"

E. Blasch (✉) • R. Kohler
Air Force Research Laboratory, Rome, NY, USA
e-mail: erik.blasch@us.af.mil; ralph.kohler@us.af.mil

Y. Chen
SUNY-Binghamton, Binghamton, NY, USA
e-mail: ychen@binghamton.edu

G. Chen • D. Shen
Intelligent Fusion Technology, Inc., Gaithersburg, MD, USA
e-mail: gchen@intfusiontech.com; dshen@intfusiontech.com

K.J. Han et al. (eds.), *High Performance Cloud Auditing and Applications*,
DOI 10.1007/978-1-4614-3296-8_4, © Springer Science+Business Media New York 2014

1 Introduction

Information fusion (IF) seeks to reduce uncertainty over data, information, and knowledge. Uncertainty comes from many sources including sensors, targets, and the environment and the subsequent processing over interpretation, context, language, and users [13]. Assessing the quality of merged and combined information requires objective and subjective measures including uncertainty representation and uncertainty reasoning as shown in Fig. 1. Information fusion (i.e., uncertainty reduction) [28] utilizes streaming data and information from a cloud-enable distributed network environment which is eventually processed by an end user [8, 11, 17].

Current advances in available processing, sensor collection, data storage, and data distribution have afforded more complex, distributed, and operational information fusion systems (IFSs). IFSs notionally consist of low-level information fusion (LLIF) and high-level information fusion (HLIF). LLIF includes data collection/-management [19], registration [64], and target tracking association in time and space [65]. HLIF includes situational awareness [25], threat assessment [6], user coordination [4], and mission control [10]. HLIF challenges [22] include resource management [14], network-centric architectures [43], and spectrum sharing [60]; which are elements of a cloud computing environment of access, storage, and retrieval. Current HLIF research is based on information management [35] and systems design [23]. HLIF and LLIF can benefit from the advances in cloud computing, but there are few reports that bring together these technologies. Topics of interest to both communities includes security [46], service oriented computing [26, 42], and integrated intelligence (such as the Open Geospatial Consortium (OGC) [18, 36]). There are examples of Google's Cloud Fusion service [34] which brings information together, but the hosting and linking of information provides a common repository that leaves the user with the goal of fusing the information. One example from Google Fusion is the linking of people to a location; however, there is little in the way of determining the quality and type of research that is needed to combine the information to make a more informed decision of the research applications.

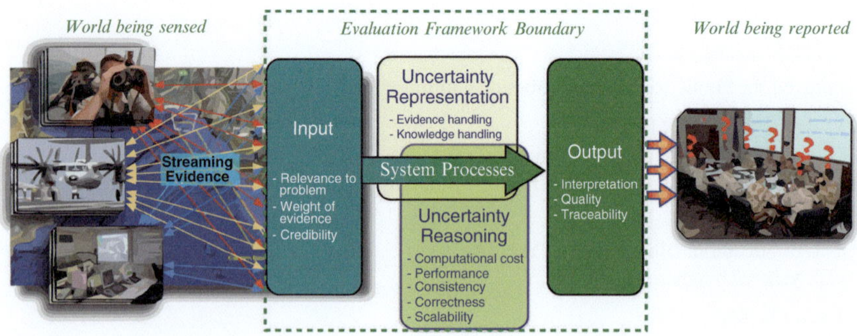

Fig. 1 Information fusion for boundaries of the uncertainty representation and reasoning evaluation (Reproduced from Costa et al. [28])

Popular methods of information fusion come from the signal processing, estimation, and control theories. Methods associated with information fusion include a host of tracking methods, Boolean logic, fuzzy logic, Markov chains, Bayesian networks, entropy/information theory, and advanced methods like Dempster Shafer [23]. Which method to use is based on the data, application, uncertainty, and assumption granularity. The developments in information fusion result from local, centralized fusion, which has proven robust in applications such as the widely used Kalman filter. However, as these methods are extended to distributed and decentralized methods, there are a host of issues such as data availability, flow, and pedigree which limit the ability of analysis. Current methods include graphical networks and nonlinear filtering for big data analysis. Many of these elements of information fusion are for LLIF, whereas the need for situation awareness requires knowledge of the environment, enterprise, and distributed users which can be enabled by cloud computing. In this chapter, *low/high-level information fusion (LLIF/HLIF)* and cloud computing are discussed and introduced the *Cloud-Enabled Bayes Network (CEBN)* for *wide area motion imagery target tracking and identification*.

The rest of this chapter is organized as follows. Section 2 discusses information fusion modeling techniques. Section 3 briefly describes information management methods. Section 4 presents information fusion aspects of ontologies. Section 5 introduces theoretical CEBN and CEBN-WAMI target tracking and identification. Section 6 presents a case study of information fusion on cloud platform. Finally, conclusions are presented in Section 7.

2 Information Fusion Modeling

2.1 High-Low-Level Distinctions

One aspect of the current trend in HLIF is to incorporate the user, which has gained in prominence and acceptance and is a feature of the Data Fusion Information Group (DFIG) model. The DFIG model [12, 13, 23], shown in Fig. 2, couples resource management (RM) functions with information fusion (IF) estimation needs.

The DFIG model supports differing control functions based on the spatial/temporal/spectral differences. The spectral needs drive sensor selection. The temporal needs are based on the user's need for timely information to afford action. Finally, the spatial needs are based on the mission goals.

The current DFIG model process definitions include:

- *Level 0 Data Assessment (DA)*: Estimation and prediction of signal/object observable states on the basis of pixel/signal level data association (e.g., information systems collections);
- *Level 1 Object Assessment (OA)*: Estimation and prediction of entity states on the basis of data association, continuous state estimation and discrete state estimation (e.g., data processing);

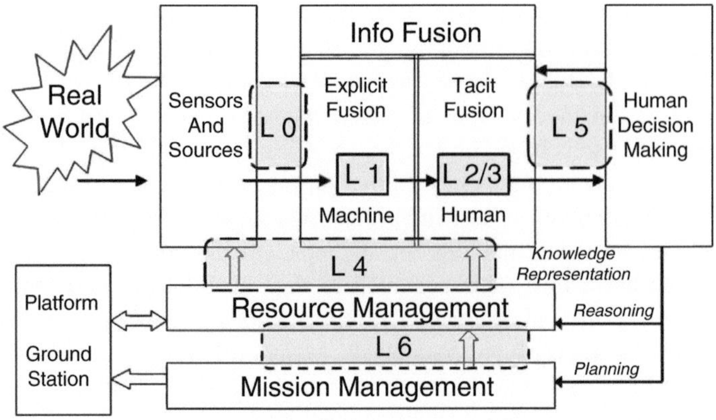

Fig. 2 Data fusion information group (DFIG) model (Reproduced from Blasch et al. [12, 13, 23])

- *Level 2 Situation Assessment (SA)*: Estimation and prediction of relations among entities, to include force structure and force relations, communications, etc. (e.g., information processing);
- *Level 3 Impact Assessment (IA)*: Estimation and prediction of effects on situations of planned or estimated actions by the participants, to include interactions between action plans of multiple players (e.g., assessing threat/intent actions to planned actions and mission requirements, and performance evaluation);
- *Level 4 Process Refinement (PR) – an element of Resource Management*: Adaptive data acquisition and processing to support sensing objectives (e.g., fusion process control and information dissemination);
- *Level 5 User Refinement (UR) – an element of Knowledge Management*: Adaptive determination of who queries information and who has access to information (e.g., information operations) and adaptive data retrieved and displayed to support cognitive decision making and actions (e.g., human systems integration (HSI));
- *Level 6 Mission Management (MM) – an element of Platform Management*: Adaptive determination of spatial-temporal control of assets (e.g. airspace operations) and route planning and goal determination to support team decision making and actions (e.g., theater operations) under social, economic, and political constraints.

In the DFIG model, the goal was to separate the IF and RM functions. RM is divided into sensor control, platform placement, and user selection to meet mission objectives. Level 2 (SA) includes tacit functions which are inferred from Level 1 (OA) explicit representations of object assessment. Since the unobserved aspects of the SA problem cannot be processed by a computer, user knowledge and reasoning is necessary. Level 3 (IA) sense-making of impacts (threats, course of actions, game-theoretic decisions, intent, etc.) helps refine the SA estimation and information needs for different actions. Resource management can be aided by cloud computing aspects of data acquisition, access, recall, and storage services. High-level information fusion (as referenced to levels beyond Level 1) is the ability of a fusion system, through knowledge, expertise, and understanding to: capture awareness and complex relations, reason over past and future events, use direct sensing exploitations and tacit reports, and discern the usefulness and intention of results to meet system-level goals. The Information Fusion community has coined the term "high-level fusion"; however, this implies that there is a low-level/high-level distinction when in reality they are coupled. Designs of real-world information fusion systems imply distributed information source coordination (network), organizational concepts (command), and environmental understanding (context). Additionally, there is a need for automated processes that provide functionality in support of user decision processes, particularly at higher levels requiring reasoning and inference. A cloud-enabled service can greatly enhance these attributes which benefit both LLIF and HLIF.

The DIFG model and cloud computing services share a common goal to provide information for situation awareness. Cloud computing can serve as a method to store outputs, access information, and support processing and dissemination over asynchronous services. Using the DFIG paradigm, Level 4 (sensor management) could use a cloud service to access information, Level 5 (user refinement) can be the end-user applications that query information, and Level 6 (mission management) can provide filtering and control of information dissemination to the correct user. Inherent in the analysis is that Level 0 (data preprocessing) exists from the information already resident in the cloud environment. Next, we discuss situation awareness and situation assessment as a basis where cloud environments can benefit information fusion architectures.

2.2 *Situation Awareness*

There are two main groups addressing situational information: the engineering IF community (i.e., Situation Assessment [SA]) and the human factors community (i.e., Situation Awareness [SAW]). SAW is a mental state while SA supports (e.g., fusion products) that state. Given the developments of SAW and SA, we combine the ideas into an integrated information fusion situation assessment (IFSA) model in which the role of SA stratifies the object/event analysis. The IFSA captures the elements of the SAW reference model with the proposed changes to the Joint Director of the Laboratories (JDL) [23] model that includes the DFIG elements of a combined Level 2/3 analysis as well as the coordination with the user in Level 5 information fusion. The IFSA model is presented in Fig. 3.

Starting with Boyd's observe-orient-decide-act (OODA) control loop [17], the right side of Fig. 3 captures the needs of the user and their ability to observe and orient themselves to the information. As the user requests information for their SAW, they must regress over the data they have, what they need, and the control actions. A cloud environment can provide these services. The information fusion

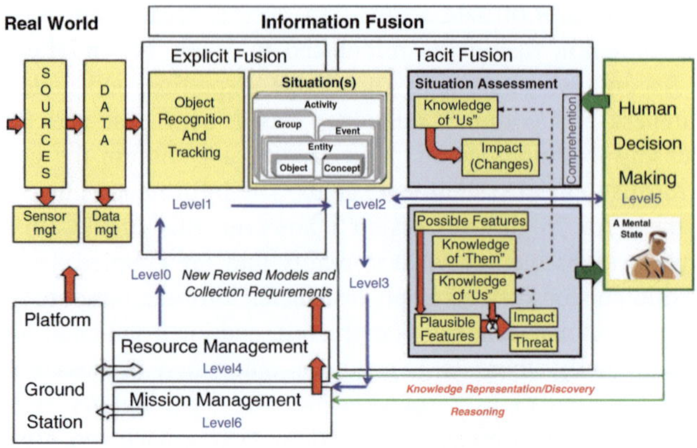

Fig. 3 Information fusion situation assessment model (Reproduced from Blasch et al. [20, 23])

system provides the elements of the information from the left side of Fig. 3, which provides alerts that call to the attention of situations of interest (denoted as explicit fusion). The user can coordinate with any of these levels to update the SAW and control the collections for SA through tacit fusion. Finally, we note that there are needs between resource (e.g., airborne assets, and web pages) management versus that of mission management (e.g., goals, policies, and doctrines) as shown in the bottom of Fig. 3. What is not detailed in the DFIG/IFSA models is access to the information about the real world (i.e., constant flux and change over political, social, and environmental contexts).

While the IFSA model captures our current assessment of the domain of SA and SAW, other issues need to be considered such as metrics for evaluation and the challenges of situation model refinement for practical use. Other cloud computing models in the book also explore representations and theories to instantiate data products that are useful to the community. The interchange between the *us* and *them* refers to an environment, such as the cloud, which requires analysis of security, access, and authentication of users to obtain the information. Cloud computing security analysis will directly impact future methods in situation awareness.

2.3 *Information Fusion Modeling in the Enterprise*

The current trends in information fusion are data mining, the enterprise architecture, and the communications. Different mission applications require coordination over (1) data (models, storage access control, process and transport flow), (2) architecture (e.g., service-oriented architecture (SOA)), and (3) enterprise (e.g., service bus, computing environment, and the cloud). Specifically, the cloud serves as a basis for access to resource information.

Most developments that imply fusion and cloud computing refer to methods of the enterprise such as an SOA [3,49,62]. One example of an SOA is an information management service.

3 Information Management

3.1 Information Management Model

The goal of Information management (IM) is to maximize the effectiveness of an enterprise by maximizing its ability to act upon information that is produced or consumed within the enterprise. There are several means by which this can be accomplished [23]:

- Reducing barriers to effective information use by providing notification, mediation, access control, and persistence services.
- Providing an information space wherein information is managed directly, rather than delegating information management responsibilities to applications that produce and consume information.
- Focusing on consumer needs rather than producer preferences to ensure that information can be effectively presented and used.
- Providing tools to assess information quality and suitability.
- Exploiting producer-provided characterization of information to support automated management and dissemination of information.

If these means can be accomplished, edge-user applications can be made less complicated, thereby enabling the enterprise to be more agile in adapting to changing requirements and environments. (By edge-user applications, we mean applications that are not part of, or controlled by, the information management infrastructure.)

There are several best practices that help achieve the goals of information management. Organizations will greatly improve the interoperability and agility of their future net-centric information fusion and command and control (C2) systems by Blasch et al. [23]:

1. Adopting dedicated information management infrastructures (e.g., cloud computing),
2. Packaging information for dissemination and management,
3. Creating simple, ubiquitous services that are independent of operating system and programming language,
4. Using a common syntax and semantics for common information attributes such as location, time and subject, and
5. Adopting interfaces among producers, consumers and brokers that are simple, effective and well-documented.

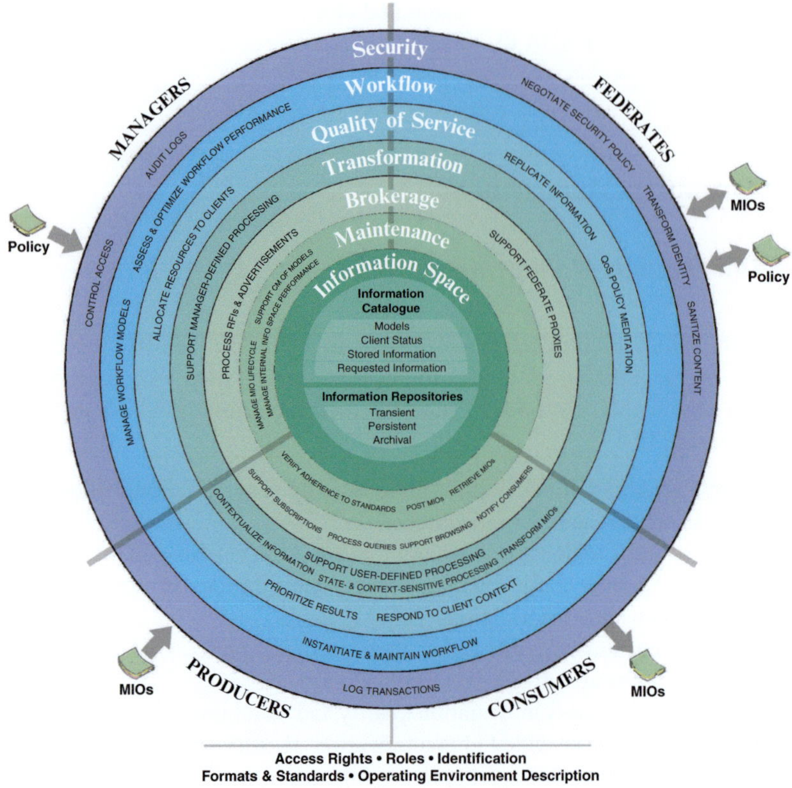

Fig. 4 Information management model (Reproduced from Blasch et al. [20, 23])

If appropriately employed, these best practices can reduce the complexity of fusion and C2 systems, allow for effective control of the information space, and facilitate more effective sharing of information over a cloud-enabled environment. Viewing data as a *managed information object*, then information fusion can be viewed as process that uses the tenets of a cloud environment. Figure 4 presents the information management model [23, 44]. The quantum of managed information is called a Managed Information Object (MIO). A MIO comprises a payload and metadata that characterize the object such as topic, time, and location. It is desirable that all of the information needed for making management decisions, such as content-based dissemination, be present in or referenced within the metadata in a form that permits efficient processing.

An important element of characterization is the concept of *type*. The type of an object (e.g., Wide Area Motion Imagery (WAMI)) is useful for determining how the information should be characterized and for setting policy on its appropriate use. A given type may be encoded in many formats, for example WAMI may be encoded in a number of image formats including say, JPEG. Conversely while the JPEG

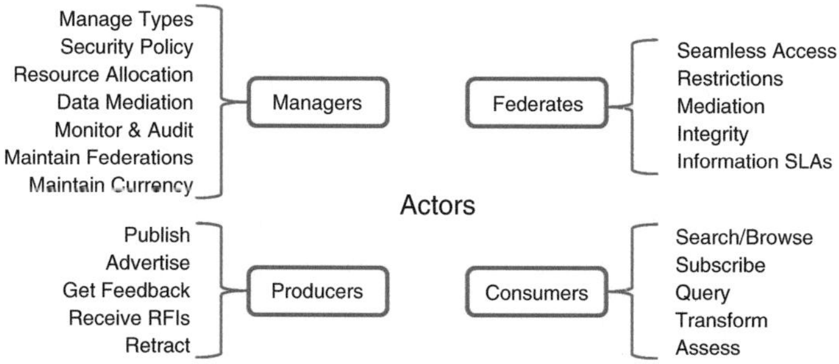

Fig. 5 Information actors (Reproduced from Blasch et al. [20, 23])

format may be used for overhead imagery it could also be used for a similar type say, hand held imagery or a substantially different type say, scanned human-observation intelligence reports. Type is distinct from *format* in that type relates to the purpose of the information, whereas the format relates to its encoding. While format is essential for processing or presenting the information, type is more important for determining management of the information.

3.1.1 Actors

People or autonomous agents interact with the managed information cloud-enabled environment by producing and consuming information or by managing it. Federated information spaces that interact with a given information space are also considered actors from the perspective of that information space, shown in Fig. 5 with a summary of the elements in Table 1. Cloud computing can enable access between the users/actors and the information fusion solutions in the information space. The illustration puts the extended relation of the actors coordinating through the cloud with the various layers and inner circles providing the protocol for information service access and dissemination. Note that in many real cases, producers may also be consumers, and vice versa.

3.1.2 Service Layers

A set of service layers are defined that use artifacts to perform specific IM activities and are inherent in cloud environments. An artifact is a piece of information that is acted upon by a service or that influences the behavior of the service (e.g., a policy). The services layers defined by the model are: Security, Workflow, Quality of Service (QoS) [48], Transformation, Brokerage, and Maintenance, as shown in Table 2. These services are intelligent agents that provide information to the architecture.

Table 1 Elements of information actors (Modified from Blasch et al. [23])

Actors	Elements
Consumers	*Search/Browse* the information space
	Subscribe to new information
	Query the information space to retrieve previously published information
	Transformation, as a part of browsing, searching or query, user may filter, set priorities and transform the information prior to delivery
	Provide Feedback on the effectiveness of the information to information producers
	Republish transformed information, so others may benefit from it
Producers	*Publish* information to the information space and provide appropriate metadata.
	Advertise their capabilities and products that may be produced if they are requested
	Get Feedback to enable them to optimize their service
	Receive Requests for Information (RFI) to service new requests
	Retract information that is subsequently deemed to be incorrect. With the exception of a security spill retracted information is not removed from the repository, it is just marked as retracted
Managers	*Manage Types* and the relationships between types and how they support the goals of the enterprise, monitor the introduction of new types
	Security Policy, including the balance between security concerns and operational effectiveness. The policy may change as a result of the operational context
	Resource Allocation must be managed to provide an optimal outcome
	Data Mediation between evolving information standards and legacy systems
	Monitor & Audit functions must be appropriate for the task
	Maintain Federations with external information spaces
	Maintain Currency by removing or achieving old information, manage the movement of information through transient, persistent and archive catalogues
Federates	*Seamless Access* is provided to consumers accessing external information spaces
	Restrictions on usage may need to be enforced as the trust relationships will be different for external consumers
	Mediation of inconsistent data standards
	Integrity of information provided by a federated partner
	Information Service Level Agreement (SLA) through formal negotiation

3.1.3 Information Spaces

The information space is a collection of catalogues and repositories that provide common functions for storage, retrieval and lifecycle management. The Information Space operates on our managed information objects and forms the heart of the managed information environment. An information space is thus a key element of future coordination between cloud computing and information fusion. Due the rapidly increasing number of sensors and monitors, the expanding coverage on physical world has necessitated creation of higher dimension information space. Within a high dimensional space it is non-trivial to tailor computing resources for a specific information fusion task while maximizing the system utility efficiency. Obviously, it is highly undesired to index through the entire space given a

Table 2 Service layers (Modified from Blasch et al. [23])

Services	Functionalities
Security	Control access, Log transactions, Audit logs, Negotiate security policy with federated information spaces, Transform identity and Sanitize content
Workflow	Manage workflow model configurations, Instantiate and maintain workflows, Assess and optimize workflow performance
QoS	Respond to client context, Allocate resources to clients, QoS policy mediation, Prioritize results, and Replicate information
Transformation	Contextualize information, Transform MIOs, Support state and context-sensitive processing, Support user defined processing functions, Support manager defined processing functions
Brokerage	Process queries, Support browsing, Maintain subscriptions, Notify consumers, Process requests for information and advertisements, Support federated information space proxies
Maintenance	Post MIOs, Verify Adherence to standards, Manage MIO lifecycle, Manage information space performance, Retrieve specific MIOs from repositories, Support configuration management of information models

simpler task. At the same time, under-estimation on task complexity will make the user suffer unexpected long delay or even the risk of task failure.

The Cloud computing platforms will meet the challenge by providing a highly elastic information fusion mechanism in the high dimensional information space. The elasticity and illusion of infinite resources make transactional tasks practically feasible, such as scalable database management systems (DBMS) [1, 29]. With a certain initial goal, a user can start with traditional light-weight data processing using a sub-set of the operations/resources. Then, extra functions and resources increase in demand later with the elastic scalability. Whether in the context of Infrastructure as a Service (IaaS), Platform as a Service (PaaS), or Software as a Service (SaaS), the capability of dynamical job reassignments enables the cloud system to match the fluctuation in resource requirements seamlessly.

3.2 Layered View of the Cloud

Using elements of the DFIG, the IFAS, the enterprise, and the information management model, these methods seek to extend LLIF to HLIF needs. Figure 6 presents the information as LLIF to HLIF where the end-user (operator or machine) desires quality information as fused products from data which requires various methods and services from sensor collections to information delivery. *Sensors/Sources* can be viewed as a general term as it relates to physical sensors, humans, and database services (e.g. data mining) that seek data from the environment and process it as a transducer for analysis.

Some of the current trends in cloud computing and information fusion share common developments. Here we discuss three: agent-based network service architectures, ontologies, and metrics [68].

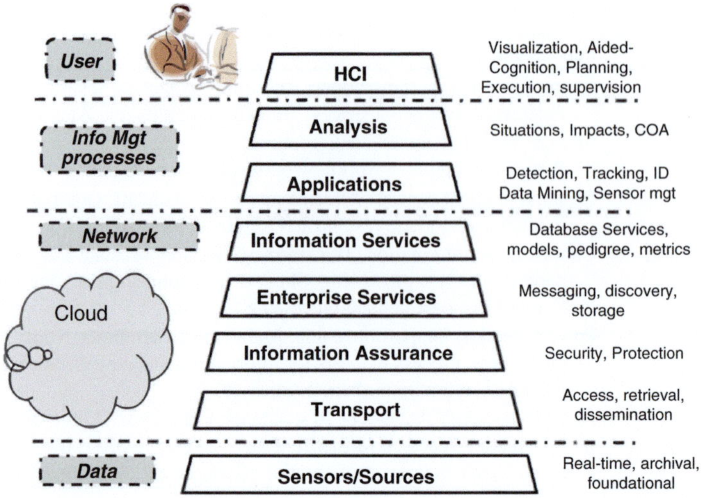

Fig. 6 Layered information services

4 Ontologies

Ontologies [27] are important for the submission and access to the cloud which consists of protocols, networks, and data characterization. The ontology supports the processing of information for measures of performance and measures of effectiveness. An ontology can be defined in many ways. In Artificial Intelligence (AI), and in computer science in general, an ontology refers to an engineering artifact, made up of a specific vocabulary used to describe a certain reality [5]. The ontology definition also includes a set of explicit assumptions regarding agreement on the intended meaning of the vocabulary words. A number of ontologies have been developed for use in information systems [2,30,37,58]; however they were primarily concerned with characterizing and tracking data associated with the systems assets and not the content of the data and streaming information. Work in the cyber forensics for security includes: *Cyber Forensic Ontology* [54], *Holistic Network Defense*, *Fusing Host and Network Features for Attack Classification*, and *Security Ontology for Annotating Resources*. These ontologies could contribute to security and targeting solutions for C2 services. Examples include processing textual reports of target identities from human observations.

Promising research in unsupervised ontology induction from text [52] has shown that semantic content can be extracted from unstructured content using a method that induces and populates a probabilistic ontology using dependency-parsed text as input. The unsupervised ontology induction approach has been found to be effective in extracting a medical knowledge base and could easily be extended into the cyber domain. In the case of unstructured textual data (or an oral stream converted to its textual representation), the transformation process consists of several stages of Natural Language Processing (NLP). These stages will differ depending on the goal of the process, and could include, among others: identification

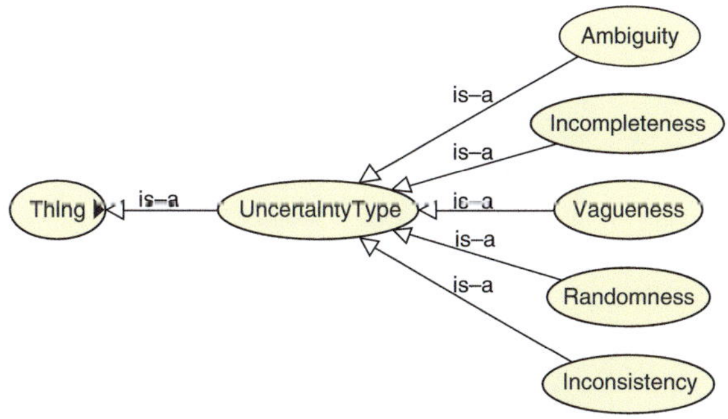

Fig. 7 URREF ontology: uncertainty type class (Reproduce from Costa et al. [28])

of word and sentences boundaries (text segmentation); assignment of labels to each word indicating its type (part-of-speech tagging); syntactic representation of each sentence according to the languages grammar (parsing); and selection of the meaning of ambiguous words (word sense disambiguation). At all these stages, and in order to make the right choices, NLP systems need to use syntactic and semantic knowledge of the language, word meanings and restrictions of use, word relations (ontology), and the document context [23].

The contemporary research in hard-soft fusion (HSF) allows for heterogeneous data to be aligned semantically, in order to provide a pragmatic course of action. Hard data includes the imagery and physical sensors, while soft data includes textual reports. HSF can be extremely helpful when data has been generated by a number of disparate sources and a cross alignment of *like* data is needed to help prepare data to be ingested by a reasoning engine. To process HSF data, uncertainty calculations are needed. An example from the uncertainty representation and reasoning evaluation framework (URREF) ontology is shown in Fig. 7 as a subset of the entire ontology for uncertainty analysis. For example, the textual classification can help reduce the ambiguity of terminology such as vehicle and car which are essentially the same words. The case for imagery analysis is the features processed for determining a vehicle type based on the randomness and probability calculations. To demonstrate the ability for uncertainty reduction using an ontology, we next give an example of how a Cloud-Enabled Bayes Network (CEBN) can instantiate information fusion analysis over a large set of data for simultaneous tracking and identification (STID).

5 Cloud-Enabled Bayes Network for WAMI Target Tracking

Information fusion developments include large data (e.g. imagery), flexible autonomy (e.g., from moving airborne platforms over communication systems), and human coordination for situation awareness [16]. Figure 8 demonstrates a layered architecture (Fig. 6) imagery data collection example using electro-optical (EO) cameras, millimeter-wave infrared (WMIR), and large area image recorder (LAIR) for WAMI. Information fusion for situation awareness based on imagery includes:

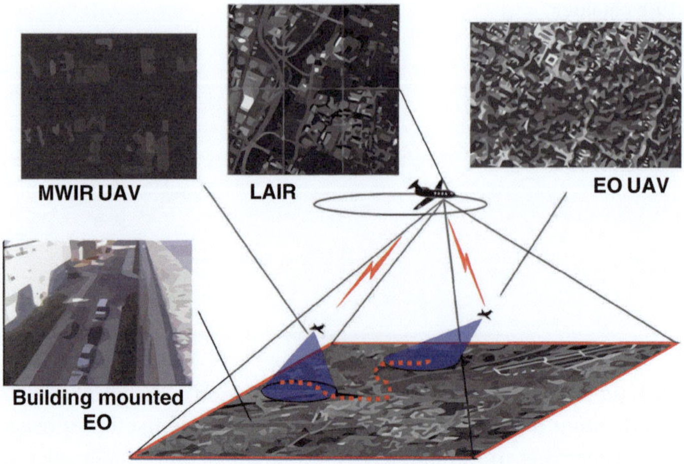

Fig. 8 WAMI information fusion (Reproduced from Mendoza-Schrock et al. [47])

(1) tracking targets in images (fusion over time) [52], (2) identifying targets using different sensors (fusion over frequency) [63], and (3) linking target measurements over wide areas (fusion over space) [45]. Demonstrated examples include the cursor on target paradigm with a limited HTML schema of target allegiance, uncertainty location, and priority. Inherent in the illustration are the collections from different sensors; however, what is needed are elements from the cloud of the physical (terrain), the resource (sensors), and the social (objects) context that come from services and resources in clouds.

5.1 Bayes Network and Cloud Computing Interactions

One recent concern for the scientific community is the ability to process large amounts of data (e.g., biological health science, social economics, and law enforcement). Examples of methods include cluster, cloud, grid, and heterogeneous computing which are compared by Schadt et al. [55].

- *Cluster computing*: Use a standard technique in information fusion of Bayes networks.
- *Cloud computing*: Search databases for relevant information so that many clusters can be transformed to work within the cloud to access relevant data.
- *Grid computing*: Build a network of distributed agents to solve a task.
- *Heterogeneous computing*: Include specific methods of speed-up such as parallelism from a graphics processing unit (GPU) [33, 45].

Likewise, there are other related methods [53] :

- *Super computing*: Enable increased computations, usually in a parallel fashion.
- *Utility computing*: Rent time and access for resources and bandwidth.

Cloud layers include:

- *Platform as a Service* (PaaS): Include basic applications (e.g., Google Maps).
- *Software as a Service* (SaaS): Host the application and data on databases at their own data center.
- *Infrastructure as a Service* (IaaS): Provide software over the Internet.

Given the various computing environments, there is an interest in high performance cloud computing which makes use of increased computations of a high-performance computer as well as a cloud-enabled environment. Information fusion applications can make use of the enterprise architecture (see Fig. 6) which could include local networking for cloud information management for image processing in the cloud. The question would be what is the value of the cloud? Cloud auditing would enable access to large data sets for a priori information (IaaS), ability to exploit streaming data in the cloud as a service (SaaS), and associating different data sets from different platform applications (PaaS). The cloud environment enables data sharing, storing, and indexing, while providing security and time scaling.

For this application, we combine elements of cluster computing (i.e., information fusion by combining relevant information for a Bayesian analysis of data and exploited features) and cloud computing (i.e., database analysis of a priori target identity information). PaaS is the STID application with SaaS maintaining the Bayesian processing and IaaS supporting the data passing and messaging.

From Fig. 9, we are interested in a cognitive-based approach to STID wherein the vehicles are to be tracked and identified in real time. Basic elements of the information fusion result from a Bayesian analysis of a family of situation events $\{T_i\}$, where T_i is the target ID for target i, $i = 1, \cdots, n$. We model the target ID as a random variable T taking its values in the set $\{T_i\}$. Let S be a fixed sensor signal that has been observed. For each possible target ID T_j belonging to the set $\{T_i\}$, we assume that $P(T_j)$, the "a priori" probability that T can have ID T_j, is known or can be accurately estimated from information in the cloud. Likewise, we assume that $P(S|T_j)$, the conditional probability that sensor signal S can occur given target ID T_j, is known or can be estimated. In order to form an estimate of T, each "a posteriori" conditional probability $P(T_i|S)$ needs to be calculated. Bayes' rule tells us that

$$P(T_i|S) = \frac{P(S|T_i)P(T_i)}{\sum_j P(S|T_j)P(T_j)} \tag{1}$$

For the given observed sensor signal S, one can then form an estimate $\hat{T}$ of the random target position T as that T_i for which $P(T_i|S)$ is a maximum; $\hat{T}$ is called the maximum a posteriori probability estimate (MAP estimate) of T based on S.

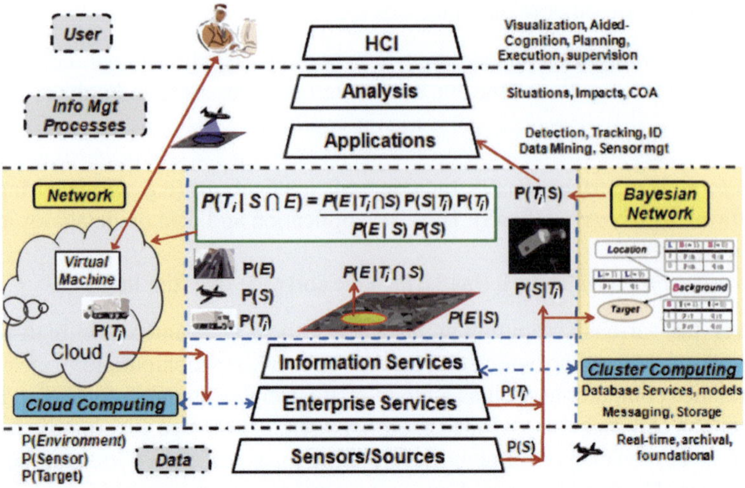

Fig. 9 Information fusion (Bayes network) and cloud computing interactions

Given that the cloud can enable more computations and analysis over a larger collection of stored information, we can bring environmental information into play in addition to sensor information in forming an estimate of the target position. Again by Bayes' rule, we have the following conditional probability that the random target T will have ID T_i given fixed sensor signal S and fixed environment E:

$$P(T_i|S \cap E) = \frac{P(E|T_i \cap S)P(S|T_i)P(T_i)}{P(E|S)\sum_j P(S|T_j)P(T_j)} \tag{2}$$

The MAP estimate $\hat{T}$ of T given S and E would then be the target ID T_i for which the left side of (2) is a maximum. To distinguish between the two types of MAP estimates, let $\hat{T}^1$ be the MAP estimate for T given S, and let $\hat{T}^2$ be the MAP estimate for T given S and E. For the first MAP estimate, $P(T \neq \hat{T}^1|S)$, the probability that the estimate is in error given S, is calculated as

$$P(T \neq \hat{T}^1|S) = 1 - \max_j P(T_j|S),$$

but then we have to average over the different possibilities S, giving us overall estimation error $P(T \neq \hat{T}^1)$ computed as

$$\begin{aligned} P(T \neq \hat{T}^1) &= \sum_S P(T \neq \hat{T}^1|S)P(S) \\ &= 1 - \sum_S \max_j P(T_j|S)P(S) \\ &= 1 - \sum_S \max_j P(T_j)P(S|T_j). \end{aligned}$$

Similarly,

$$P(T \neq \hat{T}^2) = 1 - \sum_{S,E} \max_j P(E|T_j \cap S)P(S|T_j)P(T_j).$$

The mathematics tells us that the probability of estimation error $P(T \neq \hat{T}^2)$ using estimate $\hat{T}^2$ is less than or equal to the probability of estimation error $P(T \neq \hat{T}^1)$ using estimate $\hat{T}^1$, which is intuitively clear because more information is being used in forming the estimate $\hat{T}^2$. To decrease the probability of estimation error still further, we can use multiple sensors rather than just one sensor. In this case, the formula for the conditional probability of target ID T_i given the multiple sensor readings and the environment is somewhat more complicated to derive and so we omit the computation, which can be found in [24, 57, 69].

Using the layered functional hierarchy of information fusion and management functions, Fig. 9 brings together the ideas of cloud computing and information fusion (as a Bayes network [41, 61]). Starting from the upper left, a user queries the cloud service for information (such as background a priori information on targets to include make, amount in area, etc.). The cloud can also provide contextual information on the environment such as road networks, political activity, and weather. At the same time, the user calls a sensor to make a data collection. Together both an *enterprise service* from the cloud and *information service* from the sensor are activated.

A Bayes network can be used locally to estimate the target position from on-board analysis and then delivered to the application service. In a cloud-enabled information fusion solution, a key fact is that the cloud provides storage and access to an enormous amount of information. Thus, additional details on the environment and targets are readily available to refine the map estimate $\hat{T}$ of target position in order to make the estimate more accurate. In forming these refined estimates via increased use of cloud information, there are three important points to keep in mind:

1. The information needed varies over many conditions and the cloud's storage ability affords a refined estimate of the environment and targeting information (*Enterprise Service*)
2. The computations needed also vary based on the situation and information in the cloud allow for data to be processed for information (*Information Service*)
3. The ability to connect through a cloud enables the combination of different sensors and collections for Information Fusion (*Communication Service*)

Exploiting all the services available in the cloud allow a Bayesian network to provide more accurate estimates than would be possible based on limited world knowledge. An example follows.

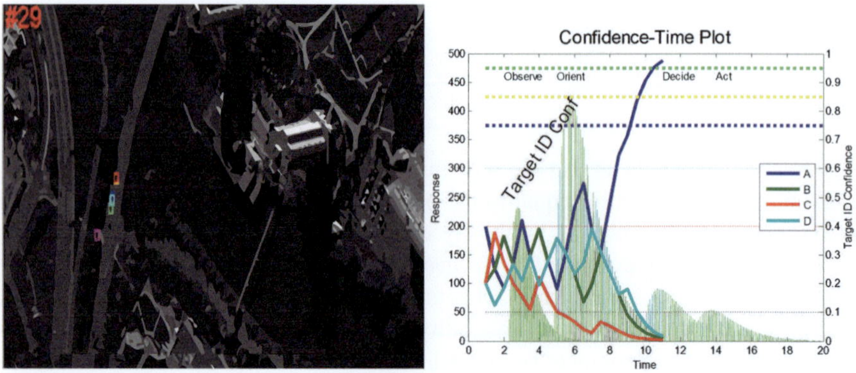

Fig. 10 (*Left*) *Annotated feature analysis* using different tracking results (Reproduced from Ling et al. [45]). (*Right*) *Decision Support Timeliness-Confidence Plot* reporting from cumulated measurements in an OODA time analysis (Reproduced from Blasch et al. [16])

5.2 CEBN-Based WAMI Tracking and Identification

Simultaneous tracking and identification (STID) require a priori information for enhanced accuracy, timeliness, and confidence in decision making. To better enable the analysis, cloud computing serves three useful aspects of future technologies to aid the contextual processing. For tracking, the environment can serve as the world model (gained from a cloud service for any area on the Earth), while the a priori is the road networks, vehicle capabilities, vehicle make, etc. that help refine the position location and target prediction. Target identification can be greatly enhanced by assessing the activity analysis of the vehicle to include the type, intention, and interaction with other vehicles which can come from the physical (imagery) and human observations (text). Persistent target tracking requires an analysis of likelihood functions extracted from the WAMI over the targets [50, 51].

Examples from the cloud include the vehicle data, the social context (e.g. rhythm of the city), and the political (rules of the road). Figure 10 demonstrates that if the cloud provides a priori information, then the timeliness, confidence, and accuracy of the objects can be assessed as related to the uncertainty for each metric. Given a WAMI image and the determination of four targets on a road, we can use a CEBN to identify targets. Using the cognitive-observe-orient-decide-act (C-OODA) paradigm, we can improve the user query for object assessment in a time-confidence plot. For example, when the user designates an area of interest (Fig. 9), the machine can then detect and track targets (observe). After a few time steps, the machine can access information through data mining (orient) from the cloud to enhance the Bayesian analysis of the situation. Together cloud computing and information fusion aid to determine (decide/evaluate) of the target type and activity. Finally, the results are used to query the sensors to get more information (act), store the results and disseminate back to the cloud.

Next, we explore a recent trend in cloud computing of Google Fusion Tables.

6 Google Fusion Tables: Information Fusion on Cloud Platform

Cloud providers offer data management, integration, and collaboration tasks in cloud computing environments. In particular, outsourcing data-intensive and compute-intensive information fusion tasks to cloud service is a natural solution for applications in which either on-site computing power is insufficient or decision making requires integrated analysis of data collected by distributed sensors or monitors. For example, many research efforts have been reported to relieve the burden of information fusion for wireless sensor networks (WSNs) to cloud service platforms [40, 59]. In this section, Google Fusion Tables [31, 32, 34] is selected to illustrate important design principles for cloud-based information fusion applications.

Initially launched in June 2009, the Google Fusion Tables service is a cloud-based data management and integration service [34], which aimed at to meet three important requirements [32]:

1. Have the capability of supporting collaborative operations among multiple users and/or organizations.
2. Provide user-friendly applications.
3. Integrate the information fusion functions to web services.

The objective of the Google Fusion Tables is to exploit the cloud computing facility to achieve high efficient data utility. Below are the small set of guiding principles the design follows [32], which enable a continuous improvement in both the user experience and the performance of Fusion Tables:

1. Provide seamless integration with web services.
2. Emphasize ease of use.
3. Provide incentives for data sharing.
4. Facilitate collaboration.

Figure 1 in [32] shows the system architecture of Google Fusion Tables service. The requests of data processing and fusion are from multiple sources connected to the front end dispatcher: the Fusion Tables website, other applications using the application programming interface (API), and visualizations that are embedded in other web pages, for example, charts. Google Maps is one of the featured services currently integrated into Fusion Table application. The front end dispatcher formats the received queries into a predefined representation and sends them to the query processing module, which generates the query plan by parsing the query and conducts optimization. At the back end, the query plan is executed. Google Maps is an application service that supports the STID methods for elements of context (e.g., road networks, path of travel, speed considerations, etc.).

Scalability and throughput are the main challenges the storage plane needs to handle. In particular, there are hundreds of thousands of tables with different schemes, sizes, and query characteristics. A two-layer storage stack is adopted in Google Fusion Tables: BigTable and Megastore. The BigTable stores information in form of tuples (key, value), which are stored and shared on the key. Writing and flexible reading operations are supported by BigTable. As a library on top of BigTable and Megastore provides higher level primitives. The library is used for three purposes: (1) maintaining property indexes, (2) providing table level transactions, and (3) replicating tables across multiple data centers.

As a cloud-based data management and information fusion service, the Google Fusion Tables focuses on collaborative data processing. It has successfully supported seamless integration of map features based on the Google Maps infrastructure. However, more work is to be conducted to provide users more experiences on other applications. While BigTable and Megastore helped to achieve a scalable and replicated data store, this two layer storage made certain type of Structured Query Language (SQL) queries trickier. Many extensions are expected to make the Fusion Tables easier for user and more powerful to process larger data sets.

Using the Google Fusion Tables paradigm, there are future trends that support CEBNs. The BigTable is the information provides the ontology of queries such as location, target identity, and sensor attributes (shown in Fig. 9). The Megastore can provide the consistency of analysis across distributed centers such as a hand off of STID results from one sensor to another (shown in Fig. 8). Elements of coordination between the Google Fusion Tables enables HLIF products from LLIF capabilities. Together the Google Fusion Tables provides enterprise services (shown in Fig. 6).

The processing of larger data sets moves from LLIF of object tracking and identification to HLIF of situation awareness and analysis [21]. New paradigms from CEBN using Google tables can enhance Sensor, User, Mission (SUM) resource management over information and data for Level 2/3 information fusion situation and threat awareness [7]. Such examples from the cloud enables linking individual targets to group behaviors [9], road contextual information [66], and net-centric sensor management [67] to extend dynamic track lifetime. With a CEBN, new elements of LLIF/HLIF are available such as high performance computing solutions [39], trust-based search over communication network systems [56], and aspects of political and cultural effects [15]. Without a cloud-enabled environment, access, storage, and recall of data from large databases, is not practical for real-time applications. Table 3 serves as an organization on on-going trends, over the big data, user coordination (metrics), and architectures for flexible autonomy.

7 Conclusion

The integration of information fusion and cloud computing can provide useful developments for both communities. The highlight is ontologies, metadata formats (protocols), metrics (e.g. security), and SOA. Future developments would require cloud standards, such as the IEEE P2301 Draft Guide for *Cloud Portability and Interoperability Profiles* [38]. Inherent in the operational deployment is the standards that exist from policy decisions of the web-based community over user, industry, and international regulations. The limitations include the accepted and approved

Table 3 Category of concerns for contemporary developments in cloud computing

Concerns	Big Data	Standards	Metrics	Architecture
Service	Ontology based access	Metadata formats	QoS, IQ, security multi-modal	SOA interoperable
Policy	Tagged	Industry standard	Information theoretic	Open source
Processing	Extraction	Compression	Confidence & timeliness	Extensible
Collaboration	Collection & management	Unified data language (schemas)	Visualization (social network analysis)	Shared architectures (Bayes networks)
Limitations	Protocols	Approval	Representations	Graph theory modeling
Performance	Database storage	Protocols	Computing analysis	Layered sensing

formats, standards, representations and graphical methods. Together performance would increase for both information fusion through database access and protocols; while cloud computing would be enhanced through standardized metrics of service and hierarchical/layered approaches to information dissemination of data through information fusion (estimation, filtering, and prediction).

Acknowledgements This material is based upon work partially supported by the Air Force Office of Scientific Research (AFOSR) and the Air Force Research Laboratory (AFRL) Visiting Faculty Research Program (VFRP) extension grant LRIR 11RI01COR. The authors appreciate the insightful directions from Dr. Frederica Darema of the Dynamic Data Driven Application System (DDDAS) concept for big data concerns. The views and conclusions contained herein are those of the authors and should not be interpreted as necessarily representing the official policies or endorsements, either expressed or implied, of the Air Force.

References

1. Agrawal, D., Das, S., El Abbadi, A.: Big data and cloud computing: current state and future opportunities. In: Proceedings of the 14th International Conference on Extending Database Technology, EDBT/ICDT'11, Uppsala, pp. 530–533. ACM, New York (2011). doi:10.1145/1951365.1951432
2. Assadi, H.: Construction of a regional ontology from text and its use within a documentary system. In: Proceedings of the 1st International Conference on Formal Ontology in Information Systems, FOIS'98, Trento (1998)
3. Baird, S.A.: Heterogeneity and interoperability at the core: SOA, virtualization, the cloud and the government role. In: Proceedings of the 4th International Conference on Theory and Practice of Electronic Governance, ICEGOV'10, Beijing, pp. 387–388. ACM, New York (2010). doi:10.1145/1930321.1930409
4. Blasch, E.: Situation, impact, and user refinement. In: Proceedings of the SPIE, Orlando, vol. 5096, pp. 463–1134 (2003). doi:10.1117/12.542897

5. Blasch, E.P.: Ontological issues in higher levels of information fusion: user refinement of the fusion process. In: Proceeding of the 6th International Conference on Information Fusion, FUSION'03, Cairns, pp. 634–641 (2003)
6. Blasch, E.: Modeling intent for a target tracking and identification scenario. In: Proceedings of the SPIE, Orlando, vol. 5428, pp. 260–851 (2004). doi:10.1117/12.542897
7. Blasch, E.: Sensor, user, mission (SUM) resource management and their interaction with level 2/3 fusion. In: Proceedings of the 9th International Conference on Information Fusion, FUSION'06, Florence (2006). doi:10.1109/ICIF.2006.301791
8. Blasch, E.: User refinement in information fusion, Chap. 19. In: Liggins, M.E., Hall, D.L., Llinas, J. (eds.) Handbook of Multisensor Data Fusion, 2nd edn. CRC, Boca Raton (2008)
9. Blasch, E., Connare, T.: Improving track maintenance through group tracking. In: Proceedings of the Workshop on Estimation, Tracking and Fusion: A Tribute to Yaakov Bar-Shalom, pp. 360–371 (2001)
10. Blasch, E., Hanselman, P.: Information fusion for information superiority. In: Proceedings of the 2000 IEEE National Aerospace and Electronics Conference, NAECON'00, pp. 290–297 (2000). doi:10.1109/NAECON.2000.894923
11. Blasch, E., Plano, S.: Level 5: user refinement to aid the fusion process. In: Proceedings of the SPIE, vol. 5099, pp. 288–735 (2003). doi:10.1117/12.486899
12. Blasch, E., Plano, S.: DFIG level 5 (user refinement) issues supporting situational assessment reasoning. In: Proceeding of the 8th International Conference on Information Fusion, FUSION'05, Philadelphia (2005). doi:10.1109/ICIF.2005.1591830
13. Blasch, E., Kadar, I., Salerno, J., Kokar, M.M., Das, S., Powell, G.M., Corkill, D.D., Ruspini, E.H.: Issues and challenges in situation assessment (level 2 fusion). J. Adv. Inf. Fusion **1**(2), 122–139 (2006)
14. Blasch, E., Kadar, I., Hintz, K., Biermann, J., Chong, C., Das, S.: Resource management coordination with level 2/3 fusion issues and challenges. IEEE Aerosp. Electron. Syst. Mag. **23**(3), 32–46 (2008). doi:10.1109/MAES.2008.4476103
15. Blasch, E., Valin, P., Bosse, E., Nilsson, M., Laere, J.V., Shahbazian, E.: Implication of culture: user roles in information fusion for enhanced situational understanding. In: Proceedings of the 12th International Conference on Information Fusion, FUSION'09, Seattle, pp. 1272–1279 (2009)
16. Blasch, E., Breton, R., Valin, P.: Information fusion measures of effectiveness (MOE) for decision support. In: Proceedings of the SPIE, Orlando, vol. 8050 (2011). doi:10.1117/12.883988
17. Blasch, E., Breton, R., Valin, P., Bosse, E.: User information fusion decision making analysis with the C-OODA model. In: Proceedings of the 14th International Conference on Information Fusion, FUSION'11, Chicago (2011)
18. Blasch, E., Deignan, Jr., P.B., Dockstader, S.L., Pellechia, M., Palaniappan, K., Seetharaman, G.: Contemporary concerns in geographical/geospatial information systems (gis) processing. In: Proceedings of the 2011 IEEE National Aerospace and Electronics Conference, NAECON'11, Dayton, pp. 183–190 (2011). doi:10.1109/NAECON.2011.6183099
19. Blasch, E., Russell, S., Seetharaman, G.: Joint data management for MOVINT data-to-decision making. In: Proceedings of the 14th International Conference on Information Fusion, FUSION'11, Chicago (2011)
20. Blasch, E., Salerno, J.J., Tadda, G.: Measuring the worthiness of situation assessment. In: Proceedings of the 2011 IEEE National Aerospace and Electronics Conference, NAECON'11, Dayton (2011). doi:10.1109/NAECON.2011.6183083
21. Blasch, E., Costa, P.C.G., Laskey, K.B., Stampouli, D., Ng, G.W., Schubert, J., Nagi, R., Valin, P.: Issues of uncertainty analysis in high-level information fusion – Fusion 2012 panel discussion. In: Proceedings of the 15th International Conference on Information Fusion, Edinburgh, FUSION'12 (2012)
22. Blasch, E., Lambert, D.A., Valin, P., Kokar, M.M., Llinas, J., Das, S., Chong, C.Y., Shahbazian, E.: High level information fusion (HLIF) survey of models, issues, and grand challenges. IEEE Aerosp. Electron. Syst. Mag. **27**(9), 4–20 (2012). doi:10.1109/MAES.2012.6366088

23. Blasch, E.P., Bosse, E., Lambert, D.A.: High-Level Information Fusion Management and Systems Design. Artech House, Norwood (2012)
24. Bruzzone, L.: An approach to feature selection and classification of remote sensing images based on the bayes rule for minimum cost. IEEE Trans. Geosci. Remote Sens. **38**(1), 429–438 (2000). doi:10.1109/36.823938
25. Chen, G., Shen, D., Kwan, C., Cruz, J., Kruger, M., Blasch, E.: Game theoretic approach to threat prediction and situation awareness. J Adv Inf. Fusion **2**(1), 1–14 (2007)
26. Chen, G., Blasch, E., Shen, D., Chen, H., Pham, K.: Services oriented architecture (SOA) based persistent isr simulation system. In: Proceedings of the SPIE, Orlando, vol. 7694 (2010). doi:10.1117/12.849783
27. Costa, P., Carvalho, R., Laskey, K., Park, C.: Evaluating uncertainty representation and reasoning in HLF systems. In: Proceeding of the 14th International Conference on Information Fusion, FUSION'11, Chicago, pp. 1–8 (2011)
28. Costa, P.C.G., Laskey, K.B., Blasch, E., Jousselme, A.L.: Towards unbiased evaluation of uncertainty reasoning: the URREF ontology. In: Proceedings of the 15th International Conference on Information Fusion, FUSION'12, Edinburgh (2012)
29. Das, S., Agrawal, D., El Abbadi, A.: ElasTraS: an elastic transactional data store in the cloud. In: Proceedings of the 2009 Conference on Hot Topics in Cloud Computing, HotCloud'09, San Diego. USENIX Association, Berkeley (2009)
30. Fenz, S., Ekelhart, A.: Formalizing information security knowledge. In: Proceedings of the 4th International Symposium on Information, Computer, and Communications Security, ASIACCS'09, Sydney, pp. 183–194. ACM, New York (2009) doi:10.1145/1533057.1533084
31. Gonzalez, H., Halevy, A., Jensen, C.S., Langen, A., Madhavan, J., Shapley, R., Shen, W.: Google fusion tables: data management, integration and collaboration in the cloud. In: Proceedings of the 1st ACM Symposium on Cloud Computing, SoCC'10, Indianapolis, pp. 175–180. ACM, New York (2010). doi:10.1145/1807128.1807158
32. Gonzalez, H., Halevy, A.Y., Jensen, C.S., Langen, A., Madhavan, J., Shapley, R., Shen, W., Goldberg-Kidon, J.: Google fusion tables: web-centered data management and collaboration. In: Proceedings of the 2010 ACM SIGMOD International Conference on Management of Data, SIGMOD'10, Indianapolis, pp. 1061–1066. ACM, New York (2010). doi:10.1145/1807167.1807286
33. Grauer-Gray, S., Kambhamettu, C., Palaniappan, K.: GPU implementation of belief propagation using CUDA for cloud tracking and reconstruction. In: IAPR Workshop on Pattern Recognition in Remote Sensing (PRRS), Tampa (2008)
34. Halevy, A., Shapley, R.: googleresearch.blogspot.com, Google fusion tables. http://goo.gl/6jX09 (2009)
35. Kessler, O., White, F.: Data fusion perspectives and its role in information processing, Chap. 2. In: Liggins, M.E., Hall, D.L., Llinas, J. (eds.) Handbook of Multisensor Data Fusion, 2nd edn. CRC, Boca Raton (2008)
36. Khan, Z., Ludlow, D., McClatchey, R., Anjum, A.: An architecture for integrated intelligence in urban management using cloud computing. J. Cloud Comput. Adv. Syst. Appl. **1**(1) (2012). doi:10.1186/2192-113X-1-1
37. Kim, A., Luo, J., Kang, M.: Security ontology for annotating resources. In: Proceedings of the 2005 OTM Confederated International Conference on the Move to Meaningful Internet Systems: CoopIS, COA, and ODBASE – Volume Part II, OTM'05, Agia Napa, pp. 1483–1499. Springer, Berlin/Heidelberg (2005). DOI 10.1007/11575801_34
38. Kowalenko, K.: ieee.org, Standards for seamless cloud computing. http://goo.gl/ajLfS (2012)
39. Kumar, P., Palaniappan, K., Mittal, A., Seetharaman, G.: Parallel blob extraction using the multi-core cell processor. In: Blanc-Talon, J., Philips, W., Popescu, D., Scheunders, P. (eds.) Advanced Concepts for Intelligent Vision Systems. Lecture Notes in Computer Science, vol. 5807, pp. 320–332, Springer, New York (2009)
40. Kurschl, W., Beer, W.: Combining cloud computing and wireless sensor networks. In: Proceedings of the 11th International Conference on Information Integration and Web-based Applications & Services, iiWAS'09, Kuala Lumpur, pp. 512–518. ACM, New York (2009). doi:10.1145/1806338.1806435

41. LarrañAga, P., Karshenas, H., Bielza, C., Santana, R.: A review on evolutionary algorithms in bayesian network learning and inference tasks. Inf. Sci. **233**, 109–125 (2013). doi:10.1016/j.ins.2012.12.051
42. Li, B., Yan, X.: Modeling of ambient intelligence based on information fusion and service oriented computing. In: Proceedings of the 5th International Conference on Ubiquitous Information Technologies and Applications, CUTE'10, Sanya, pp. 1–5 (2010). doi:10.1109/ICUT.2010.5677852
43. Liggins, M.E., Chang, K.C.: User refinement in information fusion, Chap. 17. In: Liggins, M.E., Hall, D.L., Llinas, J. (eds.) Handbook of Multisensor Data Fusion 2nd edn. CRC, Boca Raton (2008)
44. Linderman, M., Haines, S., Siegel, B., Chase, G., Ouellet, D., O'May, J., Brichacek, J.: A reference model for information management to support coalition information sharing needs. In: Proceeding of the 2005 International Command and Control Research and Technology Symposium, ICCRTS'05, Washington, DC (2005)
45. Ling, H., Wu, Y., Blasch, E., Chen, G., Bai, L.: Evaluation of visual tracking in extremely low frame rate wide area motion imagery. In: Proceedings of the 14th International Conference on Information Fusion, Chicago (2011)
46. Mazur, S., Blasch, E., Chen, Y., Skormin, V.: Mitigating cloud computing security risks using a self-monitoring defensive scheme. In: Proceedings of the IEEE 2000 National Aerospace and Electronics Conference, NAECON'11, Dayton, pp. 39–45 (2011). doi:10.1109/NAECON.2011.6183074
47. Mendoza-Schrock, O., Patrick, J.A., Blasch, E.P.: Video image registration evaluation for a layered sensing environment. In: Proceedings of the 2009 IEEE National Aerospace and Electronics Conference, NAECON'09, Dayton (2009). doi:10.1109/NAECON.2009.5426624
48. Nathuji, R., Kansal, A., Ghaffarkhah, A.: Q-clouds: managing performance interference effects for QoS-aware clouds. In: Proceedings of the 5th European Conference on Computer Systems, EuroSys'10, Paris, pp. 237–250. ACM, New York (2010). doi:10.1145/1755913.1755938
49. O'Brien, L., Brebner, P., Gray, J.: Business transformation to soa: aspects of the migration and performance and QoS issues. In: Proceedings of the 2nd International Workshop on Systems Development in SOA Environments, SDSOA'08, Leipzig, pp. 35–40. ACM, New York (2008). doi:10.1145/1370916.1370925
50. Palaniappan, K., Bunyak, F., Kumar, P., et al.: Efficient feature extraction and likelihood fusion for vehicle tracking in low frame rate airborne video. In: Proceedings of the 13th International Conference on Information Fusion, FUSION'10, Edinburgh (2010)
51. Pelapur, R., Candemir, S., Poostchi, M., Bunyak, F., Wang, R., Seetharaman, G., Palaniappan, K.: Persistent target tracking using likelihood fusion in wide-area and full motion video sequences. In: Proceedings of the 15th International Conference on Information Fusion, FUSION'12, Singapore (2012)
52. Poon, H., Domingos, P.: Unsupervised ontology induction from text. In: Proceedings of the 48th Annual Meeting of the Association for Computational Linguistics, ACL'10, Uppsala, Uppsala, pp. 296–305. Association for Computational Linguistics (2010)
53. Puig, E.J., Kwasniewksi, T.J.: Cloud computing in the government: a DACS critical review and technology assessment. In: DACS report 518136 (2011)
54. Raskin, V., Taylor, J.M., Hempelmann, C.F.: Ontological semantic technology for detecting insider threat and social engineering. In: Proceedings of the 2010 Workshop on New Security Paradigms, NSPW'10, Concord, pp. 115–128. ACM, New York (2010). doi 10.1145/1900546.1900563
55. Schadt, E., Linderman, M., Sorenson, J., Lee, L., Nolan, G.P.: Computational solutions to large-scale data management and analysis. Nat. Rev. Genet. **11**, 647–657 (2010). doi:10.1038/nrg2857
56. Shen, D., Chen, G., Pham, K., Blasch, E.: A trust-based sensor allocation algorithm in cooperative space search problems. In: Proceedings of the SPIE, Baltimore, vol. 8044 (2011). doi:10.1117/12.882904

57. Sotoca, J.M., Sanchez, J.S., Pla, F.: Attribute relevance in multiclass data sets using the naive bayes rule. In: Proceedings of 17th International Conference on the Pattern Recognition, ICPR'04, Cambridge, vol. 3, pp. 426–429. IEEE Computer Society, Washington, DC (2004). doi:10.1109/ICPR.2004.188
58. Takahashi, T., Kadobayashi, Y., Fujiwara, H.: Ontological approach toward cybersecurity in cloud computing. In: Proceedings of the 3rd International Conference on Security of Information and Networks, SIN'10, Taganrog, pp. 100–109. ACM, New York, (2010). doi:10.1145/1854099.1854121
59. Tan, K.L.: What's NExT?: Sensor + Cloud!? In: Proceedings of the 7th International Workshop on Data Management for Sensor Networks, DMSN'10, Singapore, pp. 1–1. ACM, New York (2010). doi:10.1145/1858158.1858160
60. Tian, X., Tian, Z., Blasch, E., Pham, K., Shen, D., Chen, G.: Performance analysis of sliding window energy detection for spectrum sensing. J. Comput. Netw. Commun. Special Issue Trends Appl. Cogn. Radio (2012, Submitted)
61. Wang, Z., Chan, L.: Learning bayesian networks from markov random fields: an efficient algorithm for linear models. ACM Trans. Knowl. Discov. Data **6**(3), 10:1–10:31 (2012). doi:10.1145/2362383.2362384
62. Wilde, N., Simmons, S., Pressel, M., Vandeville, J.: Understanding features in SOA: some experiences from distributed systems. In: Proceedings of the 2nd International Workshop on Systems Development in SOA Environments, SDSOA'08, Leipzig, pp. 59–62. ACM, New York (2008). doi:10.1145/1370916.1370931
63. Wu, Y., Blasch, E., Chen, G., Bai, L., Ling, H.: Multiple source data fusion via sparse representation for robust visual tracking. In: Proceedings of the 14th International Conference on Information Fusion, Chicago (2011)
64. Wu, Y., Chen, G., Blasch, E., Ling, H.: Feature based background registration in wide area motion imagery. In: Proceedings of the SPIE, Baltimore, vol. 8402 (2012). doi:10.1117/12.918804
65. Yang, C., Blasch, E.: Pose angular-aiding for maneuvering target tracking. In: Proceedings of the 8th International Conference on Information Fusion, FUSION'05, Philadelphia (2005)
66. Yang, C., Blasch, E.: Fusion of tracks with road constraints. J. Adv. Inf. Fusion **3**(1), 14–32 (2008)
67. Yang, C., Blasch, E.: Performance measures of covariance and information matrices in resource management for target state estimation. IEEE Trans. Aerosp. Electron **48**(3), 2594–2613 (2012)
68. Yu, W., Wang, X., Fu, X., Xuan, D., Zhao, W.: An invisible localization attack to internet threat monitors. IEEE Trans. Parallel Distrib. Syst. (TPDS) **20**(11), 1611–1625 (2009). doi:10.1109/TPDS.2008.255
69. Zhang, Y., Ji, Q.: Active and dynamic information fusion for multisensor systems with dynamic bayesian networks. IEEE Trans. Syst. Man Cybern. Part B **36**(2), 467–472 (2006). doi:10.1109/TSMCB.2005.859081

Part II
Cloud Auditing and Assured Information Sharing

Diagnosing Vulnerability Patterns in Cloud Audit Logs

Rui Xie, Rose Gamble, and Norman Ahmed

Abstract A service cloud architecture that allows web service compositions to answer complex requests improves the accessibility and flexibility of web services from different vendors. However, security issues exist in the service cloud, including both vulnerabilities of traditional web service communications and new issues brought by inter-cloud communications. Cloud-wide auditing to uncover security issues is a complex task due to the large scale and decentralized structure of the cloud environment. Existing security standards, protocols and auditing mechanisms can provide audit logs, but in most cases, these logs cannot pinpoint type, location, and impact of threats. Given a cloud architecture that specifies the scope of audit logs and a definition of the expected auditable events in the cloud providing evidence of potential threats, we define *Vulnerability Diagnostic Trees (VDTs)* to formally manifest vulnerability patterns across several audit trails generated within the service cloud. Our attack examples are based on the allocation of services to a web service composition that answers a client request through end-to-end round trip messaging.

"Approved for Public Release; Distribution Unlimited: 88ABW-2013-0074, 09-Jan-2013"

R. Xie (✉) • R. Gamble
University of Tulsa, Tulsa, OK, USA
e-mail: rui-xie@utulsa.edu; gamble@utulsa.edu

N. Ahmed
Air Force Research Laboratory, Rome, NY, USA
Purdue University, West Lafayette, IN, USA
e-mail: norman.ahmed@us.af.mil; ahmed24@purdue.edu

K.J. Han et al. (eds.), *High Performance Cloud Auditing and Applications*,
DOI 10.1007/978-1-4614-3296-8_5,

1 Introduction

Cloud computing has become an emerging model to provide dynamically composed Service-Oriented Architectures (SOAs) [3, 45] of web services from different vendors . This model underlies the service cloud architecture [10, 39]. In the service cloud, service vendors or owners contract with a cloud provider for their service to be hosted and deployed in response to user or client requests, possibly as part of a web service composition. Despite benefits, such as flexibility and easy-accessibility, allocating services into web compositions to create service invocation chains in the service cloud to resolve client requests can lead to a variety of security concerns. These security issues are not just the focus of existing SOA protocols and standards such as communication-based attacks [5, 42] and sensitive information disclosure [38, 39], but also contain new challenges brought by cloud environment, such as judging the trustworthiness of information from foreign services and clouds [9, 17].

Information assurance controls in multiple governing documents propose auditing requirements which can be used to reveal potential security issues within the information systems, such as the NIST SP800-53 [29], the DoDI 8500.2 [14] and related Security Technical Implementation Guides (STIGs), and the Common Criteria [12]. These controls are usually based on traditional components and component based systems, and transferring them to service-oriented cloud models is not seamless due to the lack of fundamental logging concepts for working within these environments. One reason for this adoption delay is their deployment of web services, which must often be forcibly embedded with logging mechanisms to capture the messaging, resource utilization, and session involvement activities.

In this chapter, we will use the concepts from basic control statements as the foundation of our auditing discussion, and extend the scope of auditing to more complicated SOA and cloud environments. Furthermore, we will review a threat model based on potential attacks in web compositions within a service cloud, including attacks specifically related to Simple Object Access Protocol (SOAP) [33, 34] messaging such as Extensible Markup Language (XML) rewriting attacks, denial of service (DoS) attacks aimed at compromising the communication link (e.g., packet dropping attack) or authentication entity (e.g., security token service (STS) bombardment) and threats violating message flow control policy (e.g., message tainting and resource misuse). We will discuss the related auditable events and use a previously defined abstract architecture for scoping captured events as audit records held by specific cloud entities. We will provide a diagnostic tool in the form of decision trees to show how using distinct audit assets captured at different abstraction levels can manifest security vulnerability patterns. This visual model illustrates the diagnosis process to give a more comprehensive view of certain threats that are translated to specific auditable events, while at the same time allowing the implementation of automated detection mechanisms when the records for these events are logged.

In order to state this kind of composition, we devise a *Vulnerability Diagnosis Tree (VDT)* that is structured to reflect the audit requirements of different cloud entities and the cross-associations among them to manifest the vulnerability pattern. Thus, the different branches of the decision trees monitor specific audit trails and their organization, while their root nodes can be used to indicate the purpose or result of the composition of multiple records from those trails that may include different information capture scopes. This model of diagnosis can be used as the underlying script for a *cloud-wide vulnerability detection mechanism* that automatically takes various audit trails as input, and pinpoints threats using their type and location.

2 Cloud Security and Auditing

2.1 Service Cloud and Security Concerns

SOAs often comprise web services that are designed to perform discrete processing and seamlessly interoperate with services that have compatible interfaces. Services can be discovered, integrated and used on distributed servers across a network. Communication among services is performed using multiple protocols, one of which, SOAP, has an associated web services standards specification and is typically conveyed using Hypertext Transfer Protocol (HTTP) [43].

The cloud model can provide a platform for hosting SOAs for large numbers of diverse web services in which vendors (i.e., service providers) package business functionalities and resources [4]. This cloud model platform is often defined as the service cloud architecture [39]. By registering their services with the cloud service provider, vendors can increase the computational power to execute their services with comparatively low cost. These services can be dynamically composed into a web service invocation chain within the cloud using existing, multi-criteria allocation algorithms [15, 20, 24]. The allocation of services to a composition is often based on the compatibility of services [2], the client request, the performance status and the security assurance level as defined by the Service Level Agreement (SLA) [17]. Furthermore, within an SOA, end-to-end round-trip messaging can be configured to traverse different clouds in federation [9], forming an inter-cloud invocation chain. This kind of external composition expands the capability of an individual service and cloud.

Along with the benefits brought by pushing SOA into the cloud environment, common security issues with traditional SOAs remain. Due to the properties of the cloud environment (e.g., larger scale and more geographically distributed), new security challenges have emerged [11, 18, 25, 37, 44]. Thus, threat mitigation strategies to address both traditional and emergent security concerns must adapt. These strategies also must respect the different cloud service models (i.e., Software as a Service (SaaS), Platform as a Service (PaaS), and Infrastructure as a Service (IaaS)) and the information capabilities available at each level [11,37]. For instance, the local data access control policy of a standalone web service may need to be augmented to include its involvement in an invocation chain within the cloud environment to comply with information flow control requirements that are part of managed cloud sessions [38, 39].

Using the cloud environment brings new security concerns with respect to cross-domain communication, which includes the inter-cloud service compositions. For example, a trust broker that is tasked with monitoring information propagation between several collaborative SOAs may not be the best solution for a trust broker between federated clouds, since the large amount of potential communications may overburden a standalone, intermediate trust broker. Therefore, gateways held on each cloud can be more flexible and suitable for the cloud scenario [16].

A cloud, especially a public cloud, usually has more distributed and layered structures than traditional SOA systems. Therefore, although many security mechanisms and solutions made for traditional SOA are still working and the information gathered from federated clouds can be propagated within the federation, there are new questions regarding how to perform cloud-wide auditing across the different parts of the cloud, how to scope the captured audit records, and how to recombine them to provide new explanatory analyses from different perspectives [8, 13].

There exist specific potential threats associated with the service cloud. These threats include information leakage, Man-in-The-Middle (MiTM) attacks, DoS attacks, and resource misuse. Loopholes in information flow controls can be maliciously used by attackers to cause unexpected information disclosure and tainting [38, 39]. Assume that the service cloud requires an SOA to use an identity manager in the form of an STS in order to form a trusted conversation among the services in the service invocation chain [26, 31, 32]. If a service in the *trusted conversation* invokes a service outside of that trust chain even if it is a service hosted by a federated cloud, this untrusted service may respond with tainted information that can be propagated throughout the service invocation chain. MiTM attacks toward web services can be achieved by intercepting and rewriting XML messages during transmission between two web services [5].

DoS is a type of attack aimed at compromising a service's functionality by flooding it with messages, sending overly large messages, or cutting its inbound communication channels, such as forcing packet dropping to the service [23]. In addition, resources held on the web service can also be vulnerable to unauthorized or spoofed access, which can lead to information leakage and tainting [41]. These attacks or vulnerabilities can affect different perspectives of information assurance including confidentiality, integrity and availability within a cloud, as well as within its federation [6].

2.2 Auditing and Analytical Models for Manifesting Threats

Given the known threats affecting the security assurance of a service cloud, governing documents provide a set of guidelines for performing security-related auditing [12, 14, 29]. These guidelines yield a set of goal-oriented auditing mechanisms that when properly placed can offer security-related information by recording specific auditable events. For instance, services that are required to log their inbound and outbound messages as auditable events must have a built in logging mechanism to achieve the goal of auditing message exchange. Even when mechanisms are put in place, they must guarantee that the localized audit trails produced throughout the cloud service architecture have some level of consistency in the representation and expectation of the available audit records. In addition, a localized audit trail has a single perspective that cannot provide the complete picture of the type, location, and potential impact of the threat throughout the cloud. The logging mechanisms and entities we enforce in our research are discussed in a later section in conjunction with the service cloud architecture.

For example, suppose the security token activity logged by the STS only manifests the potential risk/indexpotential risk of information tainting by revealing an anomalous or broken sequence of service communication. Additional investigation into the occurrences of communication among federated clouds is needed to judge if the broken sequence is caused by allowed inter-cloud message exchange or unauthorized external service invocation. For MiTM attacks, message attack detection handlers that have been created to build security awareness into web service [1] will only flag the fault at the local service that deploys it. Within a round-trip service invocation chain, the fault code attached to the message will be masked by intermediate services before it returns a failed communication back to the client. Thus, the client has no real knowledge about what happened. The possibility must be explored that combining an understanding of available fault codes and their translations to events that occurred along the service invocation chain will more directly manifest a potential attack.

Similarly, fault codes from local message error handlers reveal potential security issues. These handlers may be embedded in web application servers, denoting such events as the inability to communicate with the next service in the chain and possibly pointing to a DoS attack. The handlers may be related to the use of Web Services Security (WS-Security) [28] to determine potential digital signature weaknesses. With a myopic view of only the local compilation of records, the attacks cannot be pinpointed or confirmed.

Composing multiple audit trails generated from distinct goal-oriented mechanisms among different cloud entities is quite necessary for cloud-wide security auditing since it can provide broader views of threat manifestations. In order to formally describe how to assess the threat given local audit trails, we need a suitable model to perform the analysis and render a decision. One type of analytical model is an *Attack Graph* [35]. An Attack Graph can have a hybrid-dependency representation [30] in which there are two kinds of nodes: conditions and vulnerabilities. Edges in the graph represent the transfer relationship between them. Common uses of attack graphs include comparing two networks to determine which one is more secure, exposing potential threats and providing countermeasures by helping generate hypotheses of how an attacker may penetrate the system, and helping an organization minimize its security cost by providing quantitative measurement [19]. Attack Graphs perform well at visualizing the process of malicious penetration as

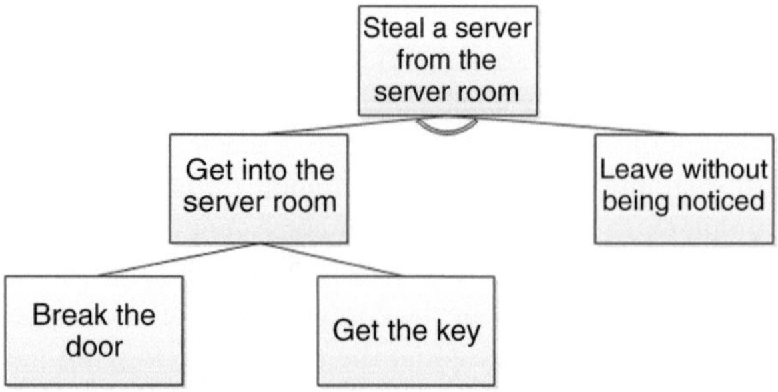

Fig. 1 A basic attack tree

a work flow chart. However, in some cases, the potential vulnerabilities or attacks are better formulated strictly hierarchically to impart a clearer sense of the distinct entities of the system that can be impacted by different security vulnerabilities and attacks.

From this perspective, *attack trees* have become a de facto model for attack analysis. Traditionally, an attack goal is the root node and intermediate nodes are the sub-goals for achieving that goal. Leaf nodes represent the specific activities an attacker can take to fulfill a malicious purpose [36]. The typical attack tree can assess the presence of a vulnerability or risk from the attacker's perspective. Logical relationships join siblings to evaluate their parent node. In most trees, the default relationship among siblings is OR, whereas the AND relationship is explicitly denoted by a curve across each branch included in the conjunction [36].

Figure 1 shows a simple instance of an attack tree found in [7]. It is important to note that some leaf nodes in this model can still be further specified. For example, a more detailed version of this model may specify the node "Leave without being noticed" t as two sub goals with OR relationship, "Cause the burglar alarm to fail" and "Impersonate the janitor".

Researchers have extended the basic attack tree to form more complex trees, such as *defense trees* [7] and *protection trees* [42]. These trees direct analysis toward the strength of an attack, as well as advocate possible countermeasures. In defense trees, countermeasures to certain attack activities are attached below the corresponding leaf node of an attack tree. Each countermeasure node contains quantitative indices for both the attack activity and its countermeasures' assessment.

The protection tree is built and analyzed similarly to a traditional attack tree, in which each node of the protection tree corresponds to an attack goal, with sub-goals or specific activities depending on its position in the tree. Protection Trees impose metrics for quantitative analysis, which align with the leaf nodes and then propagate upward to the root node. The configuration provides an understanding of analyzing countermeasures for malicious activities, both quantitatively and qualitatively.

3 Scoping Audit Logs to Specific Cloud Entities

Using the PaaS model as a backdrop, we review the cloud architecture as it was first presented in [48]. Figure 2 depicts the entities that we target within the cloud for capturing security related auditable events and storing them in local audit trails. In the subsections below, we discuss specific cloud management entities within the cloud architecture in Fig. 2.

3.1 Service Auditing Tier

At the service auditing tier, we impose a protocol on services hosted by the cloud to log specific local events such as inbound and outbound messages and local handler faults which may be due to message attack detection [1]. Each service forms a local *Service Audit Trail* of records across a timeline of its activity. Below is the structure of an inbound message record for a service S_i within a certain cloud session. The outbound messages have a similar format.

$$S_i.AT.im_z = (session\_id, timestamp, sender, receiver, payload, faults)$$

As shown in the above tuple, the inbound message record contains the ID of the corresponding cloud session (*session_id*), the time of the recorded event (*timestamp*), the *sender* and the *receiver* of the message, the actual SOAP content (*payload*) and any *faults* found.

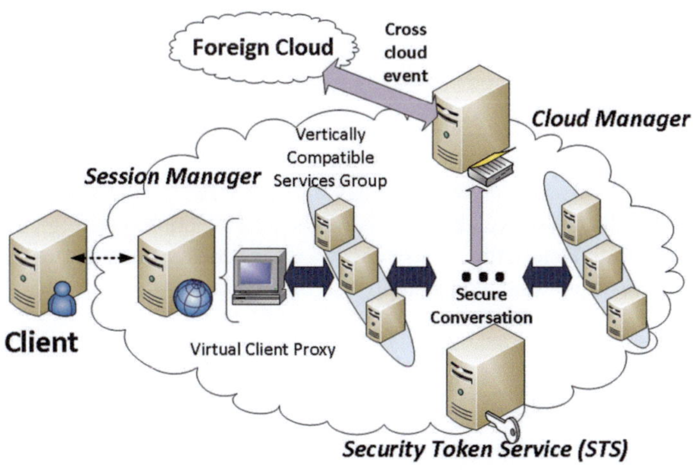

Fig. 2 Cloud architecture entities (Reproduced from Xie and Gamble [48])

Table 1 Fault codes thrown under different attacks

No.	Situation	Fault code	Type
1	Service time out	Could not send message	Protocol
2	Multiple bombardments	Error reading XMLStreamReader	Protocol
3	Packet dropping without history	Failed to create service	Protocol
4	Delegation issue-STS bombardment	Unable to communicate with the STS	Protocol
5	Packet dropping with history	Problem writing SAAJ model to stream	Protocol
6	Validation issue-STS bombardment	Problems creating SAAJ object model	Protocol
7	Modified ciphertext	The signature or decryption was invalid	Custom
8	SOAPAction spoofing	SOAP action mismatch	Custom
9	Malicious body element	Repeated body element	Custom

Note: SOAP with Attachments API for Java (SAAJ)

There are two kinds of fault codes thrown by services. The first is a fault code generated by custom attack handlers configured for the service [1]. All of the three types of XML rewriting attacks presented can have in place custom handlers that produce specific fault codes defined by the handler developer. The second kind of fault code is thrown by WS-* standards protocols. DoS attacks (e.g., packet dropping and STS bombardment) cause this kind of fault code. Depending on different invoking configurations (i.e., with/without history in packet dropping) or different attacked links (i.e., delegation/validation link in STS bombardment), this kind of protocol-based fault code can vary for the same type of attack.

Table 1 lists the possible fault codes and their reference numbers, which we will refer to later in the analytic model.

Also in the service auditing tier, a service that serves as a resource owner creates a separate *Resource Audit Trail* containing logs generated by policy-aware resources [27, 41]. These records have the following structure.

$$S_i.RAT.r_z = (session\_id, timestamp, operator, data\_location, type, status, policy)$$

The *session_id* indicates the identity of the cloud session, the *timestamp* represents when the access event occurred, the *operator* records the identity of the accessing service, the *data_location* shows the location of the resource and is still the unique identity of that data resource, the *type* shows the access action type (e.g., read, read-fail, write, write-fail, etc.), the *status* indicates if the access is legal or not, and lastly, the index of corresponding *policy* could also be recorded.

3.2 Cloud Auditing Tier

For the cloud auditing tier, we introduce a *Session Manager (SM)* to facilitate constructing a secure session of web services to process the client request, using existing allocation algorithms. The SM proxies the client request and directs the request to the first service based on an initial composition aimed at fulfilling each portion of the client request. Since it is possible for a service to be substituted within a vertically compatible group (consisting of functionally equivalent services) [2] during session proceedings, the allocation of services to securely satisfy the client request can be dynamic. Such dynamism means the services inside a composition chain will inquire with the SM to pick a next service before sending their outbound message. The SM is responsible for logging the history of service composition and the final session completion status. The format of the record as part of the Session Manager History (SMH) appears below.

$$SM.SMH = (session\_id, status, start\_time, end\_time, < service\_chain >)$$

Shown in the tuple above, the Session Manager History contains the ID (*session_id*), the completion *status*, the time period (*start_time* and *end_time*) and the basic composition (*service_chain*) of a cloud session.

Because auditing is a major part of information assurance, we also require the services to comply with a set of WS-* standards, specifically WS-Security [26], WS-Trust [32] and WS-SecureConversation [31]. Our approach relies on the presence of an STS [32] to authenticate services in the trusted conversation. The STS uses a security context token (SCT) [31] and associated encryption algorithms [26] to provide basic assurances of data integrity and confidentiality during transmission among services in the trusted conversation. The STS issues, delegates, validates and cancels tokens passed along the service invocation chain within the secure session to provide basic identity management.

The STS is a web service that is considered at the cloud auditing tier because of its interaction with the secure session of services. We use an enhanced version that logs the timeline of these token-based interactions which it stores locally. The following tuple shows the form of an *STS Audit Trail* with the identity of the session (*session_id*) and a timeline of token activities (*token passing flow*), which contains four types of token events (issues, delegates, validates and cancels) and the services that perform the request from the STS:

$$STS.AT = (session\_id, < token\_passing\_flow >)$$

A valid *target sequence* and *broken protocol* both represent a message sequence that is a portion of the STS Audit Trail of logged messages for a particular session. A valid target sequence indicates a protocol compliant sequence, while a broken protocol means the sequence has some anomaly that makes it nonstandard. For example, in a cloud session with a valid target sequence, the client first requests the issue of token, and then it delegates the token to the next service, which receives the token and requests the validation. After completing the processing at that service, the token will be delegated to the next service in the cloud session, along with the actual message. After all of the services have been reached via delegation/validation actions, the response eventually returns to the client that makes the last request to the STS for token cancellation.

On the contrary, if there exists an attack or potential threat, the token passing flow may reveal an anomaly. For instance, certain expected token activities could be missing in a sequence, or the delegation/validation could be made by the same service successively. In the analytic model discussion within Section 5, we show multiple types of broken protocol and present how they can be used to manifest the threat patterns.

3.3 Inter-cloud Auditing Tier

The inter-cloud tier houses the authority for exchanging messages and resources between clouds or services external to the cloud. Federating lets the cloud provider extend its abilities by "borrowing" processing power, storage or service functionality from another cloud which we refer to as a *foreign cloud* [9]. We advocate using a *Cloud Manager (CM)* that records communications between federated clouds, including the actual messages and requested audit assets that illustrate the activity in the foreign cloud related to its invocation by the CM. When the SM cannot find a suitable service within the cloud to be invoked next in the service chain, the last service in the chain configures a request for an external service. The CM receives the request and acts as a gateway to negotiate and invoke the external service within another cloud, while controlling the information exchange. Using the CM in this manner contrasts with allowing either a service in the chain or the SM to perform the cross-cloud communication. This design provides more assurance of integrity and confidentiality of communications between services from different clouds. While reducing the management burden through separation of duty, the CM can be responsible for passing audit requests associated with the outbound message, gathering and returns audit response for the inbound response, and recording these audit asset propagations along with the exchanged messages as *Cross Cloud Event within Inter-Cloud Audit Trail*. The record format of a *Cross-cloud event* is shown below.

$$CM.ICAT.cce = (session\_id, timestamp, home\_cloud, foreign\_cloud, direction, \\ audit\_request, audit\_response, agreement, payload)$$

The cross-cloud event contains the unique ID of the session (*session_id*), the time when the CM performed the invocation (*timestamp*), the clouds involved in the cross-cloud communication (*home_cloud* and *foreign_cloud*), the direction of message (i.e., inbound or outbound), the audit assets expected (*audit_request* and *audit_response*), the contractual agreement of the federated cloud for the message exchange (*agreement*), and the necessary contents of the SOAP message (*payload*).

4 Threats and Auditable Events

Given the potential threats toward the service cloud architecture and the defined scoping of local audit trails discussed in Section 3, Table 2 indicates certain targeted threats to the service cloud architecture (Column 1) and the security objectives related to each threat are shown in Column 2 (A: availability, C: confidentiality and I: integrity). The evidence that may assert the presence of threats is shown in Column 3. If an enforced logging mechanism is in place, Column 4 indicates what auditable events should be captured that support the evidence of a threat as well as the cloud entity responsible to record the event and maintain the logs (in parentheses).

To reduce potential message tainting or authorized disclosure brought by improper service invocation, evidence of a potentially unauthorized service invocation external to the trusted conversation or "home" cloud can be manifested by the STS forcing it to log token related communications (i.e., *issue, validation, delegation, cancel*) with the services. In some cases, even messages passed to services in federated foreign clouds can face the potential risk of being tainted or disclosed.

Table 2 Threats and auditing inside a service cloud

Threat	Assurance	Evidence	Auditable event
Message tainting	I and C	Broken trust conversation protocol	Token activity (STS)
		Cross-cloud invocation	Inter-cloud event (CM)
MiTM attack	I	Expected secure conversation	Token activity (STS)
		Expected message flow	I and O messages (Service)
		Message modification at trusted link	Message scan for attacks (Service)
Packet dropping attack	A	Client request failure	Session status (SM)
		Token cancellation failure	Token activity (STS)
		Broken message flow	I and O messages (Service)
		Service connection failure	Message scan for errors (Service)
STS Bombard -ment	A	Client request failure	Session status (SM)
		STS function tampering	Token activity (STS)
		Token cancellation failure	Token activity (STS)
		Broken message flow	I and O messages (Service)
		Service connection failure	Message scan for errors (Service)
Resource misuse or disclosure	I and C	Normal secure conversation	Token activity (STS)
		Invalid session	Session creation (SM)
		Policy-violating access	Policy-aware access control record
			(Service as the resource holder)

Note: Security Assurance (availability (A), confidentiality (C), integrity (I))
Inbound and Outbound messages (I and O messages)

This is possible since the security agreements used in cloud federation may only provide partial assurance, such as through SLAs, which dictate service level expectations between cooperative entities. SLA matching to form a federation can be used to reduce or at least manifest the potential risks brought by cross-cloud communication [17]. Information regarding federation risk brought by external communication to clouds in the federation would be revealed through the inter-cloud events recorded by the CM [16] on each cloud, and the cross-cloud events can contain the audit requests and responses propagated between federated clouds, along with the actual exchanged messages [47].

Although secure conversation and cloud federation can provide security assurance to transmitted messages by keeping them within the trusted scope, the communication links can still be attacked by MiTM attacks on SOAP messages which can bypass the security standards [22, 23, 40]. We focus on three types of XML rewriting attacks discussed in [1], including modified ciphertext attack, SOAPAction spoofing attack and malicious body element attack. They are configured to intercept, modify and resend the message to the original receiver without being noticed. Under this scenario, the evidence of data integrity compromising during message transmission will be manifested by failure of session and corresponding fault records (caused by failed message integrity checking) thrown by local message attack detection handlers deployed on individual services [1]. Since sophisticatedly deployed MiTM attacks will not explicitly impact the token authentication, secure conversation and the message flow inside it usually appear as expected.

Denial of Service (DoS) attacks [23, 40] can cause services to fail by flooding them with requests or blocking outbound transmissions. Often, on the service side, if a service is unable to contact its next service in a service chain or the STS, there should be a connection failure alert from a default service handler [21], logged locally at the service, as well as the records of redundant messages which are unable to be sent. If the attack compromises the function of STS, then the broken protocol can be revealed by the missing components in the token activity logging at STS side. Both types of DoS attacks can cause time outs by services in earlier parts of the service invocation chain that wait for the response past some threshold. In this case, there may also be a missing token cancellation within the trusted conversation that is manifested by the token activity logging provided by the STS, since the original requester for token issuing can no longer request token cancellation. To summarize, a failed session, error records indicating failure connections provided by local message error handlers plus the token activities recorded by the STS can be used as auditable events for revealing DoS attacks.

In addition to the security concerns regarding transmitted messages, resources stored by services within the cloud model can also become targets of misuse or unauthorized disclosure [38]. If the resource is policy-aware [27, 41], the evidence regarding unauthorized resource access will be revealed by its logged information. These logs provide assurance toward the integrity and confidentiality of stored resources. However, this kind of access control mechanism is usually not session-aware, meaning that the access should be associated with a secure session as dictated by the SM. Thus, if there is any mimicked service with a fake session ID access, no alarm will be raised from the policy-aware access control mechanism. In this case, the evidence of invalid sessions (still considered as secure conversations since the STS is spoofed) should be revealed by session creation history generated by the SM.

5 VDT Architecture and Decision Tree Model

The VDT decision tree model that is the focus of this section is used to reveal security threats by manifesting vulnerability patterns over audit trails generated in the service cloud. Similar to the attack trees and graphs presented in the background, the root node of a VDT indicates the result of the attack diagnosis. The logical division of branches indicates any subordinate diagnostic results to achieve. The structure differs in three meaningful ways. First, it dictates the scope of the audit assets considered. Second, it identifies the parameters for comparison across distinct audit assets in different scopes or collected by different entities (e.g., cloud management services or vendor services). The third way is the embedding of an algorithm within ordered leaf nodes to determine the satisfaction of the diagnosis predicate at the parent.

Figure 3 shows the basic VDT architecture. The root of the tree indicates the overall attack type being diagnosed. The root is tagged with any session ID (SID) in its lower right corner from which the service communication stems. The SID at the root is then scoped throughout the tree providing a direct filter of the audit assets examined. Diagnostic nodes are non-leaf nodes that direct the evaluation of audit assets to make a particular conclusion. If the children are linked by conjunction, as

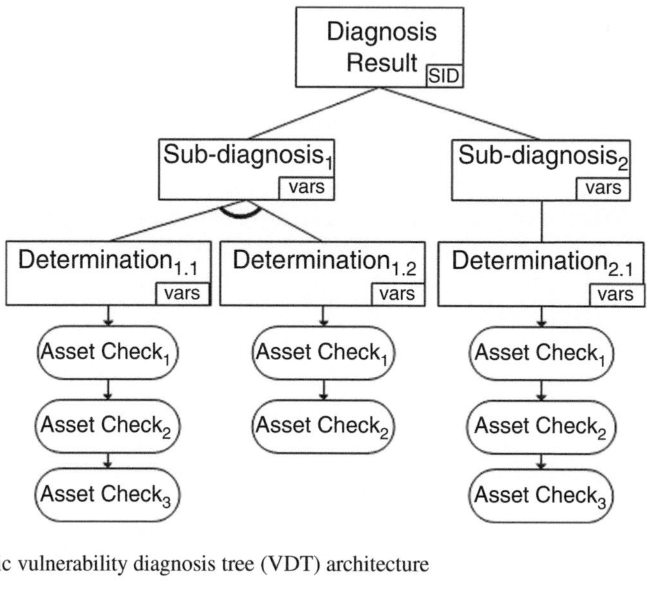

Fig. 3 Basic vulnerability diagnosis tree (VDT) architecture

shown in Subdiagnosis1, then together they verify the parent diagnosis. Otherwise, they are assumed linked via disjunction. A *Diagnostic node* may be tagged with variables in its lower right corner. These variables are scoped to all prodigies of the tagged node. In most decision and attack trees, the Determination node, as the last rectangular node in the tree on each branch, would be a leaf node. Because the goal of our decision trees is to manifest vulnerability patterns, the determination leads to a process of examining more detailed audit assets, such as single records, often taken in a sequence. These *Determination nodes* also may also be tagged with variables scoped to their prodigy.

Children of Determination nodes are called *Asset Check nodes*. The sequence of examining audit assets appears as *Asset Check nodes* with arrows directing the examination progress, where the last examination step performed in the chain returns the result to its Determination node. These checks are performed against the contents of the audit asset tagged in the Determination node or at an ancestor node, if shared across branches. The asset checks can take many forms, such as generic send/receive message identification, a fault code receipt after a certain ordering of message communication records, the absence of expected messages in the log according to the communication protocol, or an unexpected sequence of messages. The checks also rely on the protocol of communication among the cloud entities and the service composition or invocation chain as defined by the cloud architecture described in Section 3. If any one of the Asset Check nodes fails on the directed, and possibly time-ordered, path or the communication protocol is not met, then the conclusion of the Determination node is considered to be false. If the last Asset Check is satisfied, this success is propagated to the Determination node. For our current attacks and case study, only a linear ordering of checks is needed. However, these checks can be adapted to other order dependent structures.

Within a request/response round-trip message configuration, a service will time out when (1) a response to its query has not been received past a set time threshold or (2) it is unable to send its request/response message. Service time out may be a symptom of a potential attack, especially for the DoS scenarios. It may also deliver strength toward verifying an attack depending on which entities capture the auditable event information related to the time out and when this capture took place. We denote time-out as a weak contributor in a VDT using a dashed branch. Weak contribution means the sub-diagnosis node cannot fully support its parent's diagnostic result but can contribute some degree of strength toward verifying the vulnerability. Thus, fulfilling the sub-diagnosis attached to a contributor branch means strong, but inconclusive evidence regarding its parent's diagnosis. If only a weak contributor branch propagates a positive diagnosis to its parent, then the tree can claim only a potential attack. Figure 4 depicts the three branch types in the VDT.

Figure 5 illustrates a reusable sub-tree example for service time out. The hexagon attached to the root node labels the sub-tree as reusable with input parameter, service *S*. If there is a record, *r*, in the audit trail, *AT*, of *S* (*S.AT.r*) for a particular session ID (SID) that logs a fault code (i.e., 1 = "cannot send message") as shown in Table 1 related to service time-outs, then there is a verifiable time-out that may be a symptom of a larger attack. The time-out may also indicate that the STS does

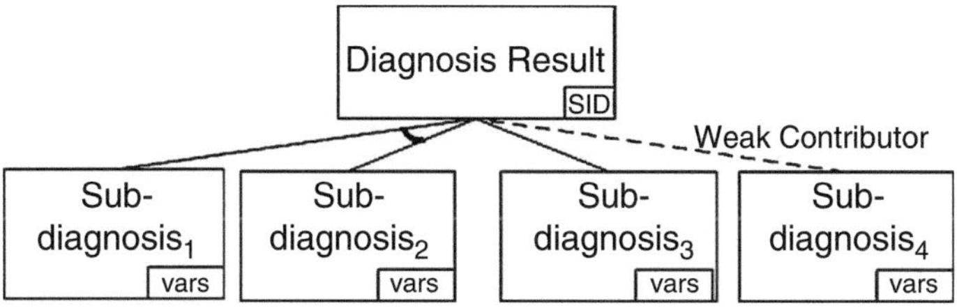

Fig. 4 Branching types with weak contribution

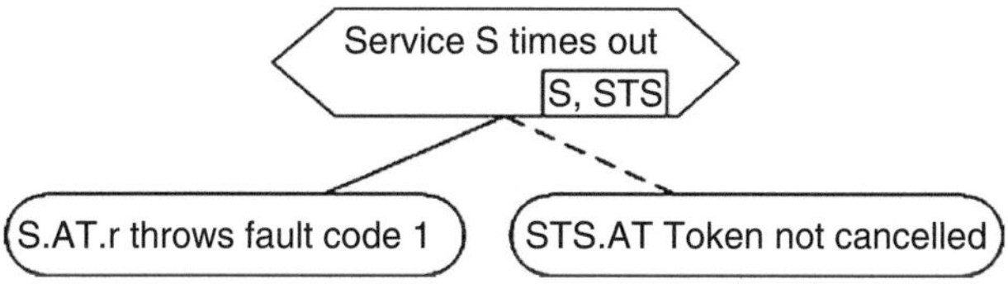

Fig. 5 A reusable sub-tree indicating a time-out scenario

not have a log of the cancelled token for the same session ID, which implies that a response was not received by the client to request cancellation. However, this weak contributing state within the STS only indicates the potential for a service time out somewhere in the trust chain. It is not verifiable. Thus, the time-out sub-tree is reused in multiple VDTs.

6 Vulnerability Pattern Diagnosis

This section shows how the VDTs use patterns within and across different logs in the cloud tiers to manifest security vulnerabilities. They can indicate the type of vulnerability as well as the endpoint(s) impacted. For each VDT event, it is assumed that the SM formed a web composition that consists of a sequence of services, $< S_1, \ldots, S_n >$ within a single cloud with a session ID, and the *SID* is scoped to all nodes. This assumption reduces the notation burden of having the service point to the session ID each time, i.e., $S_i(SID)$. Though only sub-trees focusing on specific pattern processing are shown in this section, complete trees can be found in [46].

6.1 VDT for Potential Information Tainting

Information tainting can occur if a message is altered by a MiTM attack, which will be discussed later. It can also occur if a service is invoked that is not part of the trusted conversation and provides a response to the invocation that

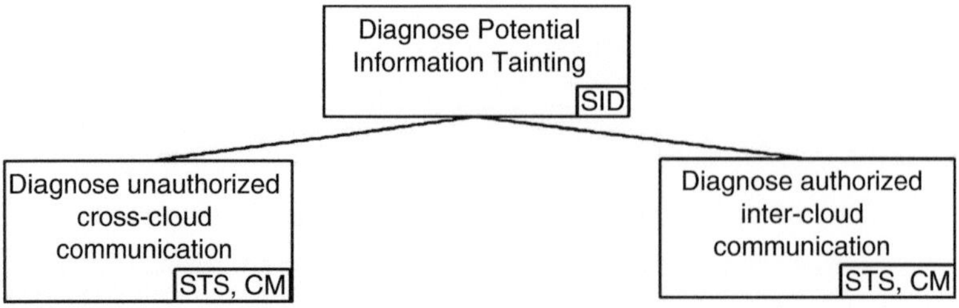

Fig. 6 High level nodes of the VDT for information tainting

Fig. 7 Sub VDT for diagnosing unauthorized cross-cloud communication

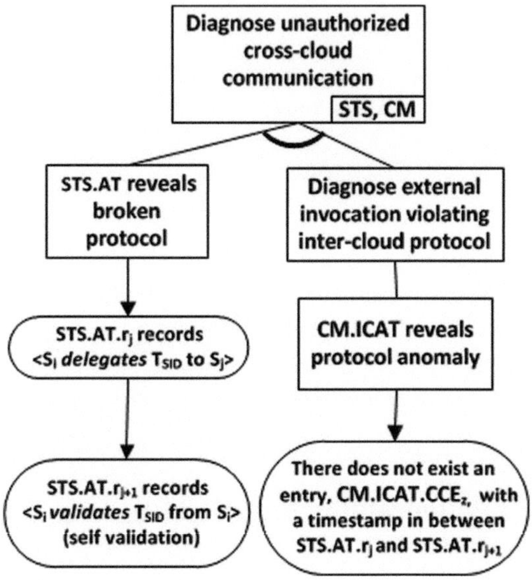

included information that is tainted. The cloud architecture trusts only services that authenticate with the STS, because it presumes that an STS has been used for client requests. There are scenarios when the cloud does not have the services needed to complete the request and must use a service external to its security domain, perhaps within another cloud. In this case, the external service request would be authorized by the cloud manager (CM).

The CM will require of an external cloud certain audit assets along with the response. The CM examines these assets to determine if there is a threat of attack from the external cloud. Thus, even though the initial external request is authorized, attacks may be revealed in the audit asset patterns returned by the external CM and must be diagnosed separately. In contrast, if a service in the trusted conversation invokes an external service without using the CM, then this is diagnosed as potential tainting. Figure 6 shows the roots of the sub-trees for diagnosing this attack. Though not shown, the rightmost sub-tree will reuse the VDTs against the audit assets provided by the external cloud. The left sub-tree is expanded in Fig. 7.

Table 3 Audit assets diagnosing information tainting

Type	Entity	Tier
Cloud Manager Audit Trail	CM.AT filtered by session ID	Inter-Cloud Auditing Tier
STS Audit Trail	STS.AT filtered by session ID	Cloud Auditing Tier

Corresponding to Row 1 of Table 2, Table 3 shows the audit assets needed within the cloud to diagnose potential information tainting due to seeking information that is external to a trusted conversation. These assets belong to the STS which reveals a communication anomaly and the CM which configures and stores a Common Configuration Enumeration (CCE) (formalized in Section 3) recorded every time it receives a request to invoke a service external to the "home" cloud. Though the request and response pass through the SM within the cloud, since it serves as the service composer, its audit assets are not needed to diagnose the attack. The left sub-tree is expanded in Fig. 7. The STS reveals an anomaly in the communication protocol for session SID. Service S_i self-validates. Thus, it appears to the STS that S_i is receiving a message from itself. But in reality, S_i is communicating with the STS after receiving a message from a service that does not have or use the token and does not communicate with the STS.

6.2 VDT for XML Rewriting Attacks

Our vulnerability patterns for XML rewriting attacks rely on the fault codes logged by the local attack handlers. Similarly, a wrapper can be deployed to log a meaningful fault code when WS-Security detects a digital signature wrapping attack, such as the Modified Ciphertext attack. These fault codes (listed in Table 1 as custom fault codes) simplify the pattern diagnosis as seen in the single attack tree that represents them. However, other attack detection methods for XML rewriting attacks could be used to log events that point to particular vulnerabilities and modification to the trees would be minimal. Fault codes in other VDTs are more generic and are used to represent the auditable event that occurs within the log, such as fault code number 1 indicating that the service could not send a message.

Figure 8 shows the two major branches of the VDT for the XML rewriting attacks. The overall diagnosis is separated into two attack diagnoses: one on inbound messages and one on outbound messages of a particular service. These branches are partitioned into diagnosing the attack location in the invocation chain and its actual type with the assistance of a local message handler. The division of the type and location diagnoses yields strong and accurate evidence of an XML rewriting attack. Whereas other analytical methods for security attack detection focus on revealing the potential for attack such as finding particular attack codes, the VDT sufficiently pinpoints the node location, even if remote from the client, given the knowledge of the communication protocol.

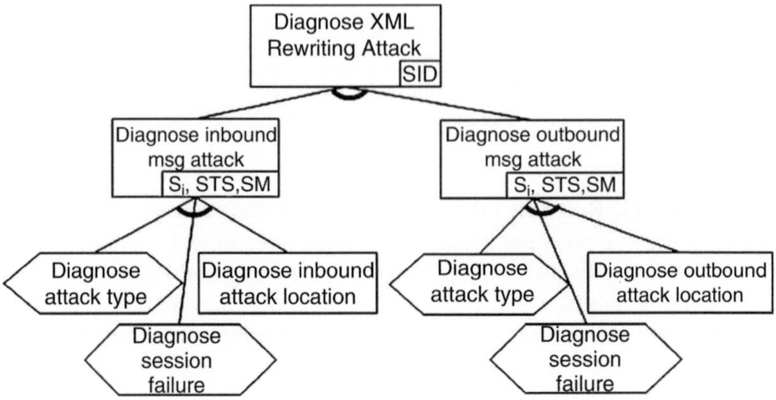

Fig. 8 High level nodes in the XML rewriting attack VDT

Table 4 Audit assets to expose XML rewriting attack

Type	Entity	Tier
Session Manager History	SM.SMH filtered by session ID	Cloud Auditing Tier
Service Audit Trail	Service S_i.AT filtered by session ID	Service Auditing Tier
Service Audit Trail	Service S_{i-1}.AT filtered by session ID	Service Auditing Tier
STS Audit Trail	STS.AT filtered by session ID	Cloud Auditing Tier

The diagnosis attack type node indicates that it is the root of a reusable sub-tree because it is used for both the inbound and outbound message evaluation. In addition, the sub-tree of diagnosing session failure through SM history is also reused in both inbound and outbound attack detection (the complete details of the tree expansion can be found in [46]).

For the targeted XML rewriting attacks, following the composition strategy shown in Row 2 of Table 2, Table 4 shows the log affected, the entity that records the auditable event, and the tier to which its log is scoped (discussed in Section 3). In the leftmost path in Fig. 8, to identify the specific attack type, service S_i's local audit trail is examined for specific fault events given the deployment of a local message handler that can detect the three forms of XML rewriting attacks on inbound messages. Determining the fault event requires finding a record in S_i.AT with one of the fault codes 7, 8 and 9 that translate to modified ciphertext attack, SOAPAction spoofing attack and malicious body element attack, respectively. The diagnosis attack type subtree is reused for identifying the type of attack on outbound messages.

Figure 9 expands the sub-tree for finding the location of the XML rewriting attack on an inbound message. It examines the audit trails, filtered by SID, from services S_i and S_{i-1}, and the STS. In the rightmost branch of Fig. 9, a reusable sub-tree uses audit records from the STS log to confirm that all communications between S_i and S_{i-1} follow the prescribed communication protocol. The leftmost branch of Fig. 9

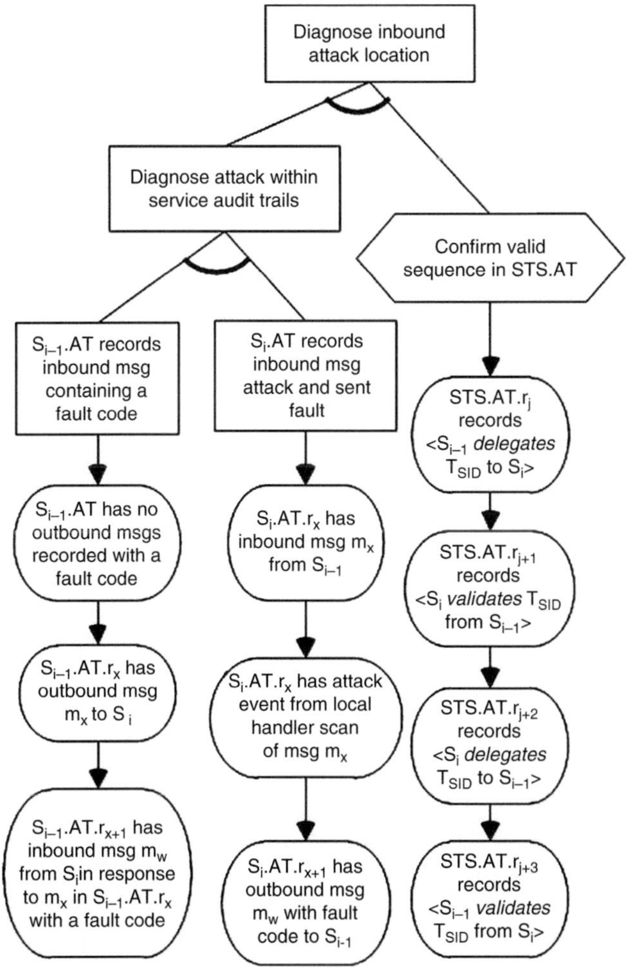

Fig. 9 Determining attack location for an inbound message

examines the audit trail of S_{i-1} filtered for its logged message events that are related to its communication with S_i. The audit trail for S_{i-1} has no record of a fault on any of its outbound messages, including those to S_i. Thus, S_{i-1} is not the culprit, should a fault be found. S_{i-1} does find that it has recorded at least one inbound response message from S_i containing a fault code, which indicates that S_i detected an anomaly of some type. This anomaly is reflected in the middle branch in Fig. 9, in which S_i logs an inbound message that includes an attack event from its local handler. S_i has similarly logged the same outbound message response recorded by S_{i-1} that indicates the attack.

A malicious insider can deploy the same XML rewriting attacks described above that are manifested as MiTM attacks on an outbound message. The VDT branch that

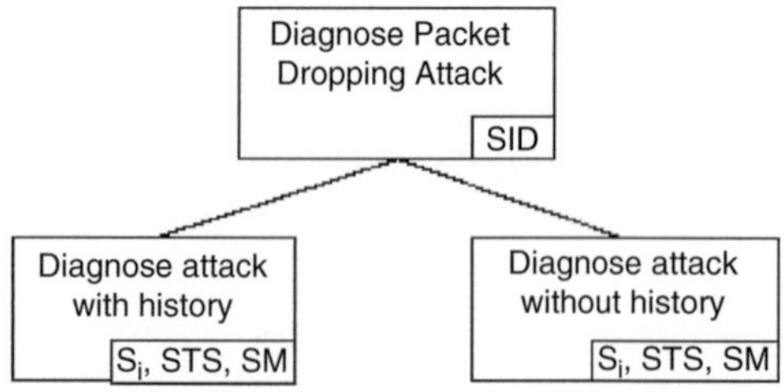

Fig. 10 High level nodes to diagnose packet dropping

Table 5 Audit assets to diagnose packet dropping and STS bombardment

Type	Entity	Tier
Session Manager History	SM.SMH filtered by session ID	Cloud Auditing Tier
Service Audit Trail	S_i.AT filtered by session ID	Service Auditing Tier
Service Audit Trail	Service S_{i-1}.AT filtered by session ID	Service Auditing Tier
Service Audit Trail	Service S_{i+1}.AT filtered by session ID	Service Auditing Tier
STS Audit Trail	STS.AT filtered by session ID	Cloud Auditing Tier

detects this anomaly follows the same decision strategy as shown in Fig. 9. However, the focus is on specific outbound messages in the service audit trails in the cloud.

6.3 *VDT for Packet Dropping*

Figure 10 shows the top three nodes of the VDT to diagnosis a packet dropping attack. Unlike the VDT for an XML rewriting attack, a packet dropping attack does not have different attack types. However, the diagnosis differentiates between whether there was a history of prior communication in a distinct session between two services involved in the attack or whether the first interaction between two services (no history) was attacked. The audit assets needed to evaluate packet dropping are listed in Table 5 (corresponding to Row 3 in Table 2).

Assuming a trusted conversation $< S_1, \ldots, S_n >$, Fig. 11 expands the left sub-tree for detecting a Packet Dropping attack occurring on the outbound communication of service S_i, given a communication history with service S_{i+1} that S_i is attempting to invoke. The sub-tree has five branches: four that are conjoined for examining the STS and local service logs and the fifth as a weak contributor branch indicating a service time out, as shown in Fig. 5. The leftmost branch determines if a prior communication history exists by examining the audit trail of S_i filtered by SID where SID$' \neq$ SID. If a record exists that embeds an outbound message to S_{i+1}, then there is a history. This history influences the interpretation of anomalies in the communication protocol followed by the services with respect to their token interchange with the STS. The second, leftmost branch with the label "Diagnose session Failure" has a shape indicating it is a reusable diagnosis. A similarly labeled branch previously appeared in Fig. 8, in which the SM history was used to confirm if the session has failed.

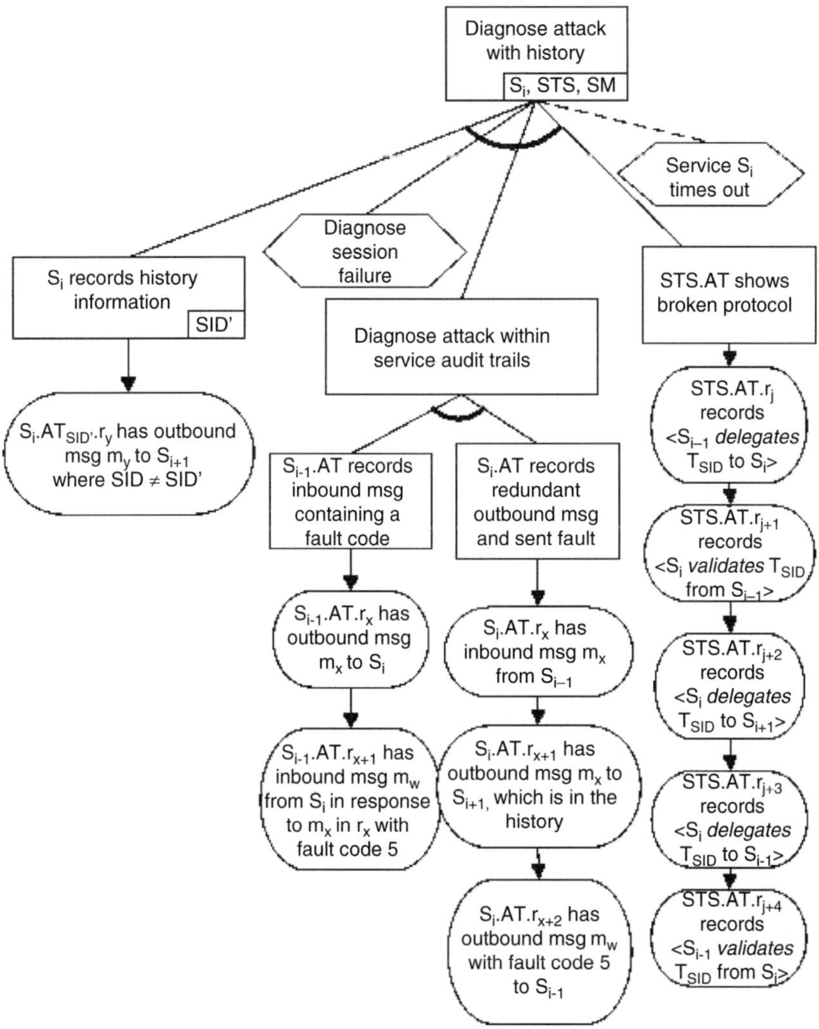

Fig. 11 Sub VDT for diagnosing packet dropping attacks

The examination of the STS audit trail indicates a communication protocol anomaly where service S_i was found to have performed successive, uninterrupted delegations. The interpretation of the anomalous pattern is that S_i delegated the token to S_{i+1}, a service that it had previously communicated with and thus has information regarding its location and invocation interface. When the first S_i delegation failed, S_i delegated its failure response to S_{i+1}. This pattern indicates that there was no intermediate validation request to the STS from S_{i+1}. This protocol anomaly contributes potential diagnosis of a DoS attack in which trusted conversations follow the established protocol. Similarly, the local audit trail of

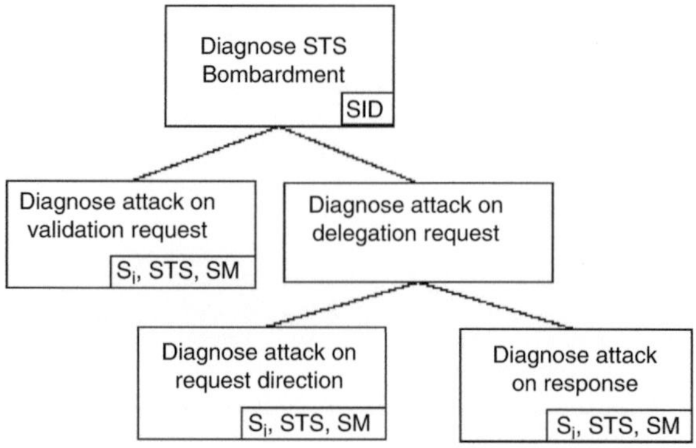

Fig. 12 VDT high level nodes for STS bombardment

service S_i shows that the message was attempted to S_{i+1}, while the audit trails of S_{i-1} and S_i show the fault code exchange on the response message indicating a Packet Dropping DoS attack.

6.4 VDT for STS Bombardment Attack

Because the STS is also a service within the cloud, it can be subjected to denial of service attacks. As discussed previously, these attacks are in the form of a bombardment of fake token issue requests. Figure 12 shows the high level nodes of the STS bombardment attack VDT. The sub-trees extending from the root node segregate the attack diagnosis into whether it occurred on a token validation request from a service to the STS or whether it occurred on a token delegation request from a service to the STS. The right delegation-related sub-tree is partitioned according to whether the attack affected a service request or response. The scoping of the sub-trees indicates that the S_i can be different depending on the direction of the message communication (i.e., request or response). Table 5 shows the audit assets needed for the diagnosis (corresponding to Row 4 in Table 2).

Figure 13 expands the left sub-tree as an example of how the attack can be diagnosed given logging and communication protocol expectations of the services in the trusted conversation and the STS. Whenever a service times out (rightmost branch) there exists the potential for a DoS attack because the service has not received the expected response from another service that it invoked within the trusted conversation. In the leftmost sub-tree in Fig. 13, service S_i confirms that it communicated with S_{i+1} in the form of a request and response, but that response included fault code number 4 that is interpreted as part of the diagnosis to be that the service logged "unable to communicate with STS". This missing communication is confirmed in the STS audit trail in which the protocol is missing an expected validation step. Similar to the previously discussed threats which cause the session failure, there is a reusable branch of detecting failed session status through SM history.

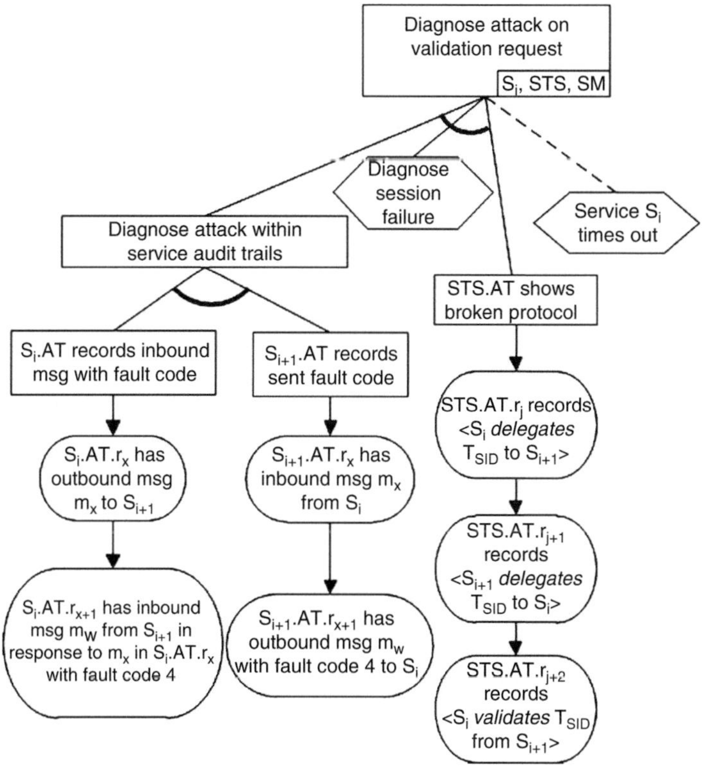

Fig. 13 Sub VDT for diagnosing STS bombardment

Table 6 Audit assets for diagnosing resource misuse

Type	Entity	Tier
Session Manager History	SM.SMH filtered by session ID	Cloud Auditing Tier
STS Audit Trail	STS.AT filtered by session ID	Cloud Auditing Tier
Service Resource Audit Trail	S_i.AT filtered by session ID	Service Auditing Tier

6.5 *VDT for Resource Misuse*

The last security vulnerability examined is resource misuse. There are multiple forms of misuse. This research focuses on the trusted conversation and how vulnerability can occur and be diagnosed given the available assets. The examined scenario is when a service within the trusted conversation can be spoofed, obtain a token from the STS, and access a service resource illegally. The audit assets needed are shown in Table 6 diagnosis (corresponding to Row 5 in Table 2).

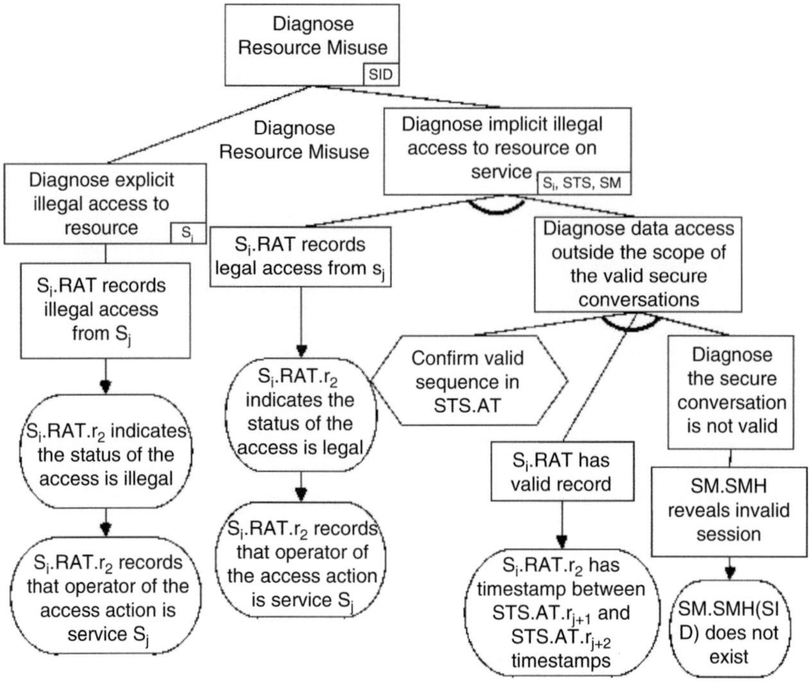

Fig. 14 VDT to diagnose resource misuse

Figure 14 shows the complete VDT to diagnose two kinds of resource misuse. The left sub-tree (only containing the leftmost branch) indicates that the access from S_j explicitly violated the access control policy (e.g. deletion of a protected file), thereby causing the alert of the access control mechanism, which recorded an illegal access record in the resource audit trail of S_i. The right sub-tree in Fig. 14 depicts the form of misuse presented earlier, in which the access control policy appears to be enforced (shown by the resource audit trail of S_i, as presented by the leftmost branch of the right sub-tree). However, the rightmost branch reveals that although the attack happened in a secure conversation (proved by the STS.AT and S_i.RAT), the secure conversation itself is not valid, since it is not recognized by the SM (the rightmost sub branch). Therefore, the access is, though implicitly, illegal.

7 Conclusion

This chapter discussed how to deploy a cloud-wide auditing solution for revealing threats in a service cloud environment. The solution focuses on providing a novel analytical model called the Vulnerability Diagnostic Tree (VDT) model. The VDTs provide a means via which audit assets captured and retained by multiple cloud entities can be examined together to determine security vulnerability patterns, especially patterns emerging from SOA communications in conjunction with service composition allocation and cross-cloud communication. There are additional vulnerabilities that arise with violating information flow control policies [38, 39] as well other communication issues. By defining the auditable events in the cloud, the audit assets needed to record those events, and the diagnosis structure that can be used to determine vulnerability patterns, the VDTs can be adapted to detect new patterns.

Acknowledgements This material is based on research sponsored in part by the Air Force Office of Scientific Research (AFOSR) grant FA-9550-09-1-0409, the Air Force Research Laboratory (AFRL) grant FA8750-10-2-0143 and the AFOSR/AFRL LRIR 11RI01COR. The U.S. Government is authorized to reproduce and distribute reprints for Governmental purposes notwithstanding any copyright notation thereon. The views and conclusions contained herein are those of the authors and should not be interpreted as necessarily representing the official policies or endorsements, either expressed or implied of the AFOSR, AFRL, or the U.S. Government.

References

1. Andrekanic, A., Gamble, R.: Architecting web service attack detection handlers. In: Proceedings of the 19th International Conference on Web Services, ICWS'12, Honolulu, pp. 130–137. IEEE Computer Society, Washington, DC (2012). doi:10.1109/ICWS.2012.69
2. Andrikopoulos, V., Plebani, P.: Retrieving compatible web services. In: Proceedings of the 2011 IEEE International Conference on Web Services, ICWS'11, Washington, DC, pp. 179–186. IEEE Computer Society, Washington, DC (2011). doi:10.1109/ICWS.2011.24
3. Beaton, J.K., Myers, B.A., Stylos, J., Jeong, S.Y., Xie, Y.: Usability evaluation for enterprise SOA APIs. In: Proceedings of the 2nd International Workshop on Systems Development in SOA Environments, SDSOA'08, Leipzig, pp. 29–34. ACM, New York (2008). doi:10.1145/1370916.1370924
4. Bell, M.: Service-Oriented Modeling: Service Analysis, Design, and Architecture. Wiley, Hoboken, New Jersey (2008)
5. Benameur, A., Kadir, F.A., Fenet, S.: XML rewriting attacks: existing solutions and their limitations. In: IADIS Applied Computing, Algarve (2008). doi:abs/0812.4181
6. Bishop, M.A.: The Art and Science of Computer Security. Addison-Wesley Longman, Boston (2002)
7. Bistarelli, S., Fioravanti, F., Peretti, P.: Defense trees for economic evaluation of security investments. In: Proceedings of the 1st International Conference on Availability, Reliability and Security, ARES'06, Algarve, pp. 416–423. IEEE Computer Society, Washington, DC (2006). doi:10.1109/ARES.2006.46
8. Bleikertz, S., Schunter, M., Probst, C.W., Pendarakis, D., Eriksson, K.: Security audits of multi-tier virtual infrastructures in public infrastructure clouds. In: Proceedings of the 2010 ACM Workshop on Cloud Computing Security, CCSW'10, Chicago, pp. 93–102. ACM, New York (2010). doi:10.1145/1866835.1866853
9. Celesti, A., Tusa, F., Villari, M., Puliafito, A.: How to enhance cloud architectures to enable cross-federation. In: Proceedings of the 3rd IEEE International Conference on Cloud Computing, CLOUD'10, Miami, pp. 337–345. IEEE Computer Society, Washington, DC (2010). doi:10.1109/CLOUD.2010.46

10. Cellary, W., Strykowski, S.: E-government based on cloud computing and service-oriented architecture. In: Proceedings of the 3rd International Conference on Theory and Practice of Electronic Governance, ICEGOV'09, Bogota, pp. 5–10. ACM, New York (2009). doi:10.1145/1693042.1693045
11. Chen, Z., Yoon, J.: It auditing to assure a secure cloud computing. In: Proceedings of the 6th World Congress on Services, SERVICES'10, Miami, pp. 253–259. IEEE Computer Society, Washington, DC (2010). doi:10.1109/SERVICES.2010.118
12. commoncriteriaportal.org: Common criteria for information technology security evaluation 3.1R4. http://goo.gl/IeC55 (2012)
13. Doelitzscher, F., Fischer, C., Moskal, D., Reich, C., Knahl, M., Clarke, N.: Validating cloud infrastructure changes by cloud audits. In: Proceedings of the 8th IEEE World Congress on Services, SERVICES'12, Honolulu, pp. 377–384. IEEE Computer Society, Washington, DC (2012). doi:10.1109/SERVICES.2012.12
14. dtic.mil: DoDI 8500.2, information assurance (IA) implementation. http://goo.gl/pwd2p (2003)
15. Esmaeilsabzali, S., Larson, K.: Service allocation for composite web services based on quality attributes. In: Proceedings of the 7th IEEE International Conference on E-Commerce Technology Workshops, CECW'05, Munich, pp. 71–82. IEEE Computer Society, Washington, DC (2005). doi:10.1109/CECW.2005.19
16. gictf.jp: Intercloud interface specification draft. http://goo.gl/SW4IS (2009)
17. Hale, M.L., Gamble, R.: Secagreement: advancing security risk calculations in cloud services. In: Proceedings of the 8th IEEE World Congress on Services, SERVICES'12, Honolulu, pp. 133–140. IEEE Computer Society, Washington, DC (2012). doi:10.1109/SERVICES.2012.31
18. Hamlen, K., Kantarcioglu, M., Khan, L., Thuraisingham, B.: Security issues for cloud computing. Int. J. Inf. Secur. Priv. **4**(2), 36–48 (2010)
19. Idika, N., Bhargava, B.: Extending attack graph-based security metrics and aggregating their application. IEEE Trans. Dependable Secur. Comput. **9**(1), 75–85 (2012). doi:10.1109/TDSC.2010.61
20. Ion, M., Pop, F., Dobre, C., Cristea, V.: Dynamic resources allocation in grid enviroments. In: Proceedings of the 11th International Symposium on Symbolic and Numeric Algorithms for Scientific Computing, SYNASC'09, Timisoara, pp. 213–220. IEEE Computer Society, Washington, DC (2009). doi:10.1109/SYNASC.2009.20
21. jboss.org: JBoss application server 7. http://goo.gl/0Hg9z (2011)
22. Jensen, M., Meyer, C.: Expressiveness considerations of XML signatures. In: Proceedings of 35th IEEE Annual Computer Software and Applications Conference Workshops, COMPSACW'11, Munich, pp. 392–397. IEEE Computer Society, Washington, DC (2011). doi:10.1109/COMPSACW.2011.72
23. Jensen, M., Gruschka, N., Herkenhoner, R., Luttenberger, N.: SOA and web services: new technologies, new standards – new attacks. In: Proceedings of the 5th European Conference on Web Services, ECOWS'07, Halle, pp. 35–44. IEEE Computer Society, Washington, DC (2007). doi:10.1109/ECOWS.2007.24
24. Jureta, I., Faulkner, S., Achbany, Y., Saerens, M.: Dynamic web service composition within a service-oriented architecture. In: IEEE International Conference on Web Services, Salt Lake City, pp. 304–311. IEEE Computer Society, Los Alamitos (2007). doi:http://doi.ieeecomputersociety.org/10.1109/ICWS.2007.79
25. Ko, R.K.L., Jagadpramana, P., Mowbray, M., Pearson, S., Kirchberg, M., Liang, Q., Lee, B.S.: TrustCloud: a framework for accountability and trust in cloud computing. In: Proceedings of the 7th IEEE World Congress on Services, SERVICES'11, Washington, DC, pp. 584–588. IEEE Computer Society, Washington, DC (2011). doi:10.1109/SERVICES.2011.91
26. Lakshminarayanan, S.: Interoperable security standards for web services. IT Prof. **12**(5), 42–47 (2010). doi:10.1109/MITP.2010.98

27. Lilien, L., Bhargava, B.: A scheme for privacy-preserving data dissemination. IEEE Trans. Syst. Man Cybern. A **36**(3), 503–506 (2006). doi:10.1109/TSMCA.2006.871655
28. Mainka, C., Somorovsky, J., Schwenk, J.: Penetration testing tool for web services security. In: Proceedings of the 2012 IEEE Eighth World Congress on Services, SERVICES'12, Honolulu, pp. 163–170. IEEE Computer Society, Washington, DC (2012). doi:10.1109/SERVICES.2012.7
29. nist.gov: NIST SP800-53: security and privacy controls for federal information systems and organizations. http://goo.gl/wBWYh (2012)
30. Noel, S., Jajodia, S.: Managing attack graph complexity through visual hierarchical aggregation. In: Proceedings of the 2004 ACM Workshop on Visualization and Data Mining for Computer Security, VizSEC/DMSEC'04, Washington, DC, pp. 109–118. ACM, New York (2004). doi:10.1145/1029208.1029225
31. oasis-open.org: WS-SecureConversation 1.3. http://goo.gl/90BYS (2007)
32. oasis-open.org: WS-Trust 1.3. http://goo.gl/Tv5OP (2007)
33. Phan, K.A., Tari, Z., Bertok, P.: A benchmark on soap's transport protocols performance for mobile applications. In: Proceedings of the 2006 ACM Symposium on Applied Computing, SAC'06, Dijon, pp. 1139–1144. ACM, New York (2006). doi:10.1145/1141277.1141548
34. Rahaman, M.A., Schaad, A., Rits, M.: Towards secure SOAP message exchange in a SOA. In: Proceedings of the 3rd ACM Workshop on Secure Web Services, SWS'06, Alexandria, pp. 77–84. ACM, New York (2006). doi:10.1145/1180367.1180382
35. Saha, D.: Extending logical attack graphs for efficient vulnerability analysis. In: Proceedings of the 15th ACM Conference on Computer and Communications Security, CCS'08, Alexandria, pp. 63–74. ACM, New York (2008). doi:10.1145/1455770.1455780
36. Schneier, B.: Attack trees – modeling security threats (1999). http://goo.gl/ZKRkz
37. Sengupta, S., Kaulgud, V., Sharma, V.S.: Cloud computing security–trends and research directions. In: Proceedings of the 7th IEEE World Congress on Services, SERVICES'11, pp. 524–531. IEEE Computer Society, Washington, DC (2011). doi:10.1109/SERVICES.2011.20
38. She, W., Yen, I.L., Thuraisingham, B., Bertino, E.: The SCIFC model for information flow control in web service composition. In: Proceedings of the 2009 IEEE International Conference on Web Services, ICWS'09, pp. 1–8. IEEE Computer Society, Washington, DC (2009). doi:10.1109/ICWS.2009.13
39. She, W., Yen, I.L., Thuraisingham, B., Huang, S.Y.: Rule-based run-time information flow control in service cloud. In: Proceedings of the 2011 IEEE International Conference on Web Services, ICWS'11, pp. 524–531. IEEE Computer Society, Washington, DC (2011). doi:10.1109/ICWS.2011.35
40. Sidharth, N., Liu, J.: Intrusion resistant soap messaging with iapf. In: Proceedings of the 2008 IEEE Asia-Pacific Services Computing Conference, APSCC'08, pp. 856–862. IEEE Computer Society, Washington, DC (2008). doi:10.1109/APSCC.2008.221
41. Sundareswaran, S., Squicciarini, A., Lin, D., Huang, S.: Promoting distributed accountability in the cloud. In: Proceedings of the 4th IEEE International Conference on Cloud Computing, CLOUD'11, pp. 113–120. IEEE Computer Society, Washington, DC (2011). doi:10.1109/CLOUD.2011.57
42. Tndel, I.A., Jensen, J., Rstad, L.: Combining misuse cases with attack trees and security activity models. In: Proceedings of the 5th International Conference on Availability, Reliability and Security, ARES'10, pp. 438–445. IEEE Computer Society (2010). doi:10.1109/ARES.2010.101
43. w3.org: SOAP version 1.2 part 1: messaging framework (second edition). http://goo.gl/A2agX (2007)
44. Wei, Y., Blake, M.B.: Service-oriented computing and cloud computing: challenges and opportunities. IEEE Internet Comput. **14**(6), 72–75 (2010). doi:10.1109/MIC.2010.147
45. Wilde, N., Simmons, S., Pressel, M., Vandeville, J.: Understanding features in soa: some experiences from distributed systems. In: Proceedings of the 2nd International Workshop on Systems Development in SOA Environments, SDSOA'08, pp. 59–62. ACM, New York (2008). doi:10.1145/1370916.1370931

46. Xie, R.: Manifesting security issues in a service cloud structure through auditing. M.S. Thesis, Tandy School of Computer Science, University of Tulsa, Tulsa (2012)
47. Xie, R., Gamble, R.: An architecture for cross-cloud auditing of service cloud. In: Proceedings of the 8th Cyber Security and Information Intelligence Research Workshop (2012)
48. Xie, R., Gamble, R.: A tiered strategy for auditing in the cloud. In: Proceedings of the 5th IEEE International Conference on Cloud Computing, CLOUD'12, pp. 945–946. IEEE Computer Society, Washington, DC (2012). doi:10.1109/CLOUD.2012.144

Exploiting Timing Side Channel in Secure Cloud Scheduling

Sachin Kadloor and Negar Kiyavash

Abstract Traditionally, scheduling policies used in event schedulers have been designed to optimize performance based metrics such as throughput and delay while maintaining some notion of fairness. In multi-tenancy cloud environments, it is important to ensure privacy of the users because a scheduler creates a timing based side channel through which malicious users can learn about the service usage pattern of the others. In this chapter, we demonstrate the existence of a timing side channel in shared schedulers and discuss the design of secure scheduling policies. When a processor is shared by multiple users, the delays experienced by jobs from one user are a function of the arrival pattern of jobs from other users, and the scheduling policy of the server. Consequently, a scheduling system creates a timing side channel in which information about arrival pattern from one user is inadvertently leaked to another. In this work, this information leakage is studied for a two user scheduling system. We first introduce a measure of privacy and then demonstrate that no scheduler can provide maximum privacy without idling/taking vacations, and consequently no policy can simultaneously be delay and privacy optimal.

1 Introduction

In multi-tasking systems where a finite resource is to be shared, a scheduler dictates how the resource is divided among competing processes. Examples of systems which have schedulers include, a computer where the CPU needs to be shared between the different threads running, a cloud computing infrastructure with shared computing resources, a network router serving packets from different streams, etc. Some of the commonly used schedulers are *first-come,first-served (FCFS)*,

S. Kadloor (✉) • N. Kiyavash
University of Illinois at Urbana-Champaign, Urbana, IL, USA
e-mail: kadloor1@illinois.edu; kiyavash@illinois.edu

K.J. Han et al. (eds.), *High Performance Cloud Auditing and Applications*,
DOI 10.1007/978-1-4614-3296-8_6,

round-robin (RR), *shortest-job-first (SJF)* and *priority* schedulers. Performance of a scheduler is measured in one of several metrics including, throughput (number of job completions per unit time), average delay (the difference between the job completion time and the job arrival time), fairness (a metric to measure if the resource is being distributed equally/fairly between the processes), etc. A scheduler often has to make a calculated trade-off among these conflicting metrics.

It has long been known that resources shared between processes lead to covert and side channels that can leak information from one process to another. In a *covert channel*, one process structures its use of the shared resource in a particular pattern so as to communicate secret information to another. Covert channels have been studied extensively in the context of multilevel secure systems, where they can be used to create forbidden information flows [20, 21, 26, 30, 50]. In a *side channel*, on the other hand, one process tries to learn something about the operation of another without the latter's cooperation. Side channels, therefore, focus on information that is leaked *incidentally* by a victim process, rather than explicitly coded by a sender. This makes side channels more challenging to exploit, but it also makes them much more broadly applicable in multi-tenant environments. If two mutually non-trusting processes share a resource, then a side channel could potentially leak confidential information from one to another. While *timing covert channels* have been studied extensively, most notably [18], there has been very little work in studying timing side channels. Most of the solutions proposed to mitigate the information leakage in side channels are system specific.

Some examples are cryptographic blinding [11, 24], program transformation to remove timing side channels from programs written in a sequential imperative programming language [1], and the Naval Research Laboratory (NRL) Network Pump for mitigating timing channels that arise in multilevel security systems (MLS) [31].

We consider the scenario when a scheduler is serving jobs from two users, when one of them is an innocuous user (Alice) and a malicious user (Bob). Bob wishes to learn the pattern of jobs sent by Alice. Bob exploits the fact that when the processor is busy serving one process and the other process experiences a delay, thus leaking information about one process to another. Specifically, Bob computes delays experienced by his jobs and uses these delays to infer about the times when Alice tried to access the processor, and possibly the sizes of jobs scheduled. Learning this traffic pattern from Alice can aid Bob in carrying out traffic analysis attacks. We would therefore want to design scheduling policies where the attacker cannot learn the traffic pattern reliably. The policy design of the scheduler involves deciding which user's job (from the queued jobs) to serve upon completion of the current job. One of the most commonly deployed scheduling policies is FCFS policy. For example, digital subscriber line (DSL) routers often employ such a policy. The FCFS policy is simple to implement and guarantees small delay to the users. The policy, however, provides excessive information leakage to an attacker by virtue of providing high correlation between the legitimate user's request pattern and the delays experienced by the attacker. Consequently, FCFS policy allows Bob to estimate the arrival pattern of Alice's requests to a high degree of accuracy.

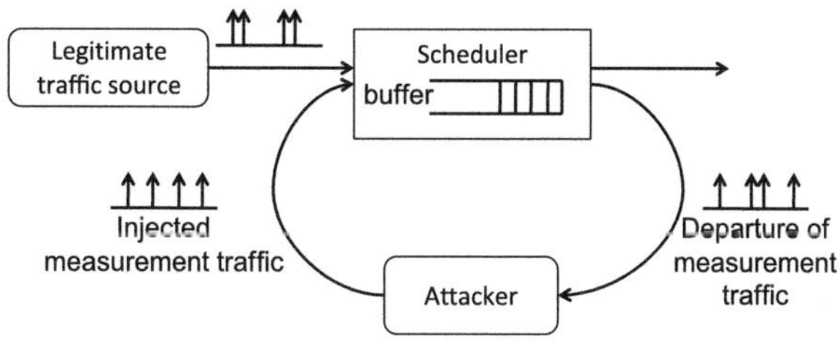

Fig. 1 An event/packet scheduler

In this chapter, we study a generic shared scheduler shown in Fig. 1. For such systems, in order to minimize the information leakage, one has to design *privacy-preserving* scheduling policies. FCFS is an example of a bad policy in this respect. An example of a good privacy preserving scheduling policy is the time division multiple access (TDMA), where a user is assigned a fixed service time regardless of whether he has any jobs that need to be processed or not. As expected, the waiting times of jobs issued by Bob are independent of Alice's arrival pattern, and consequently, the policy leaks no information. However, TDMA is a highly inefficient policy in terms of throughput and delay, especially when the traffic is varying. It is especially inefficient when the number of users using the scheduler is large. FCFS and TDMA represent two extremes of the trade-off between the information leakage and efficiency (in terms of delay or throughput). Such a trade-off is inherent to the scheduler policy design, and the goal of this chapter is to design scheduling policies that minimize this leakage without significantly compromising quality of service (QoS).

This chapter is organized as follows. Section 2 describes queuing side channel attack in cloud computing environments. Section 3 provides an overview of timing side channel threats and countermeasures. Section 4 discusses an attack strategy against FCFS policy which does leak significant information between the flows. Section 5 introduces a privacy-preserving scheduling Model model and define privacy metrics that can be used to compare the privacy of different scheduling policies. Section 6 introduces bounds on the privacy metric for some common scheduling policies. Section 7 presents job delay analysis. Finally, Section 8 presents our conclusions.

2 Queuing Side Channel Attack in Cloud Computing

While the queuing side channel can arise in any environment where a shared resource is used to serve multiple clients, we motivate this work using an example that arises in the context of packet networks. Recall our example of Bob, an attacker who is curious about Alice's activities. Bob shares a resource which is in use by Alice, namely the packet queue inside her DSL router. This resource allows him to create a side channel that will leak information about Alice's traffic. This scenario is illustrated in Fig. 2. Bob sends a low-bandwidth, but high frequency

Fig. 2 Side channel setup

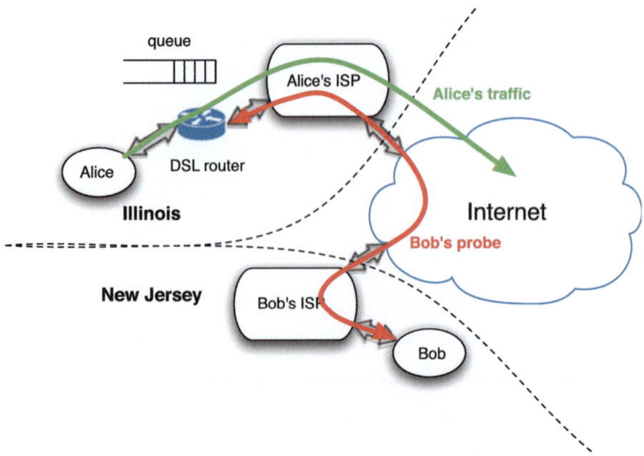

probe to the router and measures the round-trip time (RTT) of his probe packets. The DSL has an incoming and an outgoing port, for the traffics addressed to, and originating from Alice's computer. Bob's probe responses and Alice's incoming traffic share the same queue, hence the delay that Bob observes varies based on the pattern of traffic addressed to Alice. Although these pings travel through various intermediate routers, their RTT is primarily affected by Alice's traffic. This is because of two reasons. One, the intermediate routers have significantly higher bandwidth compared to the volume of the traffic flowing through them making Alice's router the bottleneck. Two, the intermediate routers carry multiple traffic flows; therefore, the delays incurred at these routers does not change with time.

To evaluate the potential for this type of attack, we observed traffic of a home DSL user in Illinois, while simultaneously sending a ping probe from a computer in New Jersey every 10 ms. For a detailed description of the setup [19]. Figure 3 shows the results: the DSL traffic is shown in Fig. 3a, and the round-trip times of the ping probe are shown in Fig. 3b, showing a clear correlation between the two. The probe traffic used less than 50 Kbps of bandwidth and is unlikely to be noticed. Although such an attack can only reveal the timing and the volume of the traffic, and not the actual contents of the packets, recent research has shown that significant inferences can be drawn by observing just the traffic pattern of a user. *The strength of this attack lies in the fact that it does not require special access or privileges for the attacker.* The attacker does not even need high computational power. This dramatically increases the attack surface available for conducting *traffic analysis*: recovery of secret data based on packet counts, timings, and sizes. Recent research has demonstrated that traffic analysis can be used for a number of attacks: identification of web sites visited [6, 27], guesses at passwords typed [42], and recovery of phrases spoken over VoIP [51]. However, it was previously assumed that such attacks required access to one of the routers along the path of Alice's traffic. We conducted an experiment wherein the attacker makes use of this side channel to identify which website the user is likely browsing. The attacker was able to classify the website correctly 70% of the time.

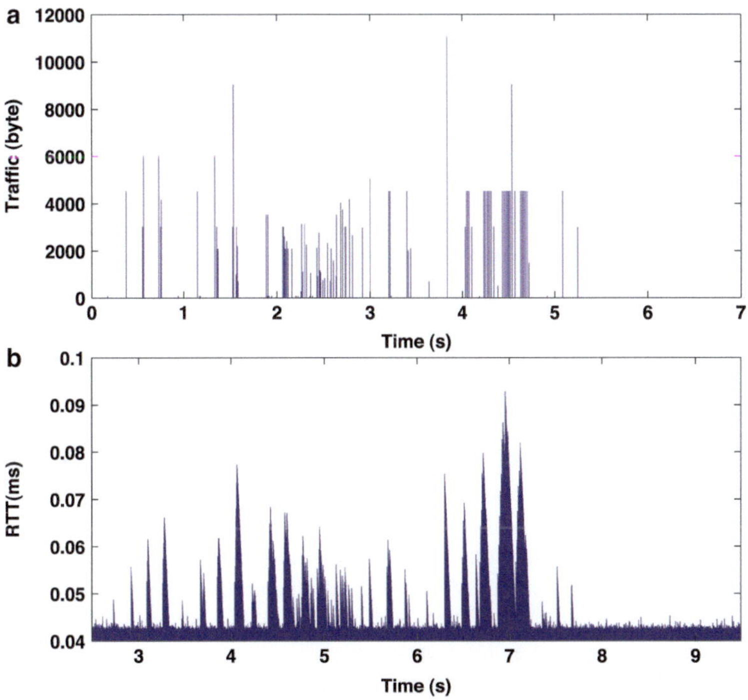

Fig. 3 Real traffic on a DSL line vs. observed probe RTTs when Alice is browsing the website (Reproduced from Kadloor et al. [22]). (**a**) DSL traffic entering Alice's router. (**b**) Round trip times of Bob's ping packets

In all these works, the attacker observes Alice's traffic and uses statistical inference techniques to carry out the attack. The assumption is that the attacker observes Alice's traffic directly (as shown in Fig. 3a). The attacker observes the traffic through a side channel (as shown in Fig. 3b), which is then used to estimate the traffic. The fact that the attacker does not observe Alice's traffic directly, but has to estimate it. The accuracy of the traffic analysis attacks will be lower than those obtained by the authors of the works mentioned. However, it must also be noted that our attack can be carried out remotely. The strength of our attack lies in the fact that it does not require special access or privileges for the attacker. It is therefore important to study these timing side channels and design schedulers to minimize the information leakage.

The increasing interest in cloud computing, wherein users issue jobs to a shared computing platform, opens up possibilities for such an attack, and is another motivation to study these timing side channels.

In [37], the authors map the internal infrastructure of Amazon Elastic Compute Cloud (Amazon EC2) computing service, and demonstrate that it is possible for an attacker to place his virtual machine (VM) on the same physical computer as the target's VM. They shown that once placed on the same physical computer, any timing channel created by sharing of the processor can be exploited by the attacker. In cryptographic side channels, the attacker aims at recovering cryptographic keys by utilizing the timing variations required for cryptographic operations [9,24]. In the following section, we give a more exhaustive survey of the related work in this area.

3 Timing Side Channel Threats and Countermeasures

Often, the timings of various events carry sensitive contextual information about the events themselves. In the context of networks, it has been recognized that encryption alone is insufficient to protect the contents of the packets [34]. The timing and sizes of packets flowing on a link carry information as well, as demonstrated in [39], wherein the identity of a video clip is guessed based on observing streaming encrypted data. In the context of cryptography, the time it takes a processor to perform a cryptographic operation reveals secret keys [8,24]. Inter-keystroke timing information reveals information on the word being typed [42]. Duration of web based transactions can potentially leak private information held at the server, such as the number of private photos in a publicly viewable gallery or the number of items in a shopping cart of an internet retailer [7].

Typical defenses to mitigate timing based leakage include ensuring that all the response times are the same [15], or adding random noise to the response times [24,40]. However, the distribution of response times of many applications is skewed [12], that is, the worst case response time is significantly longer than the average case response time. This makes the first solution to this problem infeasible in certain systems. In certain scenarios, the attacker can repeatedly carry out the attack to average out the noise added, rendering the second approach ineffective [24].

Our work falls into the broad category of study of spying channels. These are communication channels which have historically been used by spies to interact secretly among themselves. In typical scenarios, an insider with access to a secret piece of information leaks it to an outsider using covert steganographic channels or a side channel is used to spy on the activity of others. More often than not, the mode of transmission of information being transmitted or extracted is timing information. In the following, we give examples of these channels and suggested countermeasures.

In a covert channel, two users communicate with each other in a manner so that nobody is unaware of the existence of their conversation. All they share is an intermediate system, e.g., a processor or a network link. They make use of the fact that any dynamically shared resource is a channel. If one user can affect the availability of the resource to the other user, there is a channel [23]. Simultaneous multi-thread (SMT) processors run many processes concurrently, sharing most of the processor resources. Wang and Lee [46] describes a scenario in which one process either uses up all or none of the multipliers on board to covertly convey a message to another process, who receives this signal by concurrently trying to use the multipliers and observing the delay. Cabuk et al. [10] demonstrate a network timing channel over IP. The presence or absence of a packet in a timing interval is used to convey a message. Liu et al. [28] is an example of timing channel being used as a robust covert channel. These channels are usually hard to detect; nevertheless, they are important to analyze and mitigate. Under US Trusted Computer System Evaluation Criteria (TCSEC), covert channel analysis is required starting at B2 level of assurance [16]. Refer to [45] for a survey of articles pertaining to covert channels. We next discuss a few mitigation schemes.

In their seminal work, Anantharam and Verdú [2] study the capacity of a covert timing channel where Eve communicates to Bob through the timing of packets she transmits to him. An intermediate first-in-first-out (FIFO) queuing system adds noise to the timing information, by buffering each packet for a random time before transmitting it. They show that, if the delays added by the queue are independent and identically distributed, of all the distributions with the same mean D, the capacity of the channel is minimized when the delays added have an exponential distribution. Giles and Hajek [18] study a generalized version of the covert timing channel, wherein the queuing system is replaced by a jammer that can delay packets arbitrarily subject to a delay constraint. They show that the *best jammer* is one that buffers up jobs until the nearest multiple of a time D and then dumps the outputs. Such a jammer is guaranteed to delay a packet by a time at most D for any input process. The authors also develop good jamming strategies that are constrained by the average delay they add, or by the buffer size used. It is important to note a difference between this work and [2]. In [2], the authors look for the best jammer within the class of queuing disciplines with fixed mean service time (the time it takes for a packet to depart the system once it enters service) whereas in [18], the authors look for the best jammer within the class of jammers with a fixed queuing delay (this is the time spent by a packet in the queue waiting for service plus the time it takes for the server to serve the packet). A queue with a fixed service time does not guarantee a bound on the maximum or the average queuing time for arbitrary input processes.

A variant of the covert communication channel is the steganographic channel in which the adversary cannot actively generate events, but can modulate the timing of events generated by an innocuous user. In [47], Wang and Moulin consider the scenario wherein the innocuous user generates a source code according to some distribution. The malicious user then embeds a message into this source code without distorting it by much, while also ensuring that the empirical distribution of the distorted code matches the distribution of the original source code. The system then adds noise to this distorted message which is then received by the other malicious users. The authors consider the case when the mitigator is a discrete memoryless channel (DMC), and derive the maximum rate at which Eve can reliably communicate to Bob. Obviously, the data rates achievable through such channels is smaller than those achievable by covert channels, but they are also harder to detect. Note that a timing channel (e.g., the queuing channel) is typically not a memoryless channel, i.e., the current output of the channel depends not only on

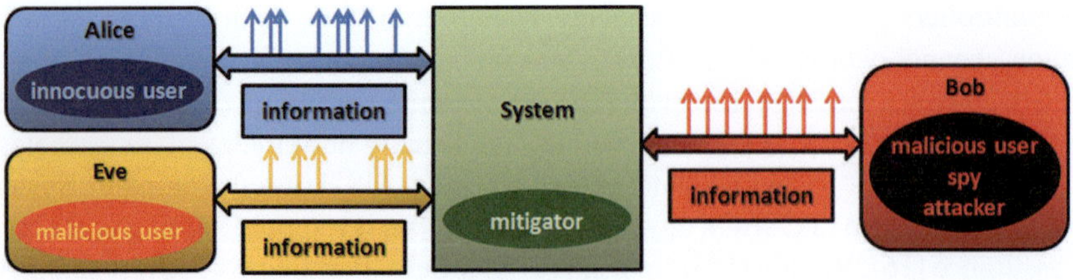

Fig. 4 A timing side channel with active adversaries

the current input to the channel, but the past inputs as well. For this reason, the results derived herein do not carry over to timing channels. Jitterbug is an example of a timing steganographic channel [41]. The authors modify a keyboard at the hardware level in a way that each time a key is pressed, instead of generating an event immediately, it delays the generation of the event. This is done in such a way as to encode the inter-keystroke timing information. This information can later be recovered and used to guess passwords typed. Steganographic channels are closely related to watermarking, wherein the cover traffic is perturbed slightly. This perturbation serves as a signature for the traffic and can be tracked. This idea has been used previously for Secure Shell (SSH) stepping stone correlation [48] and for tracking voice over internet protocol (VoIP) calls [49]. The work [53] considers a steganographic timing channel from a mitigator's perspective, and derives jamming strategies (similar to those derived in [3]) which bound the rate at which Eve can communicate to Bob. Again, delay is not explicitly considered while designing the jamming strategy, and no claim is made about its optimality. In a covert channel, the transmitter and the receiver of the secret information conspire with each other, so they can agree on a communication strategy ahead of time. This is not the case in a side channel, where the transmitter of the information is usually unaware of the fact that she is being spied upon. The shared cache memory in a CPU leads to a side channel which can be exploited to steal cryptographic keys [33, 35, 36, 43, 46]. Side-channel leakage can also occur in contexts outside of cryptographic algorithms themselves. Song et al. [42] describe a timing attack on the inter-keystroke timing of an interactive SSH connection. One of the earliest works on the study of capacity of side channels was by Wyner [52], extended by Csiszár and Körner [13]. In the area of timing side channels, a large body of literature exists on the anonymity analysis of Chaum mixes [17, 44].

In this chapter, we study a variant of this problem, a side channel with active adversaries, as shown in Fig. 4. Two malicious users Eve and Bob co-ordinate their actions to learn about the activity of an innocuous user Alice. In this scenario, a malicious user Eve can actively generate events as well, and can do so in a manner that is favorable to an attacker Bob. In [32], the authors consider the scenario where a client is connected to a rogue website using a The Onion Router (TOR) network, which is designed to protect the identity of the users. The website modulates the traffic sent to the client. The website can then try to simultaneously send data through each of the TOR nodes and measure the delay incurred. By correlating this delay with the traffic sent to the client, the website can obtain the client's identity, thus defeating the purpose of TOR. While that attack is no longer viable [14], the reason is that there are many more TOR nodes now than when [32] was published, and not because the timing based side channel has been eliminated. In [19], the authors exploit the side channel in a DSL router to infer the website being visited by the victim. A similar side channel exists within Amazon EC2 computing service, which is exploited in [37].

As mentioned earlier, FCFS is one of the most commonly deployed scheduling policies. The attack on the DSL router was feasible because FCFS policy was being used. We therefore start by describing in some detail the attack against the FCFS policy in the following section.

4 Attack Against the FCFS Scheduling Policy

The FCFS scheduling policy serves jobs in the order in which they arrive. Although this policy is easy to implement, and fares well in the metrics of throughput and mean delay, as we will demonstrate, it does not perform well in terms of preserving the privacy of a user. In order to demonstrate the fallibility of the policy, we will present one specific attack and evaluate how well an attacker is able to estimate the traffic patterns of the other users. Specifically, when the scheduler is heavily loaded, we will show that the attacker is able to estimate accurately the innocuous user's traffic pattern, implying maximum leakage. The attack involves two components such as (1) deciding on when and how many jobs Bob (the attacker) has to issue, and (2) after having observed his service times, deciding on how to use the information available to him in estimating the timing and volume of jobs issued by Alice.

4.1 Queuing Side Channel Attack Model

Let $\mathscr{X}(t)$ be the counting function denoting the total number of jobs issued by Alice until time t. Consider a related sequence $\{X_k = \mathscr{X}(kc) - \mathscr{X}((k-1)c) : k = 1,2,\cdots\}$, where $c > 0$ is a parameter. X_k is the number of jobs issued by Alice in the interval $((k-1)c, kc]$. We will consider the scenario where Bob is interested in estimating the sequence $\{X_k\}$, which will henceforth be referred to as Alice's arrival pattern. Throughout this chapter, we will use capital letters to denote the random variable, and small letters to denote its realization. The parameter c determines the resolution at which the attacker is interested in learning $\mathscr{X}(t)$. The case when c is very small corresponds to the case when the attacker wishes to learn $\mathscr{X}(t)$ exactly, for larger values of c, he is content with learning less fine grained information of the arrival pattern. For example, if the side channel corresponded to a scenario in which an attacker is snooping on an individual using a keyboard, and the attacker's intent was to measure inter-keystroke timing information accurately, he would need a resolution in tens of microseconds. On the other hand, if the goal of the attacker was to only infer if the victim was at his workstation or not by measuring the activity on his keyboard, a much coarser resolution would do.

Let λ be the average rate at which Bob issues his jobs. We consider an attack strategy where Bob issues A_k number of jobs at times kc, for $k \in \{0,1,2,\ldots\}$; the choice of A_k will be discussed shortly. Depending on whether λ is greater or smaller than $1/c$, we consider two different strategies for choosing A_k.

1. If $\lambda \geq \frac{1}{c}$, we will let $A_k = 1 + B_k$, where B_k is a Poisson random variable with mean $\lambda c - 1$ (so that $\mathbf{E}[A_k] = \lambda c$, and the average rate is equal to λ).
2. If $\lambda < \frac{1}{c}$, the attacker cannot issue a job every c time units. In this case, he issues a job every c time units with probability p and does not issue one with probability $1-p$. The value p is chosen so that the average rate at which he issues jobs is equal to λ, i.e., $p/c = \lambda$. Therefore, in this case A_k is a Bernoulli random variable with parameter λc.

By observing the departure times of the jobs he has issued, Bob estimates Alice's pattern as follows. Under the FCFS policy, the scheduler serves jobs in the strict order of their arrivals. Hence, the difference between the times when jobs sent at time $(k-1)c$, and those sent at time kc, depend only on the number of arrivals from Alice in between these two times, *viz.*, X_k. Let D_k be the random departure time of the last job issued at time kc. Let $k \geq 1$. Bob estimates X_k as $\widehat{X_k}$, defined as follows:

1. Given $D_k = d_k$, $D_{k-1} = d_{k-1}$, and $A_k = a_k$, if $d_{k-1} \geq kc$, then $\widehat{X_k} = d_k - d_{k-1} - a_k$. This is the scenario shown in Fig. 5a.
2. Given $D_k = d_k$, $D_{k-1} = d_{k-1}$, and $A_k = a_k$, if $d_{k-1} < kc$, then

$$\widehat{X_k} = \mathbf{E}[X_k | X_k \geq \lceil d_k - kc \rceil, X_k \leq \lfloor d_k - d_{k-1} - a_k \rfloor]$$

   if $\lambda \geq 1/c$, and $\widehat{X_k} = \lambda_2 c$ otherwise, where λ_2 is the long term average rate at which Alice issues her jobs (assumed known to Bob). This is the scenario shown in Fig. 5b.

Two scenarios are illustrated in Fig. 5. The upward and downward pointing arrows denote the arrival and departure times of the batch jobs, respectively. Figure 5a shows the first scenario that four jobs are issued by Alice between $(k-1)c$ and kc, and four jobs are departed after kc. Figure 5b presents the second scenario that three jobs are issued by Alice between $(k-1)c$ and kc, and two jobs are departed after kc. The scenarios indicate that the minimum number of jobs which Alice must have issued between the times $(k-1)c$ and kc are at least two Alice batch jobs to measure Alice's activities because delayed job execution and batch processing should be considered.

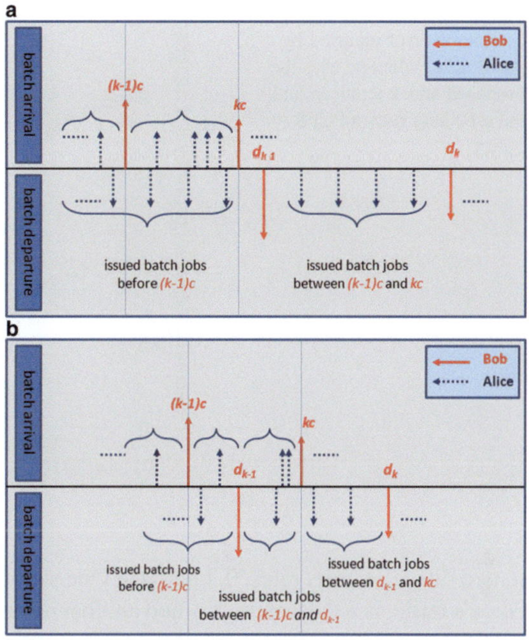

Fig. 5 Illustrations of two delay types of batch jobs for Alice and Bob (Modified from Kadloor et al. [22]). (**a**) Scenario 1: $d_{k-1} \geq kc$. (**b**) Scenario 2: $d_{k-1} < kc$

4.2 Poisson and PPB Traffic Estimation

In order to evaluate the effectiveness of the attack, we carry out two sets of simulations. In the first set of simulations, Alice's arrival process is simulated to be a Poisson traffic of rate λ_2, where $\lambda_2 = 0.2$ and then $\lambda_2 = 0.4$. This means the time between successive jobs from Alice is an independent and identically distributed (i.i.d.) exponential random variable of parameter λ_2, and the size of each job is one. We take $c = 5$. The object of the simulation is to estimate the long term average mean-squared estimation error $\lim_{n\to\infty} n^{-1}\sum_{k=1}^{n} E[(X_k - \hat{X}_k)^2]$ (the limit exists). This is closely approximated by $N^{-1}\sum_{k=1}^{N} E[(X_k - \hat{X}_k)^2]$, where we take $N = 2{,}000$. In the simulation, observing the value of X_k to be x_k and the value of $\hat{X}_k$ to be $\hat{x}_k$, there is a law of large numbers applicable to our queueing model that tells us that the simulation-based quantity $N^{-1}\sum_{k=1}^{N}(x_k - \hat{x}_k)^2$ is highly likely to be a good estimate of $N^{-1}\sum_{k=1}^{N} E[(X_k - \hat{X}_k)^2]$ and therefore a good estimate of the long term average mean-squared estimation error.

In the second set of simulations, Alice's arrival process is generated according to a Poisson Pareto Burst (PPB) traffic model [54]. Traffic in this scenario is bursty, although we chose the parameters such that the long term average rate is still the ones used earlier, $\lambda_2 = 0.2$ and 0.4. In this scenario however, Bob is assumed to know only λ_2, but not the fact that Alice's model is PPB. As a result, Bob will incur a larger estimation error at times when he uses the Poisson estimate $\widehat{X_k} = \lambda_2 c$, and consequently a higher error overall. In Fig. 6, we plot the estimation error incurred by Bob if he used the attack strategy proposed here. The sets of curves correspond

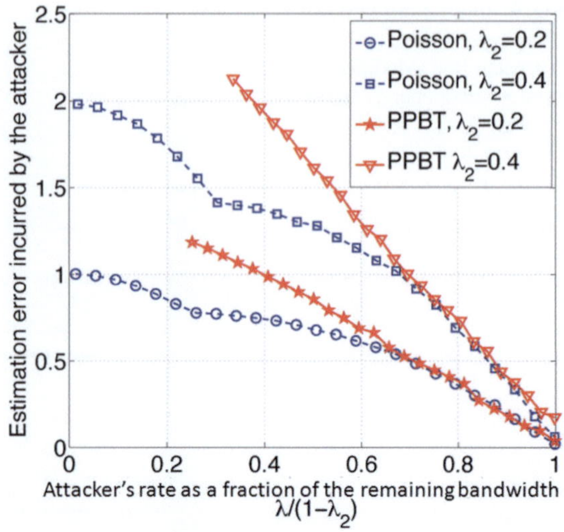

Fig. 6 A plot of the estimation error incurred by the attacker when he uses the proposed attack strategy, and the scheduler uses a FCFS policy

to two different users rates, 0.2 and 0.4. One set of curves represents the case when Alice's traffic is a Poisson traffic, and another represents the case when Alices traffic is a PPB traffic (PPBT). The resolution c is fixed at 5 [22]. Some remarks about the plot are the following:

- Even in the case when Alice's traffic is a PPB traffic, the estimation error incurred by the attacker goes to zero when he issues jobs at high rates. This means that, if a scheduler employs FCFS scheduling, an attacker can estimate Alice's traffic pattern accurately.
- When the attacker's rate is near zero, and Alice's traffic is a Poisson traffic, the estimation error incurred by the attacker converges to $\lambda_2 c$ (equal to 1 and 2 respectively for the case when λ_2 is 0.2 and 0.4). The significance of this behavior will be discussed in Section 6.

In order to design scheduling policies which eliminate, or at least reduce the potency of such attacks, we need to introduce a theoretical framework to study the information leakage. In the following section, we introduce the system model and an estimation error based metric to quantify the privacy properties of the scheduling policy.

5 Privacy-Preserving Scheduling Model

In this section, we introduce a privacy-preserving scheduling model and a metric which measures the strength of the policy in preserving the privacy of the users. The scheduler is modeled as an infinite buffer server that is serving jobs from two users. We consider the scenario when one of them is an innocuous user and the other a malicious one. The attacker, Bob, wishes to exploit the queuing side channel described earlier to learn about the pattern of jobs sent by the innocuous user, Alice. Bob is assumed to know accurately the time when his jobs are issued, and the time it took for the scheduler to process it, i.e., the difference between the completion time of the job and the time when it was issued. Knowing the delays experienced by his jobs, Bob uses this information to guess the arrival pattern of jobs from Alice (see Fig. 5).

The ability of Bob to successfully learn about Alice's arrival process depends heavily on Alice's arrival process itself. For example, on-off patterns are easier to detect reliably compared to an arrival process that is less bursty. In order to ensure that the scheduling policies we design are robust to a variety of arrival patterns, Alice's arrival process will be modeled as a Poisson process of rate λ_2, with all the jobs of unit size. We do this partly because Poisson processes are known to have maximum entropy rate among processes of a given rate [29], and hence represent a rich class of arrival processes, and also because the closed form expressions which can be derived reveal the nature of tradeoffs between privacy and delay. We will comment on the case when Alice's traffic pattern follows a general arrival pattern later. We will also consider the scenario when Bob knows the value of λ_2, the mean rate at which traffic arrives from Alice. A policy that guards the privacy of Alice in this scenario will also perform well when the attacker does not know a priori this rate.

Depending on the scenario, the attacker Bob might be interested in different aspects of Alice's arrival process. In this work, we restrict the analysis to the case when Bob is interested in learning reliably how many of Alice's jobs arrived during fixed intervals of time defined by periodically spaced clock ticks. The times at which this clock ticks is independent of all the other stochastic processes in the system. As before, define X_k to be the total number of jobs issued by Alice in between the $(k-1)$th and kth clock tick, i.e., in the time interval $((k-1)c, kc]$.

The ability of a scheduling policy to protect the privacy of a user is measured by the mean estimation error that the attacker Bob incurs in estimating the sequence $\{X_1, X_2, \ldots\}$, when he is free to choose the number of jobs and times at which he issues them, subject to an average rate constraint. For each $k \geq 1$, let $\hat{X}_k$ be Bob's estimate of X_k. More formally, let $t_1 < t_2 < \ldots < t_n$ be the arrival times of n jobs from Bob, and let $T_1', T_2', \ldots, T_n'$ be the corresponding random departure times. Bob's estimate of X_k is taken to be $\mathbf{E}[X_k | t_1, t_2, \ldots, t_n, T_1', T_2', \ldots, T_n']$, the minimum mean-square estimate. The arrival times $t_1, t_2, \ldots, t_n$ are chosen subject to the constraint that the attacker's traffic rate n/t_n is less than λ, a positive parameter. We point out that each of the random variables T_i' $(i = 1, \ldots, n)$ is a deterministic function of the sequence of arrival times of Alice's jobs, where these functions depend on the particular scheduling policy that is used.

The privacy metric of a scheduling policy is defined as

$$\mathscr{E}^{c,\lambda}_{\text{Scheduling policy}} = \limsup_{N \to \infty} \min_{n, t_1, t_2, \ldots, t_n : \frac{n}{t_n} < \lambda} N^{-1} \sum_{k=1}^{N} \mathbf{E}\left[(X_k - \mathbf{E}[X_k | t_1, \ldots, t_n, T_1', \ldots, T_n'])^2 \right].$$

In the next section, we obtain bounds on the privacy metric $\mathscr{E}^{c,\lambda}_{\text{Scheduling policy}}$ for various scheduling policies.

6 Privacy Metrics and Bounds

The privacy metrics of three different scheduling policies for TDMA, FCFS, and Accumulate-and-Serve (AS) are presented in this section.

6.1 TDMA Privacy Metric

First, note that for any scheduling policy,

$$\mathscr{E}^{c,\lambda}_{\text{Scheduling policy}} \leq \lambda_2 c. \tag{1}$$

This is because

$$\mathbf{E}[(X_k - \mathbf{E}[X_k|t_1,\ldots,t_n,T'_1,\ldots,T'_n])^2] \leq \mathbf{E}[(X_k - \mathbf{E}[X_k])^2] = \text{Variance}[X_k] = \lambda_2 c. \tag{2}$$

A scheduling policy achieving maximum privacy would then be a policy for which equality holds in (1). We show that TDMA is such a scheduling policy. If the scheduler uses the TDMA policy, it allocates dedicated time slots to process jobs from each user. For instance, when there are only two arrival processes, the scheduler could assign the odd time slots to serve jobs from the first user, and even time slots to process jobs from the other. If there are no unserved jobs from one of the users, the scheduler just idles in the corresponding time slots.

Theorem 1

$$\mathscr{E}^{c,\lambda}_{\text{TDMA}} = \lambda_2 c.$$

Proof. For the TDMA policy, the arrival pattern of one user does not influence the completion times of jobs in the other. In this scenario, the attacker can do nothing better than use his knowledge of the Alice's arrival statistics to guess the arrival pattern. That is,

$$\mathbf{E}[X_k|t_1,\ldots,t_n,T'_1,\ldots,T'_n] = \mathbf{E}[X_k],$$

and so equality occurs throughout (2). □

As we have shown, from an privacy perspective, TDMA is an optimal scheduling policy. However, TDMA may be impractical. User's job patterns are rarely periodic and would therefore incur severe delays, and TDMA also places a hard limit on each user's available processor time to half the total available time, resulting in poor resource utilization [5, 38]. The delay issues are further discussed in Section 7.

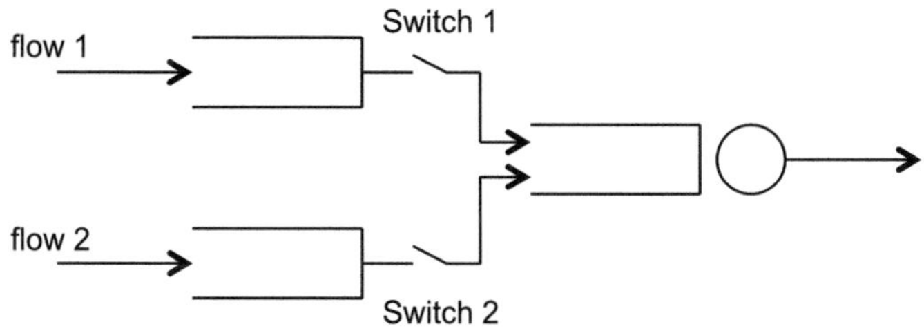

Fig. 7 Pictorial representation of the accumulate-and-serve (AS) policy

6.2 FCFS Privacy Metric

For the FCFS scheduling policy discussed earlier, we have the following theorem, which states that the estimation error incurred by the attacker under FCFS is equal to zero for most cases when the attacker issues his jobs at a high enough rate. We omit the complicated proof, which shall be given in a subsequent work.

Theorem 2 *For fixed arrival rate λ_2 from Alice, if the attacker is allowed to choose the rate at which he injects jobs to the system (while keeping the system stable), then the estimation error incurred by him is bounded by:*

$$\inf_{\lambda<1-\lambda_2} \mathscr{E}^{c,\lambda}_{\text{FCFS}} \leq \begin{cases} 0 & \text{if } \lambda_2 < 1/c \\ \lambda_2 c(1-(1-\lambda_2)^2 c^2) & \text{if } \lambda_2 \geq 1/c \end{cases}$$

Theorem 2 implies that, when Alice's arrival rate is small enough, there exists an attack by which the attacker can learn her arrival pattern accurately. For this reason, FCFS is a poor policy at protecting its user's privacy. This motivates the search for better policies.

6.3 AS Privacy Metric

In the AS scheduling policy, the scheduler accumulates all the incoming jobs in its buffer for a period of T units. We will assume that T is an integer multiple of the time it takes to serve one job. It then serves all the accumulated jobs from the first user, followed by all the accumulated jobs from the second user. The first user could be the attacker, or it could be the other user. If serving all these jobs takes M units of time and if $M < T$, then the scheduler idles for the remaining $T - M$ time slots before beginning to serve the accumulated jobs.

Figure 7 gives a pictorial representation of the working of the policy. The switches close every T time units. All the accumulated jobs from user 1 are served followed by jobs from user 2. The intuition for such a policy stems from the fact that buffering destroys the timing information of the jobs, thereby reducing the amount of information that can be extracted by the attacker. The attacker can therefore learn, at most, the total number of jobs that arrived from Alice in the accumulate period, and nothing more fine grained.

Our subsequent analysis will make use of this fact to show that this policy provides guaranteed levels of privacy to Alice, irrespective of any attack that can be launched by Bob. In [22], it is shown that the AS policy is stable when the sum of rates from all the users is less than 1, and the following result is also established.

Theorem 3 *Irrespective of the attack strategy, the AS policy guarantees a certain degree of privacy to Alice's arrival pattern. More specifically,*

$$\mathcal{E}^{c,\lambda}_{\text{AccServe}} \geq \lambda_2 c \left(1 - \frac{c}{T}\right)_+$$

Remarks:

- The AS policy guarantees a certain level of privacy regardless of the attacker strategy. Furthermore, if the accumulate period is chosen to be large enough so that $\frac{c}{T} \ll 1$, then the estimation error incurred by the attacker gets increasingly close to the maximum estimation error, which is what he would have incurred by just using the statistics of Alice's arrival. In other words, the policy achieves the saddle point in the limit. The price paid for choosing a very large value for T is of course the delay experienced by jobs. Delay analysis is carried out in Section 7. When T is chosen to be five times c or greater, the ratio is greater than 0.8, irrespective of the attacker's strategy.
- It is important to note that the error does not depend on the strategy employed by the attacker. In particular, it does not depend on the rate at which the attacker issues his jobs.

In Fig. 8, we plot the estimation error as given by the bound when c/T is 0.2. Note the large gains in privacy when the attacker's rate is large. Do note that for AS, we only have a lower bound on the estimation error. For FCFS, we plot an *upper* bound to the estimation error, as given by Theorem 2. Likewise, for the priority policy, we plot an *upper* bound to the estimation error the strongest attacker would incur. Recall that this is a bound that holds true for all non-idling policies as well (round-robin, fair queuing, etc.). We also plot the *lower* bound to the estimation error incurred by the strongest attacker against the AS policy. For this figure, the resolution c is chosen to be 5, and the accumulate period is chosen to be of duration 25.

7 Job Delay Analysis

In this section, we analyze the delay experienced by the jobs under different scheduling policies. Job delays depend on the total number of users of the system and the specific patterns of arrivals from them. For purposes of delay comparison,

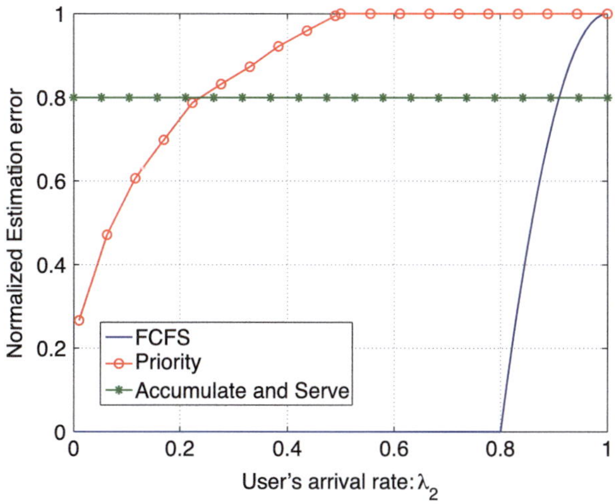

Fig. 8 A plot of the estimation error that the strongest attacker would incur when different scheduling policies are used, as a fraction of the TDMA estimation error ($= \lambda_2 c$)

we will consider a scenario where all the users of the system have time invariant arrival statistics. This would be the delay incurred when all users are legitimate innocuous users. This is a reasonable assumption, as during the *normal* mode of scheduler operation, when there are only legitimate users using the scheduler, the policy should offer as minimum delay to the jobs as possible.

We will consider the case where there are M users of the system, traffic from user i being a Poisson traffic with parameter λ_i, size of all the jobs being equal to 1 time unit. Define $\lambda = \sum_{i=1}^{M} \lambda_i$. We are interested in the mean delay of a job (averaged over all the jobs from all the users) for each of the policies discussed previously; $\mathscr{D}_{\mathrm{FCFS}}$ shall denote the mean delay for FCFS, $\mathscr{D}_{\mathrm{TDMA}}$ shall denote the mean delay for TDMA, and $\mathscr{D}_{\mathrm{Acc-Serve}}$ shall denote the mean delay for the AS.

Theorem 4

$$\mathscr{D}_{\mathrm{FCFS}} = 1 + \frac{\lambda}{2(1-\lambda)}$$

$$\mathscr{D}_{TDMA} = 1 + \frac{M}{2} + \sum_{i=1}^{M} \frac{\lambda_i}{\lambda} \frac{\lambda_i M^2}{2(1-\lambda_i M)}$$

$$\mathscr{D}_{\mathrm{Acc-Serve}} \leq 1 + \frac{T}{2} + \frac{\lambda T}{2} + \frac{\lambda + T(1-\lambda)^2}{2(1-\lambda)}$$

Proof. The FCFS system is a simple $M/D/1$ queue, and the mean delay can be derived using the celebrated formula of Polleczek and Khinchine for the $M/G/1$ system [4]. For the proof for TDMA, refer to [25]. The proof for Accumulate-and-Serve policy is given in [22]. □

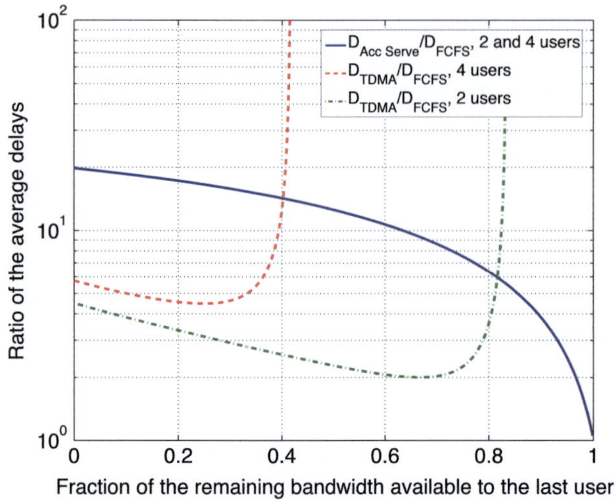

Fig. 9 A plot of the delays of the different scheduling policies as a multiple of the delay of FCFS

Remarks:

- For FCFS and AS, the mean delay is finite, equivalently the system is stable when the sum of the rates from all the users is less than 1, i.e., $\lambda = \sum_{i=1}^{M} \lambda_i < 1$. However, for TDMA, the mean delay is finite only when the rate from each user is less than $1/M$, i.e., when $\lambda_i < 1/M$. Therefore, TDMA loses out on multiplexing gains.
- $\lim_{\lambda \to 1} \frac{\mathscr{D}_{\text{Acc Serve}}}{\mathscr{D}_{\text{FCFS}}} = 1$, i.e., at high traffic loads, the mean delays offered by the AS policy is equal to that when FCFS policy is used. In the other extreme, $\lim_{\lambda \to 0} \frac{\mathscr{D}_{\text{Acc Serve}}}{\mathscr{D}_{\text{FCFS}}} = 1 + T$, which can be quite large especially if T is chosen to be large. For TDMA, the same limit would be $\lim_{\lambda \to 0} \frac{\mathscr{D}_{\text{TDMA}}}{\mathscr{D}_{\text{FCFS}}} = 1 + \frac{M}{2}$. Depending on the values of M and T, one of the two scheduling policies might offer lower delays than the other.

In Fig. 9, the ratios $\frac{\mathscr{D}_{ACC-SERVE}}{\mathscr{D}_{FCFS}}$ and $\frac{\mathscr{D}_{TDMA}}{\mathscr{D}_{FCFS}}$ are given for two scenarios. In the first scenario, $\lambda_1 = \lambda_2 = 0.1$, $\lambda_3 = 0.2$, and λ_4 is varied between 0 and 0.6. In the second scenario, $\lambda_1 = 0.4$, and λ_2 is varied from 0 to 0.6. Note that the mean delays blow up for large rates when TDMA policy is used. The accumulate period T is set to 25. In the first scenario, there are four users of the system, $\lambda_1 = \lambda_2 = 0.1$, $\lambda_3 = 0.2$, and λ_4 is varied between 0 and 0.6. In the second scenario, there are only two users, $\lambda_1 = 0.4$, and λ_2 is varied from 0 to 0.6. When there are four users using the system, the fourth user cannot issue jobs faster than rate 0.25, consequently utilizing only 40% of the bandwidth available to him (0.25/0.6). When AS is used, the fourth user can utilize all the bandwidth that is available. For the plot, T is set to 25. Recall that if the resolution at which information is to be learnt, c, is less than 5, then by choosing T to be 25, we can guarantee that the estimation error incurred by an attacker will be more than 80% of the maximum error that he can incur.

8 Conclusion

We developed an analytical framework to study the effectiveness of scheduling policies with respect to privacy preservation, arrival statistics, and scheduling delays. A key component of this framework is a quantifiable criterion to evaluate the performance of the policy with respect to the information retrieval. The performance of a scheduling policy is measured by how accurately an intelligent attacker is able to estimate the innocuous user's arrival traffic. Based on this framework, this chapter allows us to conclude that

- An estimation error based metric can be used to measure how secure a scheduling policy is against side channel attacks.
- There is an attack against the FCFS scheduling policy under which significant information is leaked between the processes when the scheduler is heavily loaded.
- TDMA offers the highest security, but at a significant performance cost.
- The AS scheduling policy *guarantees* any desired degree of privacy to the innocuous user's arrival pattern regardless of the attacker's policy. The policy achieves high levels of privacy with comparatively lower delay than the TDMA scheduler.

Acknowledgements This material is based upon work partially supported by the Air Force Office of Scientific Research (AFOSR) grants Sub TX 0200-07UI, FA9550-11-1-0016, FA9550-10-1-0573 and FA9550-10-1-0345.

References

1. Agat, J.: Transforming out timing leaks. In: Proceedings of the 27th ACM SIGPLAN-SIGACT Symposium on Principles of Programming Languages, POPL'00, Boston, pp. 40–53. ACM, New York (2000). doi:10.1145/325694.325702
2. Anantharam, V., Verdu, S.: Bits through queues. IEEE Trans. Inf. Theory **42**(1), 4–18 (2006). doi:10.1109/18.481773
3. Askarov, A., Zhang, D., Myers, A.C.: Predictive black-box mitigation of timing channels. In: Proceedings of the 17th ACM Conference on Computer and Communications Security, CCS'10, Chicago, pp. 297–307. ACM, New York (2010). doi:10.1145/1866307.1866341
4. Asmussen, S.: Applied Probability and Queues. Wiley, Hoboken (1989). doi:10.1002/asm.3150050208
5. Bertsekas, D.P., Gallager, R.G.: Data Networks. Prentice-Hall, Englewood Cliffs (1987)
6. Bissias, G.D., Liberatore, M., Jensen, D., Levine, B.N.: Privacy vulnerabilities in encrypted HTTP streams. In: Proceedings of the 5th International Conference on Privacy Enhancing Technologies, PET'05, Cavtat, pp. 1–11. Springer, Berlin/Heidelberg (2006). doi:10.1007/11767831_1
7. Bortz, A.,Boneh, D.: Exposing private information by timing web applications. In: Proceedings of the 16th International Conference on World Wide Web, WWW'07, Banff, pp. 621–628. ACM, New York (2007). doi:10.1145/1242572.1242656

8. Brumley, D., Boneh, D.: Remote timing attacks are practical. In: Proceedings of the 12th Conference on USENIX Security Symposium, SSYM'03, Washington, DC, pp. 1–1. USENIX Association, Berkeley (2003)
9. Brumley, B.B., Tuveri, N.: Remote timing attacks are still practical. In: Proceedings of the 16th European Conference on Research in Computer Security, ESORICS'11, Leuven, pp. 355–371. Springer, Berlin/Heidelberg (2011)
10. Cabuk, S., Brodley, C.E., Shields, C.: IP covert timing channels: design and detection. In: Proceedings of the 11th ACM Conference on Computer and Communications Security, CCS'04, Washington, DC, pp. 178–187. ACM, New York (2004). doi:10.1145/1030083.1030108
11. Chaum, D.: Blind signatures for untraceable payments. In: Proceedings of the 1982 CRYPTO: Advances in Cryptology, CRYPTO'82, Santa Barbara, vol. 82, pp. 199–203. Plenum, New York (1983)
12. Crosby, S.A., Wallach, D.S., Riedi, R.H.: Opportunities and limits of remote timing attacks. ACM Trans. Inf. Syst. Secur. **12**(3), 17:1–17:29 (2009). doi:10.1145/1455526.1455530
13. Csiszár, I., Korner, J.: Broadcast channels with confidential messages. IEEE Trans. Inf. Theory **24**, 339–348 (1978)
14. Evans, N.S., Dingledine, R., Grothoff, C.: A practical congestion attack on tor using long paths. In: Proceedings of the 18th Conference on USENIX Security Symposium, SSYM'09, Montreal, pp. 33–50. USENIX Association, Berkeley (2009)
15. Felten, E.W., Schneider, M.A.: Timing attacks on web privacy. In: Proceedings of the 7th ACM Conference on Computer and Communications Security, CCS'00, Athens, pp. 25–32. ACM, New York (2000). doi:10.1145/352600.352606
16. Froscher, J., Payne, C.: The Handbook for the Computer Security Certification of Trusted Systems. Naval Research Laboratory, Washington, DC (1992)
17. Ghaderi, J., Srikant, R.: Towards a theory of anonymous networking. In: Proceedings of the 29th Conference on Information Communications, INFOCOM'10, San Diego, pp. 686–694. IEEE, Piscataway (2010)
18. Giles, J., Hajek, B.: An information-theoretic and game-theoretic study of timing channels. IEEE Trans. Inf. Theory **48**(9), 2455–2477 (2002). doi:10.1109/TIT.2002.801405
19. Gong, X., Borisov, N., Kiyavash, N., Schear, N.: Website detection using remote traffic analysis. In: Proceedings of the 12th International Conference on Privacy Enhancing Technologies, PETS'12, Vigo, pp. 58–78. Springer, Berlin/Heidelberg (2012). doi:10.1007/978-3-642-31680-7_4
20. Hu, W.M.: Reducing timing channels with fuzzy time. In: Proceedings of the 1991 IEEE Computer Society Symposium on Research in Security and Privacy, Oakland, pp. 8–20 (1991). doi:10.1109/RISP.1991.130768
21. Hu, W.M.: Lattice scheduling and covert channels. In: Proceedings of the 1992 IEEE Symposium on Security and Privacy, SP'92, Oakland, pp. 52–61. IEEE Computer Society, Washington, DC (1992)
22. Kadloor, S., Kiyavash, N., Venkitasubramaniam, P.: Mitigating timing based information leakage in shared schedulers. In: Proceedings of the 2012 IEEE INFOCOM, INFOCOM'12, Orlando, pp. 1044–1052 (2012). doi:10.1109/INFCOM.2012.6195460
23. Kemmerer, R.A.: A practical approach to identifying storage and timing channels: twenty years later. In: Proceedings of the 18th Annual Computer Security Applications Conference, ACSAC'02, Los Alamitos, p. 109. IEEE Computer Society, Washington, DC (2002). doi:10.1109/CSAC.2002.1176284
24. Kocher, P.C.: Timing attacks on implementations of diffie-hellman, rsa, dss, and other systems. In: Proceedings of the 16th Annual International Cryptology Conference on Advances in Cryptology, CRYPTO'96, Santa Barbara, pp. 104–113. Springer, London (1996)
25. Lam, S.: Delay analysis of a time division multiple access (TDMA) channel. IEEE Trans. Commun. **25**(12), 1489–1494 (1977). doi:10.1109/TCOM.1977.1093784
26. Lampson, B.W.: A note on the confinement problem. Commun. ACM **16**(10), 613–615 (1973). doi:10.1145/362375.362389

27. Liberatore, M., Levine, B.N.: Inferring the source of encrypted HTTP connections. In: Proceedings of the 13th ACM Conference on Computer and Communications Security, CCS'06, Alexandria, pp. 255–263. ACM, New York (2006). doi:10.1145/1180405.1180437
28. Liu, Y., Ghosal, D., Armknecht, F., Sadeghi, A.R., Schulz, S., Katzenbeisser, S.: Hide and seek in time: robust covert timing channels. In: Proceedings of the 14th European Conference on Research in Computer Security, ESORICS'09, Saint-Malo, pp. 120–135. Springer, Berlin/Heidelberg (2009)
29. McFadden, J.A.: The entropy of a point process. SIAM J. Appl. Math. **13**(4), 988–994 (1965). doi:10.1137/0113066
30. Millen, J.K.: Covert channel capacity. In: Proceedings of the 1987 IEEE Symposium on Security and Privacy, SP'87, Oakland, pp. 60–66 (1987)
31. Moskowitz, I.S., Miller, A.R.: The channel capacity of a certain noisy timing channel. IEEE Trans. Inf. Theory **38**(4), 1339–1344 (1992). doi:10.1109/18.144712
32. Murdoch, S.J., Danezis, G.: Low-cost traffic analysis of tor. In: Proceedings of the 2005 IEEE Symposium on Security and Privacy, SP'05, Oakland, pp. 183–195. IEEE Computer Society, Washington, DC (2005). doi:10.1109/SP.2005.12
33. Osvik, D.A., Shamir, A., Tromer, E.: Cache attacks and countermeasures: the case of AES. In: Proceedings of the 2006 The Cryptographers' Track at the RSA Conference on Topics in Cryptology, CT-RSA'06, San Jose, pp. 1–20. Springer, Berlin/Heidelberg (2006). doi:10.1007/11605805_1
34. Padlipsky, M.A., Snow, D.W., Karger, P.A.: dtic.mil,ESD-TR-78-158: limitations of end-to-end encryption in secure computer networks. http://goo.gl/ujLfa (1978)
35. Page, D.: Theoretical use of cache memory as a cryptanalytic side-channel. Techincal Report CSTR-02-003, Department of Computer Science, University of Bristol (2002)
36. Percival, C.: Cache missing for fun and profit. In: Proceedings of the 2005 BSDCan, BSDCan'05, Ottawa (2005)
37. Ristenpart, T., Tromer, E., Shacham, H., Savage, S.: Hey, you, get off of my cloud: exploring information leakage in third-party compute clouds. In: Proceedings of the 16th ACM Conference on Computer and Communications Security, CCS'09, Chicago, pp. 199–212. ACM, New York (2009). doi:10.1145/1653662.1653687
38. Rom, R., Sidi, M.: Multiple Access Protocols: Performance and Analysis. Springer, New York (1990)
39. Saponas, T.S., Lester, J., Hartung, C., Agarwal, S., Kohno, T.: Devices that tell on you: privacy trends in consumer ubiquitous computing. In: Proceedings of the 16th USENIX Security Symposium, SS'07, Boston, pp. 5:1–5:16 (2007)
40. Schinzel, S.: An efficient mitigation method for timing side channels on the web. In: Proceedings of the 2nd International Workshop on Constructive Side-Channel Analysis and Secure Design (2011)
41. Shah, G., Molina, A., Blaze, M.: Keyboards and covert channels. In: Proceedings of the 15th USENIX Security Symposium, USENIX-SS'06, Vancouver. USENIX Association, Berkeley (2006)
42. Song, D.X., Wagner, D., Tian, X.: Timing analysis of keystrokes and timing attacks on SSH. In: Proceedings of the 10th Conference on USENIX Security Symposium, SSYM'01, Washington, DC, pp. 25–25. USENIX Association, Berkeley (2001)
43. Tsunoo, Y., Saito, T., Suzaki, T., Shigeri, M.: Cryptanalysis of DES implemented on computers with cache. In: Proceedings of the 2003 Cryptographic Hardware and Embedded Systems Workshop, CHES'03, Cologne, pp. 62–76. Springer, Berlin/Heidelberg (2003)
44. Venkitasubramaniam, P., Anantharam, V.: On the anonymity of chaum mixes. In: Proceedings of the 2008 IEEE International Symposium on Information Theory, Toronto (2008). doi:10.1109/ISIT.2008.4594929
45. Wagner, A.B., Anantharam, V.: NATO/ASI Workshop on Network Security and Intrusion Detection: Information Theory of Covert Timing Channels (2005)

46. Wang, Z., Lee, R.B.: Covert and side channels due to processor architecture. In: Proceedings of the 22nd Annual Computer Security Applications Conference, ACSAC '06, Miami Beach, pp. 473–482. IEEE Computer Society, Washington, DC (2006). doi:10.1109/ACSAC.2006.20
47. Wang, Y., Moulin, P.: Perfectly secure steganography: capacity, error exponents, and code constructions. IEEE Trans. Inf. Theory **54**(6), 2706–2722 (2008). doi:10.1109/TIT.2008.921684
48. Wang, X., Reeves, D.S.: Robust correlation of encrypted attack traffic through stepping stones by manipulation of interpacket delays. In: Proceedings of the 10th ACM Conference on Computer and Communications Security, CCS'03, Washington, DC, pp. 20–29. ACM, New York (2003). doi:10.1145/948109.948115
49. Wang, X., Chen, S., Jajodia, S.: Tracking anonymous peer-to-peer voip calls on the internet. In: Proceedings of the 12th ACM Conference on Computer and communications security, CCS'05, Alexandria, pp. 81–91. ACM, New York (2005). doi:10.1145/1102120.1102133
50. Wray, J.C.: An analysis of covert timing channels. In: Proceedings of the 1991 IEEE Symposium on Research in Security and Privacy, Oakland, p. 2. IEEE Computer Society, Los Alamitos (1991). doi:10.1109/RISP.1991.130767
51. Wright, C.V., Ballard, L., Coull, S.E., Monrose, F., Masson, G.M.: Spot me if you can: uncovering spoken phrases in encrypted voip conversations. In: Proceedings of the 2008 IEEE Symposium on Security and Privacy, SP'08, Oakland, pp. 35–49. IEEE Computer Society, Washington, DC (2008). doi:10.1109/SP.2008.21
52. Wyner, A.: The wiretap channel. Bell Syst. Tech. J. **54**, 1355–1387 (1975)
53. Zhang, D., Askarov, A., Myers, A.C.: Predictive mitigation of timing channels in interactive systems. In: Proceedings of the 18th ACM Conference on Computer and Communications Security, CCS'11, Chicago, pp. 563–574. ACM, New York (2011). doi:10.1145/2046707.2046772
54. Zukerman, M., Neame, T., Addie, R.: Internet traffic modeling and future technology implications. In: Proceedings of the 22nd Annual Joint Conference of the IEEE Computer and Communications, INFOCOM'03, San Francisco, pp. 587–596 (2003). doi:10.1109/INFCOM.2003.1208709

Federated Cloud Security Architecture for Secure and Agile Clouds

Weiliang Luo, Li Xu, Zhenxin Zhan, Qingji Zheng, and Shouhuai Xu

Abstract Cyber threats against clouds have evolved rapidly. Traditional reactive cyber defense technologies are not effective and sufficient to protect federated clouds. This chapter introduces the novel federated cloud security architecture that includes proactive cloud defense technologies for secure and agile cloud development. The federated security architecture consists of a set of seamlessly integrated systematic security mechanisms at the application layer, the network layer and the system layer in federated cloud computing environments. Features of the architecture include: (1) it is centered on proactive cyber defense; (2) it facilitates to detect early warning cyber attacks against at one layer and deploy early warning signs of attacks to other layers for countermeasures; (3) it uses command and control (C2) to coordinate both in-cloud and cross-cloud defense activities via federated cloud security centers.

1 Introduction

When cloud consumers decide whether or not to adopt cloud computing, cloud security is the most important factor [5, 26, 39, 40]. This is reasonable because the problem of securing cloud systems is much more challenging than the problem of securing an individual enterprise Information Technology (IT) system. First, a cloud system can be seen as the "merger" of many enterprise IT systems that share the same hardware and possibly software infrastructure. As a consequence, traditional perimeter isolation between the physical enterprise IT systems may not be competent for isolating the cloud-hosted virtual enterprise IT systems. Second, a cloud system is harder to defend than a physical enterprise IT system because it

W. Luo (✉) • L. Xu • Z. Zhan • Q. Zheng • S. Xu
University of Texas at San Antonio, San Antonio, TX, USA
e-mail: wluo@cs.utsa.edu; lxu@cs.utsa.edu; zzhan@cs.utsa.edu; qzheng@cs.utsa.edu; shxu@cs.utsa.edu

K.J. Han et al. (eds.), *High Performance Cloud Auditing and Applications*, DOI 10.1007/978-1-4614-3296-8_7, must be open to all users of the cloud-hosted enterprise IT systems. In other words, a cloud system has a much larger attack surface, including a larger population of authorized users. As a consequence, a successful attack launched by an authorized user against the cloud can cause damage to all users of the cloud.

Given the scale and complexity of cloud systems, purely reactive cyber defense is far from sufficient. Moreover, threats against clouds will evolve rapidly. In order to adequately defend cloud systems against dynamic cyber threats, proactive defense should be widely deployed because it can offer defenders some inherent situational awareness and early warning capabilities. Toward this goal, we propose a cloud security architecture that can seamlessly integrate together various proactive defense mechanisms. To our knowledge, this challenging and important problem has not been systematically explored.

In this chapter, we propose a novel federated cloud security architecture to detect dynamic cyber threats (i.e., new and possibly zero-day attacks) against clouds *as early as possible* so that the defender can have enough time to deploy countermeasures. The architecture is centered on proactive defense, and consists of a set of seamlessly integrated systematic security mechanisms at the application layer, the network layer, and the system layer. More specifically, the architecture can be characterized as follows:

- *Application-Layer Situation Awareness*: We describe how application-layer proactive cyber defense mechanisms can be deployed to defend the dynamic abuse of clouds. Although some of the integrated mechanisms were known, the others are proposed here for the first time.
- *Network-Layer Situation Awareness*: We describe how network-layer proactive defense mechanisms can be deployed to detect new cyber attacks against clouds. In particular, we illustrate how to blend honeypots into the IP address space of production cloud systems, which will make it harder for the attackers to identify and bypass the honeypots.
- *System-Layer Situation Awareness*: We describe how proactive defense mechanisms can be deployed deep in the software stack to defend stealthy attacks against clouds. This includes the use of honeytokens to detect stealthy attacks.
- *Cloud C2 Security Center Situation Awareness*: The architecture uses a cloud C2 security center to coordinate the in-cloud and cross-cloud defense activities. As a result, new attacks detected at one layer may lead to countermeasures that can be deployed at the other layers, and clouds can rapidly share information about newly detected attacks as well as the countermeasures.

To our knowledge, this is the first work that systematically explores cloud security architectures, which will play a crucial role in the cloud computing era.

The rest of the chapter is organized as follows. Section 2 presents an overview of cloud security. Section 3 describes cloud system and cloud security threats. Section 4 introduces the novel federated security architecture that consists of cross-layer and cross-cloud security components. Section 5 describes the federated cloud C2 security modules. Section 6 discusses the usefulness of the federated cloud C2 centers. Finally, Section 7 presents our conclusions.

2 Cloud Security

We briefly review cloud security [40] and related prior work based on layers at which the defense mechanisms are deployed. For detecting malicious websites as early as possible, there have been studies on combating the abuse of cloud resources to host botnet command and control (C&C) severs [22] and to crack passwords [37]. The C&C is used for the botnet operation in this chapter. There have been studies on combating spam campaigns in cloud-like settings [10, 13, 20, 32, 43, 49, 52]. The problem of detecting malicious websites is relatively new and has not been thoroughly investigated, despite some interesting initial studies [12, 14]. A competent cloud security architecture should adequately address the problem of malicious websites.

The network-layer proactive defense presented in this chapter advocates the use of honeypots in some novel ways. Previous studies related to honeypots mainly focused on analyzing honeypot-observed probe activities [25] and characterizing honeypot-observed attacks [1–3, 15, 16, 28, 35, 36, 47].

For dealing with APT-like sophisticated and stealthy attacks, there have been studies such as how to cope with the exploitation of VM image management to distribute malicious VM images [30]. More advanced studies have focused on enhancing VMM security [8, 42, 44, 50]. Overall, proactive defense at the system layer is an important problem that is not yet well understood.

Despite the aforementioned studies on various aspects of cloud security, there is no prior work that systematically explores comprehensive security architectures for securing clouds against dynamic cyber attacks. The proposed security architecture fills the void.

3 Cloud System and Cloud Security Threats

In this section we discuss the cloud system model and the associated threat model. The cloud system model is general enough to accommodate the so-called public clouds, private clouds, community clouds, and social clouds [6, 54]. This means that the threat model and the security architecture that we will present later are equally applicable to these kinds of clouds.

3.1 Cloud System Model

As shown in Fig. 1, a cloud provides both computational (including hardware and possibly software) infrastructure and storage services to its users. Cloud users may include enterprise users, home users, and mobile users. Enterprise users may outsource their data and computing tasks to the cloud, while allowing their

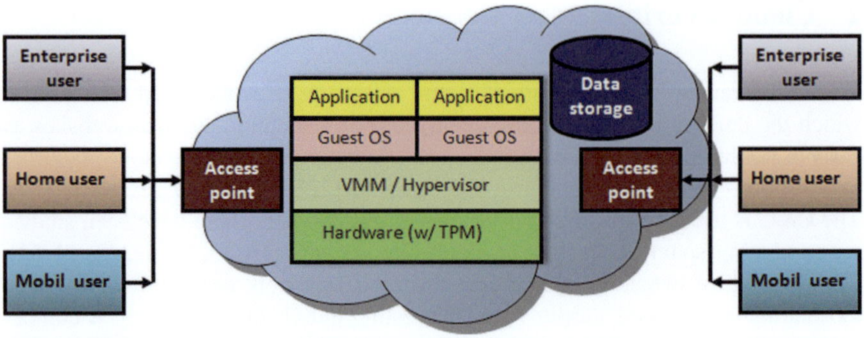

Fig. 1 Cloud system model

employees and business partners to access their data. Home users may outsource their data (e.g., emails, files and documents) to the cloud. Mobile users may access the cloud-hosted data and exploit the cloud to accomplish some compute-intensive and data-intensive tasks. Cloud users utilize cloud resources and services through some *access points*, which are managed and maintained by the cloud in question.

Cloud resource management [4] is centered on the concept of virtual machine (VM), which is an abstraction that eases the allocation of physical computing resources to cloud users. Each VM runs its own guest Operating System (OS), and may run one or more software applications of a cloud user. Multiple VMs, which may be allocated (or rented) to one or more cloud users, run on top of a Virtual Machine Monitor (VMM) or Hypervisor that manages resources of a physical computer. It is also possible that some physical computers may not need to run VMM at all because the entire physical computer is rented to a single cloud user. Each physical computer may have an associated Trusted Platform Module (TPM) [48], which may or may not be further virtualized so that the virtual machines can use the virtualized TPMs [11] (one virtual TPM per VM). A cloud can offer data storage services to its users. The outsourced data can be unstructured, semi-structured or structured. Unstructured data can be simply bitstrings or files. Semi-structured data can be emails. Structured data can be a relational database.

3.2 *Cloud Security Threats*

Threats against clouds may be launched by anyone, who may or may not be a legitimate participant in the above cloud system model. We assume that the cloud vendors (including their employees) are trusted not to attack the cloud users. While orthogonal to the focus of the present chapter, we mention that there have been studies on assuring cloud data storage integrity [7, 21, 55] and cloud data query integrity [24, 33, 56]. These mechanisms can be incorporated into our security architecture, in a plug-and-play fashion, to deal with potentially malicious cloud vendors. We assume that the cloud users may be compromised or malicious. For example, a cloud user may launch a malicious VM instance to attack the VMM or other VMs running on top of the same VMM, or may abuse a cloud VM to conduct malicious activities such as hosting malicious websites. More specifically, we focus on the following classes of powerful threats against clouds.

- *Abusing clouds to host malicious websites to launch drive-by download attacks*: Cloud resources have been, and will continue to be, abused to host malicious websites, which can launch drive-by download attacks against the vulnerable browsers and therefore the user-end computers. Cloud resources have been, and will continue to be, abused to facilitate botnet C&C operations.
- *Spreading new malwares*: These attacks often exploit software vulnerabilities that may not have been publicly available, and are powerful because it can be very difficult to detect these attacks.
- *Stealthy Advanced Persistent Threats (APT)*: These attacks do not attempt to infect as many computers as possible, but rather attempt to compromise targeted computers through vulnerabilities deep in the software stacks. These attacks are powerful because they often can bypass application-layer and network-layer defense, and may only be dealt with at the lower system layer.

Our primary goal is to detect new and possibly zero-day attacks belonging to the above classes as early as possible so that the defenders have enough time to deploy countermeasures.

4 Federated Agile Cloud Security Architecture

Now we present the cloud security architecture, which can offer the defender inherent capabilities of situational awareness and early warning against dynamic cyber threats in federated cloud computing environments. The architecture is centered on proactive defense.

4.1 Architecture Description

Figure 2 highlights the security architecture, which consists of (1) defense components at the application layer, (2) defense components at the network layer, (3) defense components at the system layer, and (4) the C2 center that coordinates both the defense within a single cloud and the defense crossing multiple clouds. In what follows we briefly discuss the functions of these defense components, which are then elaborated in the following subsections.

Application-layer defense can naturally deploy traditional reactive defense mechanisms. For example, there have been many proposals for enhancing security of execution environments [17, 19, 46], which can be deployed at the cloud-

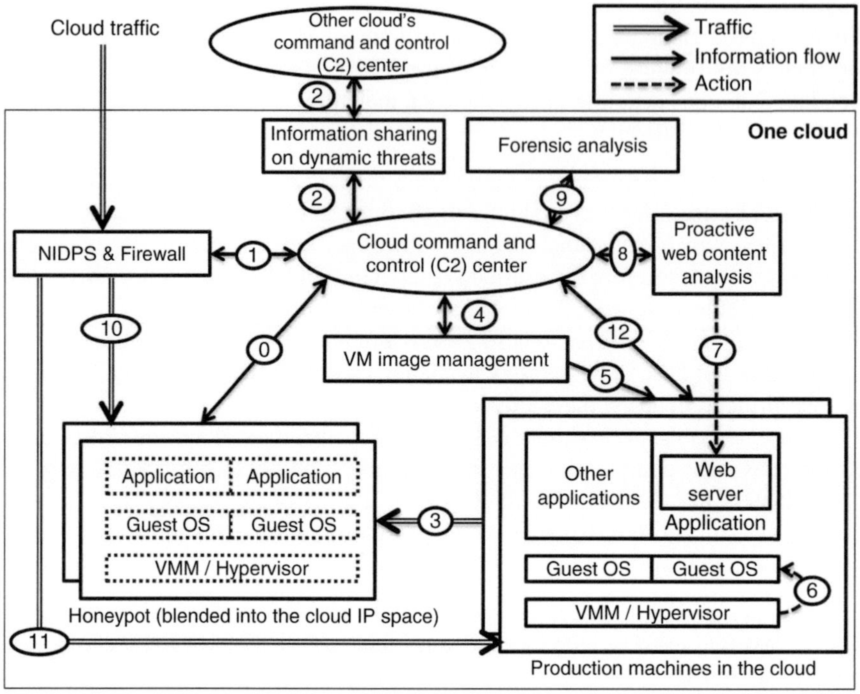

Fig. 2 Cloud security architecture with C2 security centers for coordinating in-cloud and cross-cloud defense activities in federated cloud computing environments

end to enhance the isolation between the applications. Nevertheless, we advocate application-layer proactive defense that can significantly benefit from the rich semantics of application-layer contents. Since websites have become a popular attack channel, we will focus on mitigating the abuses of websites to launch attacks. For this purpose, we propose using a *proactive web content analysis* component to identify malicious/compromised websites (Arrow 7). We elaborate application-layer proactive defense in Section 4.2.

It is well known that Network Intrusion Detection/Prevention System (NIDPS) & Firewall can possibly filter malicious traffic before it reaches the production machines (Arrow 11). However, effectiveness of NIDPS & Firewall depends on the accuracy of the detection and filtering algorithms and on the length of early warning time. We advocate fulfilling network-layer proactive defense by exploiting honeypots (http://www.honeynet.org/). Honeypots are vulnerable computers that can be used to monitor attacks [41] and collect malicious attack traffic and learn the evolution of cyber attacks (Arrow 10). The newly identified and learned attack vectors can be disseminated to NIDPS & Firewall (Arrow 0) for filtering attacks. Honeypots can also detect attack traffic originating from the cloud production machines (Arrow 3). We elaborate network-layer proactive defense in Section 4.3.

System-layer defense aims at detecting APT-like attacks against guest OS or even VMM as early as possible. System-layer defense is crucial because even network- and application-layer defenses together are not sufficient to defeat APT-like sophisticated attacks. We propose using enhanced *VM image management* component (Arrow 5) to enhance VM image security [51]. We also advocate taking advantage of VMMs to inspect guest OS kernels (Arrow 6), and taking advantage of recent progress in exploiting architectural features to enhance security of VMMs. We elaborate system-layer proactive defense in Section 4.4.

We propose letting each cloud run a C2 center. The C2 center shares updated threat information via the channel called *information sharing on dynamic threats* (Arrow 2), receives updated information from the defense components in the cloud (Arrows 0, 1, 4, 8, 9, 12), and coordinates activities of the components within the cloud (Arrows 0, 1, 4, 8, 9, 12). For example, the C2 center learns new attacks from the *proactive web content analysis* component (Arrow 8), and instructs the *NIDPS & Firewall* component (Arrow 1) and the *VM image management* component (Arrow 4) to proactively adjust their defense. We elaborate the federated cloud C2 security modules in Section 5.

4.2 *Application-Layer*

Web servers have been abused to attack users' computers. Since hosting malicious websites will hurt the reputation of cloud vendors, the vendors have the incentive to rapidly identify malicious websites they host. A cloud vendor may choose to attain situational awareness and early warning by a third-party or by itself.

4.2.1 Third-Party Monitoring Cloud-Hosted Websites

A cloud vendor can outsource the monitoring of the websites it hosts to a third party. There are three kinds of proactive mechanisms for this purpose, each of which has its own strengths and weaknesses.

- *Static approach*: Detect malicious websites via static analysis of website contents or even URLs themselves [18, 27]. This approach is very efficient, but often has limited success against sophisticated attacks such as obfuscation.
- *Dynamic approach*: Detect malicious websites by analyzing their runtime behaviors. This can be done by using various client honeypots, such as Capture-HPC [38] and PhoneyC [31]. This approach can cope with obfuscation but also resource consuming.
- *Hybrid approach*: Achieve the best of the above two approaches [12]. It uses a front-end static analysis tool to identify suspicious websites and a back-end dynamic analysis tool to further examine the websites.

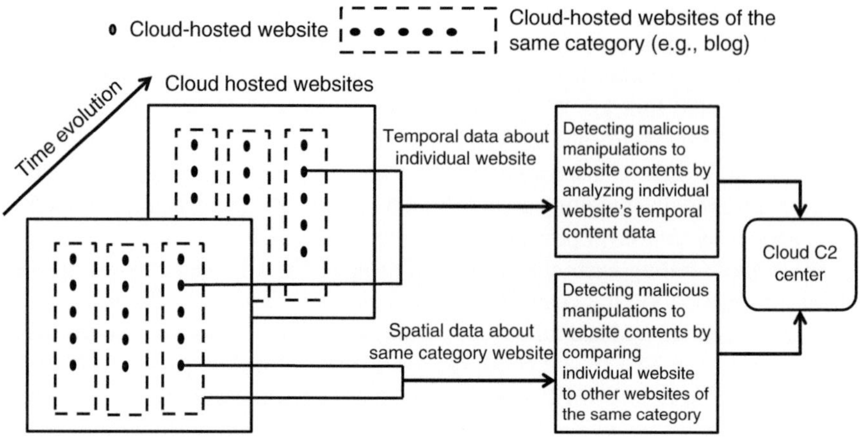

Fig. 3 Cloud vendor proactively monitors malicious websites it hosts

4.2.2 Cloud Vendor-Side Websites Monitoring

A cloud vendor can analyze the website contents by using the aforementioned static, dynamic, and hybrid methods. This approach has the drawback that the analysis must be conducted as frequently as possible. Unfortunately, it is not practical to analyze each website every minute or even every few seconds for near real-time auditing that is necessary in order to detect attacks as early as possible. We propose the following new method by which the cloud vendor (more precisely, the C2 center) can be triggered to analyze the website contents (i.e., analysis on demand rather than periodically analyzing the website contents).

We consider two cases that depend on how websites hosted in clouds become malicious. In the first case, malicious websites are set up by malicious cloud users, which are accommodated in our threat model. In the second case, websites are set up by honest cloud users but later compromised by attackers for malicious purposes. To detect malicious websites in the first case, the vendor can, as illustrated in Fig. 3, classify websites into categories such as news, blog, searching, shopping etc. The websites belonging to an individual category would exhibit similar characteristics; whereas different categories of websites would exhibit different characteristics. This allows the cloud vendor to build patterns for each category of websites. The vendor could use the pattern of a specific category to detect an anomaly of a specific website belonging to the category. This would allow the vendor to detect malicious websites that were set up by honest cloud users but compromised by the attacker to launch further attacks. To detect malicious websites in the second case, the vendor can, as illustrated in Fig. 3, analyze the temporal data associated with individual website, possibly including the IP addresses from which accesses are initiated and the IP addresses the website redirects access to.

Compared with the aforementioned proactive defense based on third parties, the above proactive defense by the cloud vendor has the following advantages. First, attacks could be detected significantly earlier because deeper analysis can be automatically triggered when certain system calls are invoked to inject malicious code into the website content. This can be done more efficiently by the third-party approach that blindly analyzes all websites in a "brute-force" fashion. To be specific, let us look at the following concrete example. Suppose an attacker attempts to embed <iframe src="www.malicioussite.com"> into an innocent website www.innocentsite.com, which will cause the browsers pointing to www.innocentsite.com to automatically access www.malicioussite.com and therefore get compromised. Since the code injection will invoke some file I/O system calls, these invocation operations can automatically alert the cloud vendor with the potential attack. Similarly, a malicious website set up by a legitimate but malicious user may be detected at the setup time because it exhibits different characteristics from other non-malicious websites belonging to the same category. Therefore, cloud vendors can enforce defense that is impossible for third parties to implement because they do not have access to the data. Second, newly detected attacks can be more easily communicated to the cloud C2 center, which can take appropriate actions upon receiving reports on malicious websites.

4.2.3 Effectiveness of Application-Layer Proactive Defense

Effectiveness of application-layer proactive defense depends on several factors. First, proactive website content analysis should scale up to the large number of websites. This is why we advocate that each cloud monitors its own websites. Second, proactive website content analysis should be able to trace redirections automatically because attackers have been abusing redirections to hide the locations of the actual web servers that host malicious malwares. This problem is not trivial because redirections can be made sophisticated enough to defeat most static tracing methods.

4.3 Network-Layer

Compromised computers/VMs can attack the vulnerable computers/VMs in the same or different cloud. Network-layer proactive defense can offer situational awareness and early warning against such dynamic threats. Honeypots are vulnerable computers that can be used to monitor attacks [41]. Although the idea of using honeypots for detecting new attacks is not new, we propose enhancing this proactive defense by blending the honeypot IP addresses into the production network IP address space, rather than running honeypots in a separate unused address space, to make it more difficult for the attackers to figure out IP addresses of the honeypot. We also propose enhancing the effectiveness of this proactive defense with more personalized or diversified honeypots.

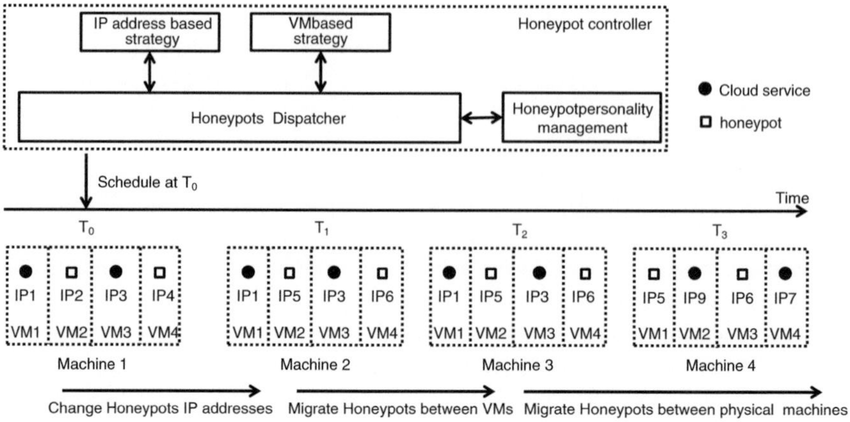

Fig. 4 Dynamic honeypots blended into the IP address space of production cloud

4.3.1 Exploiting Honeypots to Rapidly Learn New Attacks

Low-interaction honeypots can provide limited interactions with attackers, and may be easily recognized and bypassed by sophisticated attackers. In order to detect new and possibly zero-day attacks as early as possible, we advocate the use of high-interaction honeypots that use real but vulnerable operating systems to trap cyber attacks. High-interaction honeypots can collect sophisticated malicious attack traffic that would allow the C2 center to generate countermeasures for other components. For example, patterns of the observed malicious code can be extracted as NIDPS signatures. Since high-interaction honeypots often consume much more resources than low-interaction honeypots, we advocate the hybrid deployment of low-interaction and high-interaction honeypots so as to achieve better trade-off between effectiveness and efficiency. Basically, low-interaction honeypots can serve as a front-end tool for rapidly filtering known attacks, while high-interaction honeypots can serve as a back-end tool for interacting with unknown attacks that are not recognized by low-interaction honeypots.

As mentioned above and highlighted in Fig. 4, we propose the following enhancements to honeypot-based detection of new attacks. "IPx" stands for IP address x, "VMy Machine z" stands for the y-th VM on the z-th physical computer. First, we propose blending honeypot IP addresses into the production cloud network IP address. Traditionally, honeypots are set up in unused IP address space, which however can be easily recognized and bypassed by sophisticated attackers. We propose using a honeypot dispatcher to manage the dynamic IP addresses and VMs that are used for honeypot purposes. For example, at time T_0, VM2 of IP address 2 corresponding to physical computer 1 is used for honeypot purpose, which is migrated at time T_1 to VM2 of IP address 5 corresponding to physical computer 2, migrated at time T_2 to VM4 of IP address 6 corresponding to physical computer 3, and migrated at time T_3 to VM1 of IP address 7 corresponding to physical computer 4. Second, we propose to diversify the "personalities" of honeypots to achieve high fidelity in detecting new attacks. This is important because without diversifying honeypot personalities, a sophisticated attacker can recognize honeypots by analyzing the characteristics of the honeypots. This capability is independent of, and therefore can be incorporated with, the capability of recognizing honeypots via their IP addresses. Without the above enhancements, honeypots can be relatively easily evaded and cannot effectively detect new attacks.

4.3.2 Effectiveness of Network-Layer Proactive Defense

Effectiveness of network-layer proactive defense depends on the following factors. First, how well are the honeypots blended into the IP address space of the production cloud system? Traditional honeypots are often deployed at consecutive, unused IP spaces, which are relatively easy to be identified and bypassed by sophisticated attackers. Second, how often are the honeypots re-allocated or re-deployed? Even if high-interaction honeypots' IP addresses are well blended into the cloud system, attackers may still identify them because, after all, even high-interaction honeypots may not run semantically meaningful applications. This highlights the importance of re-deploying high-interaction honeypots periodically. The re-deployment operation should be conducted before attackers figure out the dynamic honeypot IP address spaces.

4.4 System-Layer

APT-like attacks might be able to bypass the application-layer and network-layer proactive defense. In order to counter such sophisticated attacks, we need system-layer proactive defense that is deployed deep into the software stack, namely at the levels of guest OS or even VMM. For this purpose, we not only need to enhance guest OS and VMM security, but also need to use new proactive approaches to detect such powerful attacks as early as possible.

4.4.1 Enhancing Guest OS and VMM Security Proactively

In order to accelerate the creation of software environment in clouds, guest OS is launched from VM image. This unfortunately also expands the attack surface of guest OS because malicious VM image directly leads to malicious guest OS. For example, a malicious VM image intentionally crafted by an attacker can cause a guest OS to run with rootkits, backdoor or manipulated kernel [30]. Therefore, cloud vendors should implement a VM image management system that enforces VM image access control, image provenance information tracking, and security patch [51]. Provenance data can be collected during the lifetime of VM images to establish a chain of trust starting at a root (e.g., Amazon or IBM). If the VMM is trusted, secure loading of guest OS is straightforward because the VMM can measure the code executed in the bootstrapping process of guest OS.

Once successfully launched, it is important to ensure runtime security of guest OS. This remains an open problem [34, 53]. There have been two promising approaches to assure runtime security of guest OS: *in-guest* protection and *VMM-supported* protection. The in-guest protection approach is similar to traditional OS protection, and depends on some defense mechanisms (e.g., anti-malware software) that run inside the guest OS itself. As long as the defense mechanisms can deal with dynamic attacks, this approach is still useful. This is possible because, for example, the anti-malware software can be behavior-based rather than signature-based. The VMM-supported protection approach is more powerful because it exploits VMM to inspect the states of the VMs. Although complete inspection and verification of system state is infeasible, it is possible to verify specific properties such as control-flow integrity in the guest OS kernel [34].

In the above, we assumed that VMMs are trusted. Security of VMMs is somewhat debatable and it is ideal not to trust VMMs. Nevertheless, there have been studies how to weaken (if not eliminate) the trust in VMMs, for example, by running VMs or applications on (almost) naked hardware [9, 44] and exploiting hardware to protect VMs from malicious VMMs [45]. There also have been studies on enhancing VMM security [8,23,29,42,50]. Future progress in these aspects can be incorporated into the cloud security architecture in a plug-and-play fashion.

4.4.2 Using Honeytokens to Detect APT-Attacks Against Guest OS

APT-like attacks may still be able to penetrate the application-layer and network-layer proactive defense and the enhanced guest OS defense. We therefore propose using the following methods to detect such attacks as early as possible. For example, attackers may exploit zero-day vulnerabilities in the guest OS to launch stealthy attacks, which can be very difficult to observe from inside of the guest OS or even from inside of the VMM. This suggests that we need to detect such powerful attacks from outside of the guest OS or even from outside of the VMM. One approach is to run a guest OS on top of a VMM, where both guest OS and VMM do not have any known vulnerabilities (from the perspective of the cloud vendor or defender). The guest OS can set up a set of user accounts, and the VMM can incorporate the inspection mechanisms mentioned above. Each user account has access to a set of honeytokens such as: (1) an email address (e.g., nobody@nowhere.com) that is never published or supposed to receive any emails; (2) a username/password for another account at possibly another computer that is never supposed to be accessed; (3) fake data containing specific keywords that may be of interest to APT-like attacks.

Even if the attacks bypassed all of the defense mechanisms mentioned above, they could still be detected by this system-layer proactive defense as follows. When the email account nobody@nowhere.com receives any email, possibly spam or email with a malicious attachment, we can conclude that the guest OS has been compromised. Similarly, when the specific fake account is accessed, we can conclude that the guest OS has been compromised.

Finally, if an APT-like attack does not encrypt the data it steals from the compromised guest OS, the VMM could detect the compromise by inspecting whether the special keywords appear in the outgoing packets; if an APT-like attack does encrypt the data it steals from the compromised guest OS, this fact of encrypted outgoing packets could still trigger the VMM's inspection mechanism for further investigation. The latter is possible because the vendor can make the outgoing traffic encrypted using some cryptographic keys that are known to the defender. When the VMM inspection mechanism cannot decrypt the outgoing packets that were encrypted by the attacker using its own cryptographic key, the VMM can conclude that the guest OS has been compromised.

4.4.3 Effectiveness of System-Layer Proactive Defense

System-layer proactive defense has a great potential in detecting APT-like stealthy attacks. The effectiveness of system-layer proactive defense depends on the following factors. First, how effective can the runtime attack-detection mechanisms be? This is because modern guest Oss are often complex and it is challenging to detect runtime security violations. Second, how effective can VMM-based inspection of guest OS state be? This largely depends on the degree at which the semantic gap (i.e., VMM may not understand the semantics of guest OS objects and operations) can be bridged. Third, what is the optimal trade-off between the power, security, and performance of VMM? This will affect the services VMMs can provide to support guest Oss and applications, and the trustworthiness of VMMs in terms of their capability in detecting sophisticated attacks.

5 Federated Cloud C2 Security Modules

The security architecture uses cloud C2 security centers to coordinate in-cloud and cross-cloud defense activities in federated cloud computing environments. In particular, it facilitates rapid deployment of countermeasures against newly detected attacks and rapid sharing of such information between clouds. In this section we elaborate the C2 center component and discuss its usefulness in fulfilling situational awareness and early warning.

The cloud C2 security center has the following four modules: information aggregation, information fusion and analysis, information feedback, and information sharing. These modules are pictorially illustrated in Fig. 5, which also highlights the usefulness of federated cloud C2 security centers.

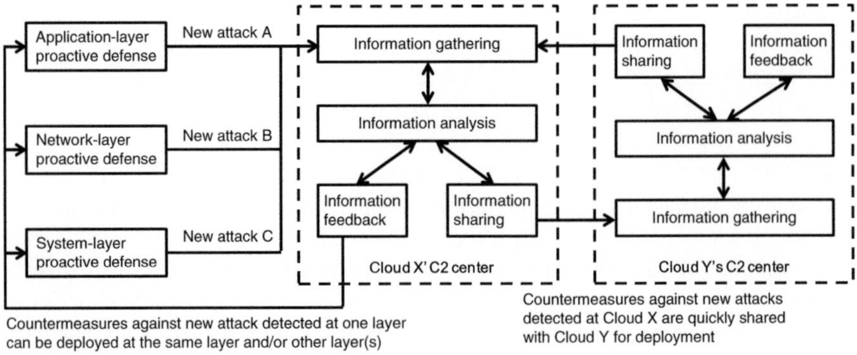

Fig. 5 One scenario of cloud C2 security centers coordinating in-cloud and cross-cloud defense activities

5.1 *Information Aggregation Module*

This module collects dynamic attack information from the NIDPS & Firewall, the honeypot, the VM image management component, the system-layer (guest OS and VMM) security monitoring mechanisms, the proactive website analysis component, and the forensic analysis component. The gathered information may have different structures and semantics. For example, application-layer information may be specific to guest OS processes (e.g. image path, creator, I/O operations, network operations); network-layer information may identify the communication peers (e.g. IP addresses, port number, protocol, application payloads); system-layer information may be specific to guest OS and VMM operations. There might be some semantic gaps between the information gathered at different layers. This highlights the importance of identifying effective cross-layer methods to bridge the potential semantic gaps so as to maximize the utility of the gathered information.

5.2 *Information Fusion and Analysis Module*

This module learns and characterizes dynamic cyber threats. This module is the core of C2 centers, and interacts with the other modules by possibly instructing them to adapt their functions. This module can instruct the information aggregation module to collect more information about some specific events. Also, this module can adopt countermeasures received from other cloud C2 centers, and can adapt them before actually instructing the information feedback module to deploy them. Various (adversarial) machine learning methods can be applied. Ideally, the information analysis engine can not only extract effective detection methods against the newly identified attacks, but also can help pin down the root cause of the newly identified attacks. For example, application-layer analysis may lead to effective methods for detecting malicious websites despite the fact that they may evolve rapidly (e.g., a website may be malicious only for the period of minutes); network-layer analysis may lead to the identification of dynamic malicious IP addresses or attack signatures that can be deployed at the NIDPS & Firewall.

5.3 Information Feedback Module

This module instructs the defense mechanisms at appropriate defense points within the cloud to deploy the newly learned countermeasures. The countermeasures may be deployed at the NIDPS & Firewall to filter malicious attack traffic, may be accommodated to adjust the configuration of honeypots, and may be incorporated to enhance the security of VM images. The countermeasures can instruct the system-layer security monitor mechanisms to pay special attention to some activities, can order the proactive web content analysis component to enhance its analysis methods, and may ask the forensic analysis engine to provide more information about certain attacks. Note that the above information gathering, information analysis, and information feedback formulate a proactive defense control loop within a single cloud.

5.4 Information Sharing Module

This module shares newly learned countermeasures with other cloud C2 centers. This can be fulfilled through the component called *information sharing on dynamic threats* that is responsible for sharing information between the C2 centers of different clouds. The rapid sharing of newly detected attacks can substantially mitigate the damage because, for example, a zero-day attack can be detected after it compromises VMs in one cloud, rather than after it compromises VMs in many clouds. The C2 centers effectively formulate a higher-layer of defense network, which is necessary for defending dynamic cyber threats.

6 Usefulness of Federated Cloud C2 Security Centers

Figure 5 elaborates the usefulness of federated cloud C2 security centers. The usefulness can be understood from the following perspectives. First, as illustrated in Fig. 5, countermeasures newly developed at Cloud X can be rapidly disseminated to Cloud Y through their respective cloud C2 security centers via cross-cloud information sharing. The C2 centers that receive the new countermeasures can coordinate their deployment within their own clouds through the information feedback module, which receives instructions from the information analysis module.

Second, as illustrated in Fig. 5, a new attack detected at a particular layer can lead to countermeasures that not only can be deployed at the application layer, but also can be deployed at the network layer and/or the system layer. For example, proactive website content analysis can trace to (perhaps after redirection hops) some malware that exploits a zero-day vulnerability, which can be quickly analyzed using an automated dynamic analysis tool. Suppose, for example, new attack *A* is detected at the application layer by proactive web content analysis. The countermeasures can be naturally deployed at the application layer to, for example, blacklist not only the websites that host the malware, but also the websites that automatically redirect to the malicious websites that host the malwares. The latter is important because even if the malicious websites that host the malwares are shut down by the defender, the attacker can quickly use the same intermediate websites to redirect to newly compromised websites for hosting malwares. Moreover, the ways attack *A* exploits the operating system vulnerabilities could lead to detection methods at the system layer to block attack *A* from success. This is possible even before the vulnerability is patched because the patch may become available only after a significant amount of time. Third, although not illustrated in Fig. 5, it is possible that C2 center's information analysis module can detect new attacks that are not detected by the defense at individual layers. For example, some new attacks may only be detected after correlating information obtained at the application layer defense mechanisms, the information obtained at the network layer defense mechanisms, and even the information obtained at the system layer defense mechanisms. Extremely sophisticated attacks may only be detected by such cross-mechanisms defense techniques.

It is important to note that effectiveness of C2 centers depends on the following factors. First, how can accurate and detailed information about attacks be gathered? This directly affects the quality of the collected raw data and ultimately bounds the usefulness of the analysis results. Second, how good are the information analysis algorithms? The better the machine learning algorithms, the more useful the extracted detection models. Third, how quick can it be to extract the patterns for detecting new and possibly zero-day attacks and how quick can it be to deploy the countermeasures? This ultimately determines the effectiveness of both in-cloud proactive defense and cross-cloud proactive defense. The above open problems should be adequately addressed in future studies.

7 Conclusion

We have developed a federated cloud security architecture and cloud C2 security modules to mitigate cloud threats against federated clouds. Cyber situational awareness and early warning feature selection functionalities are included as security components that are controlled by the cloud C2 security centers. Since cloud users access cloud-based applications through web browsers, web-security related data are collected at the application-layer, the network-layer and the system-layer.

The cloud C2 security centers aggregate cyber situational awareness data to extract early warning signs. The cloud C2 security center coordinates both in-cloud and cross-cloud defense activities. The proactive cloud security mechanisms may mitigate cloud threats and enhance cloud security. Therefore, the proposed cloud defense technologies may be appropriate for secure and agile clouds.

Acknowledgements This material is based upon work partially supported by the Air Force Office of Scientific Research (AFOSR) under Grant No. FA9550-09-01-0165 and the Air Force Research Laboratory (AFRL) Visiting Faculty Research Program (VFRP) extension grant LRIR 11RI01COR. Any opinions, findings, and conclusions or recommendations expressed in this publication are those of the authors and do not necessarily reflect the views of the funding agency.

References

1. Almotairi, S.I., Clark, A.J., Dacier, M., Leita, C., Mohay, G.M., Pham, V.H., Thonnard, O., Zimmermann, J.: Extracting inter-arrival time based behaviour from honeypot traffic using cliques. In: Proceedings of the 5th Australian Digital Forensics Conference, Perth, pp. 79–87 (2007)
2. Almotairi, S., Clark, A., Mohay, G., Zimmermann, J.: Characterization of attackers' activities in honeypot traffic using principal component analysis. In: Proceedings of the 2008 IFIP International Conference on Network and Parallel Computing, NPC'08, Shanghai, pp. 147–154. IEEE Computer Society, Washington, DC (2008)
3. Almotairi, S., Clark, A., Mohay, G., Zimmermann, J.: A technique for detecting new attacks in low-interaction honeypot traffic. In: Proceedings of the 4th International Conference on Internet Monitoring and Protection, ICIMP'09, Venice, pp. 7–13. IEEE Computer Society, Washington, DC (2009)
4. An, K.: Resource management and fault tolerance principles for supporting distributed real-time and embedded systems in the cloud. In: Proceedings of the 9th Middleware Doctoral Symposium of the 13th ACM/IFIP/USENIX International Middleware Conference, MIDDLE-WARE'12, Montreal, pp. 4:1–4:6. ACM, New York (2012). doi:10.1145/2405688.2405692
5. Anderson, T.E.: weforum.org, Exploring the future of cloud computing: riding the next wave of technology-driven transformation. http://goo.gl/BeR45 (2010)
6. Armbrust, M., Fox, A., Griffith, R., Joseph, A.D., Katz, R., Konwinski, A., Lee, G., Patterson, D., Rabkin, A., Stoica, I., Zaharia, M.: A view of cloud computing. Commun. ACM **53**(4), 50–58 (2010). doi:10.1145/1721654.1721672
7. Ateniese, G., Burns, R., Curtmola, R., Herring, J., Kissner, L., Peterson, Z., Song, D.: Provable data possession at untrusted stores. In: Proceedings of the 14th ACM Conference on Computer and Communications Security, CCS'07, Alexandria, pp. 598–609. ACM, New York (2007). doi:10.1145/1315245.1315318
8. Azab, A.M., Ning, P., Wang, Z., Jiang, X., Zhang, X., Skalsky, N.C.: Hypersentry: enabling stealthy in-context measurement of hypervisor integrity. In: Proceedings of the 17th ACM Conference on Computer and Communications Security, CCS'10, Chicago, pp. 38–49. ACM, New York (2010). doi:10.1145/1866307.1866313
9. Azab, A.M., Ning, P., Zhang, X.: Sice: A hardware-level strongly isolated computing environment for x86 multi-core platforms. In: Proceedings of the 18th ACM Conference on Computer and Communications Security, CCS'11, pp. 375–388. ACM, New York (2011). doi: 10.1145/2046707.2046752
10. Benczur, A.A., Csalogany, K., Sarlos, T., Uher, M., Uher, M.: Spamrank – fully automatic link spam detection. In: Proceedings of the 1st International Workshop on Adversarial Information Retrieval on the Web, AIRWeb'05, Chiba (2005)

11. Berger, S., Cáceres, R., Goldman, K.A., Perez, R., Sailer, R., van Doorn, L.: vtpm: virtualizing the trusted platform module. In: Proceedings of the 15th Conference on USENIX Security Symposium – Volume 15, USENIX-SS'06, Vancouver. USENIX Association, Berkeley (2006)
12. Canali, D., Cova, M., Vigna, G., Kruegel, C.: Prophiler: a fast filter for the large-scale detection of malicious web pages. In: Proceedings of the 20th International Conference on World Wide Web, WWW'11, pp. 197–206. ACM, New York (2011). doi:10.1145/1963405.1963436
13. Chellapilla, K., Maykov, A.: A taxonomy of javascript redirection spam. In: Proceedings of the 3rd International Workshop on Adversarial Information Retrieval on the Web, AIRWeb'07, Banff, pp. 81–88. ACM, New York (2007). doi:10.1145/1244408.1244423
14. Choi, H., Zhu, B.B., Lee, H.: Detecting malicious web links and identifying their attack types. In: Proceedings of the 2nd USENIX Conference on Web Application Development, WebApps'11, Portland, pp. 121–132. USENIX Association, Berkeley (2011)
15. Clark, A., Dacier, M., Mohay, G., Pouget, F., Zimmermann, J.: Internet attack knowledge discovery via clusters and cliques of attack traces. J. Inf. Assur. Secur. **1**(1), 21–32 (2006)
16. Conti, G., Abdullah, K.: Passive visual fingerprinting of network attack tools. In: Proceedings of the 2004 ACM Workshop on Visualization and Data Mining for Computer Security, VizSEC/DMSEC'04, Washington DC, pp. 45–54. ACM, New York (2004). doi:10.1145/1029208.1029216
17. Dai, W., Jin, H., Zou, D., Xu, S., Zheng, W., Shi, L.: Tee: A virtual drtm based execution environment for secure cloud-end computing. In: Proceedings of the 17th ACM Conference on Computer and Communications Security, CCS'10, Chicago, pp. 663–665. ACM, New York (2010). doi:10.1145/1866307.1866390
18. Garera, S., Provos, N., Chew, M., Rubin, A.D.: A framework for detection and measurement of phishing attacks. In: Proceedings of the 2007 ACM Workshop on Recurring Malcode, WORM'07, pp. 1–8. ACM, New York (2007). doi:10.1145/1314389.1314391
19. Garfinkel, T., Pfaff, B., Chow, J., Rosenblum, M., Boneh, D.: Terra: A virtual machine-based platform for trusted computing. ACM SIGOPS Oper. Syst. Rev. **37**(5), 193–206 (2003). doi:10.1145/1165389.945464
20. Gyongyi, Z., Garcia-Molina, H.: Web spam taxonomy. In: Proceedings of the 1st International Workshop on Adversarial Information Retrieval on the Web, AIRWeb'05, Chiba (2005)
21. Juels, A., Kaliski, B.S., Jr.: Pors: proofs of retrievability for large files. In: Proceedings of the 14th ACM Conference on Computer and Communications Security, CCS'07, Alexandria, pp. 584–597. ACM, New York (2007). doi:10.1145/1315245.1315317
22. Kartaltepe, E.J., Morales, J.A., Xu, S., Sandhu, R.: Social network-based botnet command-and-control: emerging threats and countermeasures. In: Proceedings of the 8th International Conference on Applied Cryptography and Network Security, ACNS'10, Beijing, pp. 511–528. Springer, Berlin/Heidelberg (2010)
23. Klein, G., Elphinstone, K., Heiser, G., Andronick, J., Cock, D., Derrin, P., Elkaduwe, D., Engelhardt, K., Kolanski, R., Norrish, M., Sewell, T., Tuch, H., Winwood, S.: sel4: formal verification of an OS kernel. In: Proceedings of the 2009 ACM SIGOPS 22nd Symposium on Operating Systems Principles, SOSP'09, Big Sky, pp. 207–220. ACM, New York (2009). doi:10.1145/1629575.1629596
24. Li, F., Hadjieleftheriou, M., Kollios, G., Reyzin, L.: Dynamic authenticated index structures for outsourced databases. In: Proceedings of the 2006 ACM SIGMOD International Conference on Management of Data, SIGMOD'06, Chicago, pp. 121–132. ACM, New York (2006). doi:10.1145/1142473.1142488
25. Li, Z., Goyal, A., Chen, Y., Paxson, V.: Towards situational awareness of large-scale botnet probing events. IEEE Trans. Inf. Forensics Secur. **6**(1), 175–188 (2011). doi:10.1109/TIFS.2010.2086445
26. Luna Garcia, J., Langenberg, R., Suri, N.: Benchmarking cloud security level agreements using quantitative policy trees. In: Proceedings of the 4th ACM Workshop on Cloud Computing Security Workshop, CCSW'12, Raleigh, pp. 103–112. ACM, New York (2012). doi:10.1145/2381913.2381932

27. Ma, J., Saul, L.K., Savage, S., Voelker, G.M.: Beyond blacklists: learning to detect malicious web sites from suspicious urls. In: Proceedings of the 15th ACM SIGKDD International Conference on Knowledge Discovery and Data Mining, KDD'09, Paris, pp. 1245–1254. ACM, New York (2009). doi:10.1145/1557019.1557153
28. Mahoney, M.V., Chan, P.K.: Learning nonstationary models of normal network traffic for detecting novel attacks. In: Proceedings of the 8th ACM SIGKDD International Conference on Knowledge Discovery and Data Mining, KDD'02, Edmonton, pp. 376–385. ACM, New York (2002). doi:10.1145/775047.775102
29. McCune, J.M., Li, Y., Qu, N., Zhou, Z., Datta, A., Gligor, V., Perrig, A.: Trustvisor: efficient TCB reduction and attestation. In: Proceedings of the 2010 IEEE Symposium on Security and Privacy, SP'10, Oakland, pp. 143–158. IEEE Computer Society, Washington, DC (2010). doi:10.1109/SP.2010.17
30. Meer, H., Arvanitis, N., Slaviero, M.: defcon.org, Clobbering the cloud. http://goo.gl/42hRL (2009)
31. Nazario, J.: usenix.org, PhoneyC: a virtual client Honeypot. http://goo.gl/euYt0 (2009)
32. Niu, Y., Chen, H., Hsu, F., Wang, Y.M., Ma, M.: A quantitative study of forum spamming using context-based analysis. In: Proceedings of the 2007 Network and Distributed System Security Symposium, NDSS'07, San Diego (2007)
33. Pang, H., Zhang, J., Mouratidis, K.: Scalable verification for outsourced dynamic databases. Proc. VLDB Endow. **2**(1), 802–813 (2009)
34. Petroni Jr., N.L., Hicks, M.: Automated detection of persistent kernel control-flow attacks. In: Proceedings of the 14th ACM Conference on Computer and Communications Security, CCS'07, Alexandria, pp. 103–115. ACM, New York (2007). doi:10.1145/1315245.1315260
35. Pham, V.H.: eurecom.fr, Honeypot traces forensics by means of attack event identification. http://goo.gl/wGPlV (2009)
36. Pouget, F., Dacier, M.: Honeypot-based forensics. In: Proceedings of the 2004 AusCERT Asia Pacific Information Technology Security Conference, AusCERT'04, Gold Coast (2004)
37. securityfocus.com, Zeus botnet finds hold in Amazon cloud. http://goo.gl/rFjzF (2009)
38. Seifert, C., Steenson, R.: honeynet.org, Capture – Honeypot Client (Capture-HPC). http://goo.gl/u7qJZ (2006)
39. Sherry, J., Hasan, S., Scott, C., Krishnamurthy, A., Ratnasamy, S., Sekar, V.: Making middleboxes someone else's problem: network processing as a cloud service. In: Proceedings of the 2012 ACM SIGCOMM Conference on Applications, Technologies, Architectures, and Protocols for Computer Communication, SIGCOMM '12, Helsinki, pp. 13–24. ACM, New York (2012). doi:10.1145/2342356.2342359
40. Somorovsky, J., Heiderich, M., Jensen, M., Schwenk, J., Gruschka, N., Lo Iacono, L.: All your clouds are belong to us: security analysis of cloud management interfaces. In: Proceedings of the 3rd ACM Workshop on Cloud Computing Security Workshop, CCSW'11, Chicago, pp. 3–14. ACM, New York (2011). doi:10.1145/2046660.2046664
41. Spitzner, L.: Honeypots: Tracking Hackers. Addison-Wesly Longman, Boston (2002)
42. Steinberg, U., Kauer, B.: Nova: a microhypervisor-based secure virtualization architecture. In: Proceedings of the 5th European Conference on Computer Systems, EuroSys'10, Paris, pp. 209–222. ACM, New York (2010). doi:10.1145/1755913.1755935
43. Stone-Gross, B., Holz, T., Stringhini, G., Vigna, G.: The underground economy of spam: a botmaster's perspective of coordinating large-scale spam campaigns. In: Proceedings of the 4th USENIX Conference on Large-scale Exploits and Emergent Threats, LEET'11, Boston, pp. 25–32. USENIX Association, Berkeley (2011)
44. Szefer, J., Keller, E., Lee, R.B., Rexford, J.: Eliminating the hypervisor attack surface for a more secure cloud. In: Proceedings of the 18th ACM Conference on Computer and Communications Security, CCS'11, Chicago, pp. 401–412. ACM, New York (2011). doi: 10.1145/2046707.2046754
45. Szefer, J., Lee, R.B.: Architectural support for hypervisor-secure virtualization. In: Proceedings of the 7th International Conference on Architectural Support for Programming Languages and Operating Systems, ASPLOS'12, London, pp. 437–450. ACM, New York (2012). doi:10.1145/2150976.2151022

46. Ta-Min, R., Litty, L., Lie, D.: Splitting interfaces: making trust between applications and operating systems configurable. In: Proceedings of the 7th Symposium on Operating Systems Design and Implementation, OSDI'06, Seattle, pp. 279–292. USENIX Association, Berkeley (2006)
47. Thonnard, O., Dacier, M.: A framework for attack patterns' discovery in honeynet data. Digit. Investig. **5**, S128–S139 (2008). doi:10.1016/j.diin.2008.05.012
48. trustedcomputinggroup.org, TPM specifications version 1.2. http://goo.gl/0IWyy (2011)
49. Wang, Y.M., Beck, D., Jiang, X., Roussev, R., Verbowski, C., Chen, S., King, S.T.: Automated Web patrol with strider HoneyMonkeys: finding Web sites that exploit browser vulnerabilities. In: Proceedings of the 2006 Network and Distributed System Security Symposium, NDSS'06, San Diego. The Internet Society, San Diego (2006)
50. Wang, Z., Jiang, X.: Hypersafe: a lightweight approach to provide lifetime hypervisor control-flow integrity. In: Proceedings of the 2010 IEEE Symposium on Security and Privacy, SP'10, Oakland, pp. 380–395. IEEE Computer Society, Washington, DC (2010). doi:10.1109/SP.2010.30
51. Wei, J., Zhang, X., Ammons, G., Bala, V., Ning, P.: Managing security of virtual machine images in a cloud environment. In: Proceedings of the 1st ACM Workshop on Cloud Computing Security, CCSW'09, Chicago, pp. 91–96. ACM, New York (2009). doi:10.1145/1655008.1655021
52. Wu, B., Davison, B.D.: Cloaking and redirection: a preliminary study. In: Proceedings of the 1st International Workshop on Adversarial Information Retrieval on the Web, AIRWeb'05, Chiba, pp. 7–16. Chiba (2005)
53. van Doorn, L.: Trusted computing challenges. In: Proceedings of the 2007 ACM Workshop on Scalable Trusted Computing, STC'07, Alexandria, pp. 1–1. ACM, New York (2007). doi:10.1145/1314354.1314356
54. Xu, S., Yung, M.: Socialclouds: concept, security architecture and some mechanisms. In: Proceedings of the 1st International Conference on Trusted Systems, INTRUST'09, Beijing, pp. 104–128. Springer, Berlin/Heidelberg (2010). doi:10.1007/978-3-642-14597-1_7
55. Zheng, Q., Xu, S.: Fair and dynamic proofs of retrievability. In: Proceedings of the 1st ACM Conference on Data and Application Security and Privacy, CODASPY'11, San Antonio, pp. 237–248. ACM, New York (2011). doi:10.1145/1943513.1943546
56. Zheng, Q., Xu, S., Ateniese, G.: Efficient query integrity for outsourced dynamic databases. In: Proceedings of the 2012 ACM Workshop on Cloud Computing Security Workshop, CCSW'12, Raleigh, pp. 71–82. ACM, New York (2012). doi:10.1145/2381913.2381927

Trust-Based Access Control for Secure Cloud Computing

Indrajit Ray and Indrakshi Ray

Abstract Multi-tenancy, elasticity and dynamicity pose several novel challenges for access control in a cloud environment. Accessing subjects may dynamically change, resources requiring protection may be created or modified, and subject access requirements to resources may change during the course of the application execution. Users may need to acquire different permissions from different administrative domains based on the services in cloud computing environment. Traditional identity-based access control models such as attribute-based access control (ABAC), role-based access control (RBAC), discretionary access control (DAC), or mandatory access control (MAC) cannot be applied directly in clouds. In this chapter, we explore challenges of cloud access control, identify desirable properties of access control models, and introduce the novel graph-theoretic semantics of access control model. We specify how authorization occurs in the proposed model, and present how to incorporate features such as separation of duty (SoD).

1 Introduction

Cloud computing is an emerging paradigm that allows on-demand network access to a shared pool of resources without requiring extensive management efforts on behalf of the clients that require these resources. It can be differentiated from other classical computing models by five major characteristics such as *on-demand self-service*, *broad network access*, *resource pooling*, *rapid elasticity* and *measured service* [41].

Since a cloud environment involves multiple resources belonging to multiple clients interacting in complex manners, proper access control to these resources is very important. As depicted in Fig. 1, access control requires addressing three non-orthogonal sub problems of *authorization*, *authentication* and *enforcement* [47, 51].

I. Ray (✉) • I. Ray
Colorado State University, Fort Collins, CO, USA
e-mail: indrajit@cs.colostate.edu; ray@cs.colostate.edu

K.J. Han et al. (eds.), *High Performance Cloud Auditing and Applications*,
DOI 10.1007/978-1-4614-3296-8_8, © Springer Science+Business Media New York 2014

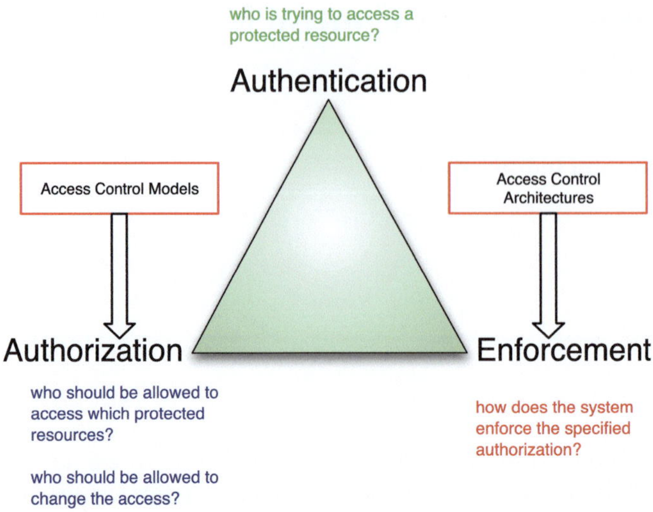

Fig. 1 Scope of the access control problem

(1) Authorization, which is expressed in terms of an access control model, specifies the resources that need to be protected, what types of accesses (operations) are possible on these resources, and who is allowed access to these resources. (2) Authentication, which is specified in terms of protocols, addresses the problem of determining who is trying to access a protected resource. (3) Enforcement involves making sure that an entity that is allowed to access certain resources, gets to access these resources when requested, and that an entity that is not allowed is denied access. In clouds, these three aspects of access control are achieved by the cooperation of four functional modules:

1. *Policy Administration Point (PAP)*: PAP is a repository for the authorization policies that are expressed in terms of the actions. Subjects (i.e., human users, devices, processes, organizations, etc.) can take on various objects in the system. The authorization policies are essentially an instantiation of the access control model tailored towards the organization. It is the main component for the authorization portion of access control.
2. *Policy Information Point (PIP)*: PIP is the module that aggregate the information to evaluate an authorization policy. It is the main component responsible for achieving authentication.
3. *Policy Decision Point (PDP)*: PDP gets relevant information from the PIP and consults the PAP to arrive at a decision whether to grant or deny an access request.
4. *Policy Enforcement Point (PEP)*: PEP receives access requests from subjects in the external world, sends them to the PDP for evaluation, and after receiving the grant or deny response from the PDP, ensures that the appropriate action is taken.

The unique characteristics of cloud computing introduce novel challenges to each of the three aspects of access control, in particular to authorization. To begin with, multi-tenancy results in the co-residency of machines (e.g., virtual machines (VMs) and database engines) and other resources (e.g., hardware or storage) owned by different clients in the same privileged position in the cloud. *Tenant*, *customer* or *user* is often used in cloud computing parlance to mean *client*. We will use these three terms interchangeably. This makes security vulnerabilities at the cloud infrastructure level particularly critical to the cloud environment.

A guest operating system (OS) can exploit vulnerabilities in the hypervisor [31] and run processes on other guests or the host. One such exploit was demonstrated in a recent work [46] where the internal cloud infrastructure was mapped to identify where a target virtual machine (VM) is likely to reside and then that information was used to mount cross-VM side-channel attacks to extract sensitive information from the target VM. In short, security breaches such as unauthorized connections, unauthorized leakage of information, unmonitored login attempts, malware propagation etc., can arise in one client and potentially propagate to another easily. It is, therefore, vitally important that proper authorization policies be employed for the *protection* of tenant resources from un-authorized disclosure and modification, *segregation* of tenants from one another, and *isolation* of computation, storage and network resources of the cloud provider from tenants.

A cloud-computing environment being massively scalable and elastic, is inherently very dynamic. This makes the problem of providing proper access control in clouds even more challenging. The accessing entities may change, resources requiring protection may be created or modified, and an entity's access to resources may change during the course of an application execution. Users may need to dynamically acquire different permissions from different domains based on the services they need. Cloud collaboration can potentially spread across enterprise boundaries where interactions among entities often occur in ad hoc manners. The access-requesting entity may not be known in advance by the access-granting entity.

In such situations, traditional identity-based access control models such as role-based access control (RBAC), discretionary access control (DAC) and mandatory access control (MAC) [23, 26, 50, 55], that rely on the access-granter knowing the identity of access-requester from beforehand and authenticating the requester, can no longer be readily applied.

It also makes the issue of *delegation* difficult. Delegation of authorization is needed to manage and mediate interactions with distributed resources having distributed administrators. Since authentication of the client may be difficult, such delegation is difficult to implement and enforce. Researchers have proposed Credential-based or Attribute-based Access Control (ABAC) models [8, 9] to address the challenges of unknown users in access control. In these models, access decisions are based on properties or attributes of requester that are specified within credentials, instead of the requester's identity. The approach provides flexibility and scalability. Unfortunately, these models do not allow access control decisions to be dynamically updated or revoked based on the history of the requester, or based on changing requirements of the requester. Such an ability is important in light of the rapid elasticity and measured-service feature of cloud computing. It appears that decisions about who to trust and how much to trust, hold the key towards satisfying these major challenges.

In this chapter, we look into the requirements of access control for cloud environments in more detail. In particular, we look into two aspects: authorization requirements and enforcement requirements. We identify the features that are required and features that are desirable. We then look into some of the more recent advanced access control models beyond DAC, MAC and RBAC. Our objective is to determine (1) if these newer models are directly applicable to cloud environments and if not, (2) whether there are strengths in these models that can be adapted to propose a more comprehensive access control model for the cloud. Recent research indicates that no comprehensive access control model exists yet, but there are promising models that can be adapted for cloud computing. We then discuss the broad outlines of a new trust-based access control model. We conclude this chapter by discussing challenges that must be met in bringing the trust-based access control model to full fruition.

2 Authorization and Enforcement Requirements

Proper access control is fundamental for achieving confidentiality, integrity and availability in clouds. A flexible authorization framework can increase security in clouds by allowing one to achieve segregation and isolation among the different resources shared between providers and customers. To understand the authorization requirements for cloud computing we begin by identifying the various resources that need to be protected, the layers in the cloud architecture to which these resources belong, and the parties responsible for these resources. Then, follows a discussion of specific features that are important in cloud systems.

2.1 Cloud Resources

The NIST definition of cloud computing [41] describes three *cloud service models*: Software as a Service (SaaS), Platform as a Service (PaaS), and Infrastructure as a Service (IaaS). It also describes four *cloud deployment models*: public cloud, private cloud, hybrid cloud, and community cloud. Depending on the service model under consideration, the major resources that need to be protected in the cloud environment can be of different types [35].

- *Software as a Service (SaaS)*: The major resources to be protected are software and users' stored/generated data. Users should be able to (1) specify all external entities with which users' data can be shared, (2) make their data available to composite applications, and (3) transfer some of their privileges to other users.

- *Platform as a Service (PaaS)*: The major resources to be protected are common services (e.g., file services, database services, directory services and web services), application development, testing and production environments, and the code repository. The access control model should enable a user proper access to the common services without allowing others to access or modify the current user's entries or interactions with these services. Each customer's development, testing and production environment should be kept separated and segregated from other customer's environments. The access control model should facilitate yet protect the movement of code between a customer's development, testing and production environments without compromising these even when they coexist on the same VM.
- *Infrastructure as a Service(IaaS)*: The major resources to be protected in the IaaS model are VMs, guest operating systems, virtualized storage and computing environments, network interfaces, firewall rule sets, load balancers, auditing and accounting facilities, etc. The main access control service is to keep these resources isolated from the tenants' applications. In addition, the resources should be dynamically provisionable for measured-service ensuring, in the process, that all tenants have a fair share of resources and that no single IaaS customer is able to affect the availability of resources to other customers.

2.2 Access Control at Different Levels

Access control needs to be provided at different granularity levels in the cloud. Further, there are different types of protection that are expected from the access control services. For example, in the SaaS model the major service is protection of user data, in the PaaS model the major service is segregation of tenant resources, and in the IaaS model the major service is isolation of provider resources from tenants. This does not mean that the other services are any less significant in a specific service model; it is just that the PDP and PEP need to be identified and designed so that the corresponding important access control service is most effective. Last but not least, since in a cloud environment resources are controlled by many different entities, access control needs to be effectuated at different levels by different entities.

Figure 2 shows the relative positioning of the major resources in a cloud environment as a stack. It identifies the placement of the major resources in the stack. The stack comprises four layers that are the infrastructure layer, the platform layer, the software layer and the service delivery layer. Relating the stack to the cloud service models, the IaaS model relies mostly on activities at the infrastructure layer, the SaaS model relies most on activities at the software layer, and the PaaS model relies mostly on the platform layer. All three models use activities at the service delivery layer to various extents.

We can broadly divide cloud resource owners into two classes: (1) providers who offer their services, and (2) tenants or customers. The provider is responsible

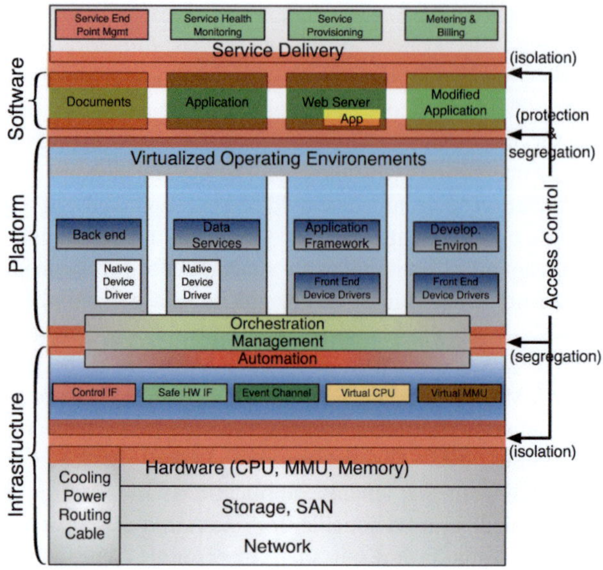

Fig. 2 Access control positioning within a cloud architecture

Fig. 3 Example illustrating need for delegation

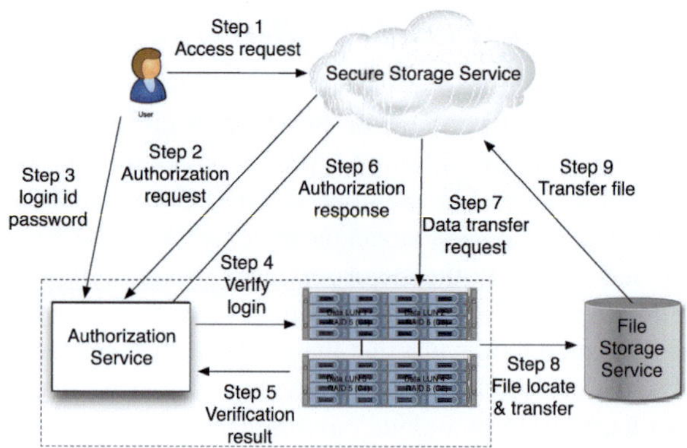

for resources specific to the infrastructure layer and the service delivery layer. The customer is primarily responsible for resources in the software layer although some of the responsibilities also rests on the provider. The customer and the provider share the responsibility almost evenly in the platform layer. Note that an entity may be a tenant to a provider of some services (the first level provider) and at the same time be a provider (second level provider) of these services to other entities. An example of such a scenario is shown in Fig. 3. In such cases, the first level provider may have to delegate some of its responsibilities to the second level provider.

2.3 Need for Delegation

In a cloud environment, resources are often distributed and managed by different administrators. To support a specific service there is frequently a need for coordination and interaction between these different administrators. A client of one service provider may need to access other providers for getting the relevant service. Consider, for example, the cloud application shown in Fig. 3.

The user is a client of the Secure Storage Service Cloud and uses its services to store user's files in a secure manner. The Secure Storage Service Cloud, in turn, uses the Authorization and File service provider to provide the storage service. The latter uses physical disk space on a File Storage Service to store bits and bytes of its clients. User's request to access some data at the Secure Storage Service Cloud will be converted to the workflow shown in Fig. 3. Delegation is the principle that allows the Secure Storage Service Cloud to act on behalf of the user by making the user's access rights available to the other providers. A delegation operation temporarily changes the access control state so as to allow the Secure Storage Service Cloud to use the initial user access privileges and thus allows the authorization of the entire workflow depicted in Fig. 3.

A naive approach to supporting delegation is to place the framework on top of the access control module, and let the delegation module handle delegation operations and manipulate access control configuration. The underlying access control module consults the access control configuration without concerning delegation. However, such an approach can be easily compromised by colluding users. For example, if the user and the Authorization File Service provider collude, the user can bypass the Secure Storage Service provider to access the File Storage Service, thereby circumventing security policies in the access control system. This suggests the need to make delegation services an integrated feature of the access control model.

2.4 Policy Integration for Secure Interoperability

In the cloud computing paradigm, information and other resources may not always be under the control of a single authority. This is typically the case when different private clouds interact with each other in a workflow to provide some functionality. In such situations, the different authorities controlling the various resources have their own individual policies. Thus, access control enforcement needs to consider the different policies independently, and compose/integrate these policies together to form a single cohesive policy ensuring that conflicts, if any, are resolved.

Policy integration is a challenging problem by its own right. It is complicated further in clouds where different administrative tasks are managed by different entities. However, policy integration is an orthogonal aspect to access control policy models and is not the topic of discussion in this article. Nonetheless, adopting a proper access control model can facilitate policy integration considerably. We identify a core set of model requirements that would facilitate such policy integration in clouds.

- *Formal semantics*: The access control policy specification language should be declarative and implementation independent. Moreover, it should have a formal semantics to avoid ambiguity across tenants and providers.
- *Expressiveness*: The access control model should allow one to express fine-grained policies. It should support combining permissions without introducing ad hoc extensions to authorizations.
- *Support for incomplete policies*: The access control framework should support policies that are not fully defined when the composition strategy is enacted. However, a complete and unambiguous access control decision should be presented to access requests at the control time.
- *Support for heterogenous policies*: Ideally, the model should support different types of policies possibly expressed in different languages and enforced by different mechanisms.

2.5 Separation of Duty (SoD)

SoD refers to the access control principle that no user should be given enough privileges to misuse the system. The purpose of SoD is to ensure "that failures of omission or commission with an organization are caused only by collusion among individuals, making such failures riskier and less likely" [30].

SoD plays a very important role in cloud access control. Refer once more to Fig. 3. Assume that the Secure Storage Service receives the user's data and applies the relevant protection functions. The resulting process changes the size of the data that the user stores. The user needs to pay for the secure storage services based on two factors: (1) The disk space to store the user's files. (2) The computation overhead to protect the user's files. An auditing and accounting facility within the cloud gathers relevant information to perform the accounting operation. Clearly, if this facility and the secure storage service can collude, the user can be easily overcharged. SoD constraints ensure that these two facilities can never collude. The ability to prevent collusion in conflict-of-interest situations is particularly important in the hybrid cloud model where multiple private clouds may be hosted on the same public cloud and compete with each other for services.

SoD minimizes the chances of collusion by assigning users to separated tasks. Separating duties applies for both the cloud provider and the customer. The cloud provider should ensure that only authorized administrators have access to resources they manage. The cloud service provider (CSP) should also have a mechanism available to the customer for giving internal administrators access to necessary resources. The SoD in conjunction with delegation of access control administration strengthens security by preventing the same administrator from modifying data and audit logs. SoD also needs to be ensured whenever there is a conflict of interest that may arise in the assignment of activities. Within a cloud this is of great importance because of the existence of numerous service level agreements between various tenants and providers. Separation of duties can be enforced statically as part of the access control policy specification or dynamically at run-time during policy enforcement.

2.6 Dynamically Adaptable Policies

Cloud environments, unlike traditional systems, are not static. They are elastic and constantly evolve in response to various factors including the need to support pay-as-you-go in the measured-service model. This dynamic nature implies that access control policies need to be adaptable to changing circumstances. In addition, often in the real world operational needs arise that require access to be granted under less than ideal security situations. These decisions are made because the benefits of allowing access outweighs the potential risks.

Consider once more the cloud based application shown in Fig. 3. The user keeps the user's sensitive documents in the cloud including a set of financial documents. The user, while preparing year-end tax returns, needs the services of a tax preparing organization (tax-service provider) and decides to let the tax-service provider have access to documents related to the user's current tax year and previous year in the Secure Storage Service provider. Thus, the tax-service provider gets permission to these specific documents on the cloud. During the execution of the tax preparation application the tax-service provider needs to have access to some documents from other years. In this case, the tax-service provider's access permissions need to be dynamically adapted to the changing circumstances.

We now investigate if there is any single existing access control model that satisfies these requirements.

3 Advanced Access Control Models

RBAC is used extensively within organizations for administering and managing privileges and is often considered the de-facto standard for access control. RBAC facilitates security administration and management of privileges by establishing permissions based on functional roles in an enterprise rather than on user's identity. Access decisions are based on roles that different users have in the organization. Users are granted membership into roles based on their responsibilities and/or competencies in the organization. RBAC was first introduced by Ferraiolo and Kuhn in [24] to address the limitations of discretionary access control model (DAC). Sandhu et al. [50] introduce four reference models to provide a systematic approach to understand RBAC model. Their framework separates administration of

RBAC from its use for controlling access. They also categorize the implementation of RBAC in different systems. Finally, after a series of modifications, the NIST standard for RBAC was proposed in [25]. The standard specifies RBAC reference model that defines the scope of features that comprise the standard and provides terminologies to support specification. The standard also specifies system and administrative functional specifications which define functional requirements for administrative operations and system level functionalities.

While the advantages of RBAC are numerous, researchers are increasingly identifying shortcomings in the model for newer and emerging applications. One of the major shortcomings is that RBAC assumes that all permissions needed to perform a job can be perfectly encapsulated within the concept of roles. For increased security, each role should be associated with the minimum number of privileges needed which invariably results in multiple roles per user. This is problematic because it goes against SoD as a fundamental security principle. Careful allocation of privileges can solve this problem. However, for security administration it is better to have fewer roles. It is becoming increasingly difficult for emerging application areas to determine a suitable tradeoff to address this dichotomy. This is particularly challenging in cloud environments where the allocation of different resources may be incompatible with organization structures and many access decisions need to be made on an ad hoc basis.

Researchers are proposing extensions to the core RBAC model as well as newer models to overcome these limitations in various manners. In the following, we review some of important research in these new access control models. We discuss their features and try to identify their suitability for cloud computing. We do not discuss eXtensible Access Control Markup Language (XACML) [42] in these sections. XACML is more of an access control specification language and a framework to evaluate authorization requests rather than an access control model. XACML is typically used to express RBAC access control policies.

3.1 Simple Extensions to RBAC

3.1.1 Team-Based Access Control

The traditional RBAC model was proposed to address the access control needs of single organizations. Thomas [52] introduces the notion of Team-based Access Control (TMAC) as an approach to applying RBAC in collaborative environments where multiple organizations collaborate with each other to perform some task. Unlike traditional RBAC where groups are defined based on users performing the same role, teams are a more natural way of organizing users into groups. A "team" is an abstract container that encapsulates a set of users with specific roles. A team is formed with the objective of accomplishing a specific task. For this purpose, the model introduces two important concepts such as user context and object context. User context allows one to identify specific users playing a single role in a team.

Object context identifies objects required for collaboration purposes. The TMAC model allows role-based permissions across object types as well as fine-grained, identity-based control on individual users in certain roles and to individual object instances. Georgiadis et al. [28] extend the TMAC model by integrating it with RBAC and using it for general contextual information like time of access, location from which access is requested, and location of the object for which access is requested etc.

3.1.2 Group-Oriented Access Control

A somewhat similar idea to TMAC is presented in the YGuard access control model [54]. YGuard employs set-based access control lists, and XML where a group of subjects is authorized to access sets of objects. Every member of a group has same access privileges. Another similar approach is Coalition-based Access Control model [17] where a coalition (group of members) shares data with their partners while ensuring that their resources are safe from inappropriate access. The authors define the protection state of a system that provides the semantics of CBAC-based access policies.

While team or group-oriented access control models are useful in collaborative environments, they do not address the challenges of cloud system identified earlier. They lack the self-administration of assignment relations between entities. These models are not yet well-developed to address ad hoc collaborations.

3.2 Access Control Models Integrating Contextual Information

Cloud computing has become very important in the mobile setting. Most mobile systems take help from clouds to perform many of the computing functionalities. We next look into extensions of RBAC that have been proposed in mobile settings.

3.2.1 Access Control for Active Spaces

Sampemane et al. [48] extends classical RBAC to formalize a new access control model for active spaces. Active space denotes the computing environment integrating physical spaces and embedded computing software and hardware entities. The active space allows interactive exchange of information between the user and the space. Environmental aspects are adopted into the access control model for active spaces, and the space roles are introduced into the implementation of the access control model based on RBAC. The model supports specification of MAC policies in which system administrator maintains the access matrix and DAC policies in which users create and update security policies for their devices.

3.2.2 Generalized RBAC (GRBAC)

Covington et al. [18] introduce environment roles in a Generalized RBAC model (GRBAC) to help control access to private information and resources in ubiquitous computing applications. The environment roles differ from the subject roles in RBAC but do have similar properties including role activation, role hierarchy, and SoD. In the access control framework enabled by environment roles, each element of permission assignment is associated with a set of environment roles, and environment roles are activated according to the changing conditions specified in environmental conditions; in this way, environmental properties like time and location are introduced into the access control framework. In a subsequent work [19], Covington et al. describe the Context-Aware Security Architecture (CASA) that is an implementation of the GRBAC model. The access control is provided by the security services in the architecture. In the CASA, policies are expressed as roles and managed by the security management service, and authentication and authorization services are used to verify user credentials and determine access to the system resources. The environmental role activation services manage environmental role activation and deactivation according to the environment variables collected by the context management services.

3.2.3 Temporal RBAC (TRBAC)

Other extensions to RBAC include the Temporal Role-Based Access Control Model (TRBAC) proposed by Bertino et al. [3]. This work adds the time dimension to the RBAC model. The authors in this paper introduce the concept of *role enabling* and *role disabling*. Temporal constraints determine when the roles can be enabled or disabled. A role can be activated only if it has been enabled. Joshi et al. [32, 33] extend this work by proposing the Generalized Temporal Role Based Access Control (GTRBAC) Model. In this work the authors introduce the concept of time-based role hierarchy and time-based SoD.

3.2.4 Location-Based RBAC

Researchers have also extended RBAC to incorporate spatial information [21, 45]. One work in this regard is the GEO-RBAC model [21]. In this model, role activation is based on the location of the user. The model supports role hierarchies but does not deal with separation of duties. Another work incorporating spatial information is by Ray et al. [45]. The authors proposed how each component of RBAC is influenced by location and defined the formal model using *Z specification language*. Role hierarchy and SoDs are not addressed in this chapter.

3.2.5 Spatio-Temporal RBAC

Incorporating both time and location in RBAC is addressed by several researchers [14, 15, 43, 44, 49, 53]. Chandran's work combines the main features of GTRBAC and GEO-RBAC [14]. Here again, role is enabled by time constraints. The user can activate the role if the role is enabled and the user satisfies the location constraints associated with role activation. Samuel et al. [49] propose GST-RBAC which incorporates topological spatial constraints into the existing GTRBAC model. The authors do this by augmenting GTRBAC operations, namely, role enabling, user-role assignment, role-permission assignment, and role-activation with spatial constraints. The operations are allowed only if the spatial and temporal constraints are satisfied. The model also introduces the notion of Spatial Role Hierarchy and Spatial SoD (spSoD) constraints. Ray and Toahchoodee [43] propose Spatio-Temporal RBAC which is an extension of RBAC model supporting the temporal and spatial constraints. In this model, user can activate the role only during the specific time at the specific location. After the user's role is activated, he can access the resource only if his role has the permission to access the object at that specific spatio-temporal point and the object is located at the specific location during the specific time. The model also supports the spatio-temporal role hierarchies and SoD constraints.

Chen and Crampton [15] develop the graph based representation for the spatio-temporal RBAC. All RBAC components are represented by vertices while the assignment and hierarchical relationships are represented by the edges of the directed graph. The model can be categorized into three model types such as standard, strong and weak models. For the standard model, component v_1 is said to be authorized to component v_n if all vertices along the authorization path satisfy the spatio-temporal constraints. For the strong model, component v_1 is said to be authorized to component v_n if all vertices together with the edges along the authorization path satisfy the spatio-temporal constraints. And in the weak model, component v_1 is said to be authorized to component v_n if both vertices satisfy the spatio-temporal constraints. The model has a well-defined semantics. However, it does not address the effects of SoD constraints.

3.3 Attribute-Based Access Control (ABAC)

Notwithstanding the success of the RBAC model and its different extensions, researchers have often found these models to be inadequate for open and decentralized multi-centric systems where the user population is dynamic and the identity of all users is not known in advance. Examples of such systems are service providers operating over open systems like the Internet and cloud environments. It is almost impossible to know beforehand all the users that will request services in these systems. Assigning appropriate roles to these users thus becomes an irrational and ad-hoc exercise.

Researchers have proposed ABAC models or credential-based access control models [6, 7, 36] to address the problem of not knowing the identity of the access requester. Attributes are specific characteristics of users or resources, conditions defined by authority, or aspects of the environment. The basic idea is to use characteristics or attributes of the access requesters as a basis for access decision.

In these models, the access requester has to produce a predetermined set of credentials (e.g., proof of membership in certain groups) to gain specific access privileges. The credentials provide information about the rights, qualifications, responsibilities and other characteristics attributable to its bearer by one or more trusted authorities. This information forms the attributes of the access requester. In addition, credentials provide trust information about the authorities themselves. Researchers have also integrated ABAC with RBAC to facilitate security administration [12, 37, 38].

Bonatti and Samarati [9] propose a uniform formal framework to regulate service access and information disclosure on the Internet. The regulation is based on credentials. The framework includes a mechanism to treat information which is not in the form of a certificate and is needed for required access. It comprises a language for expressing access and release policies. The framework also has a policy filtering mechanism that helps the entities involved in the communication to exchange their requirements in a concise and privacy-preserving way. In [2] Bauer et al. describe the design, implementation, and performance of a system to control access on the web.

Their system is based on proof-carrying authorization (PCA). The access control model locates and uses pieces of the security policy that have been distributed across hosts, and keeps the policies hidden from unauthorized clients. It also provides iterative authorization by which a server can require a browser to prove a series of challenges. In [29], the authors address the problems of access control in large open systems where the authenticated identity of an entity does not provide any information regarding the likely behavior of that entity. Their scheme, called cryptographic access control, is based on cryptography to guarantee confidentiality and integrity of objects stored in potentially untrusted servers in the system.

ABAC can provide more complex access control than allowed by RBAC and its different extensions. ABAC also scales well for large enterprises and cross-domain or federated situations assuming that a scalable approach to managing attributes and credentials is available. However, the ability to express complex policies in ABAC comes with additional administrative overhead. Moreover, strictly speaking, a credential that is needed to support ABAC does not bind a user to its purported behavior or actions. It does not guarantee that its bearer really satisfies the claims encoded in the credential. It does not convey any information about the behavior of the bearer between the time the credential was issued and its use. A credential does not reveal whether it was obtained via devious means. In a cloud environment, such information may play a crucial role in access control. Additionally, attribute based access control does not keep track of the user's behavior history. Access per-session is given on the basis of the credential presented in that particular session without any sense that a user can potentially misuse the privileges. Either the user's credentials are accepted and required privileges are allowed, or the credentials are rejected, and the user does not get the access rights. Thus, good behavior by the user cannot be rewarded with enhanced privileges nor bad behavior be punished. In light of the "pay-as-you-go" model of use in a cloud environment, this appears to be severely restrictive.

Cloud environments unlike traditional organizations are not static. They are elastic and constantly evolve in response to various factors including the need to support a pay-as-you-go model. This dynamic nature implies that their access control policies need to be adaptable to changing circumstances. Often operational needs may arise that require access be granted under less than ideal security situations. Attribute-based access control cannot adequately model this dynamism. Researchers have investigated incorporating notions of risk levels into the access control decision process. This goes beyond the traditional reliance on static attributes or credentials of users to perform access control and instead rely on the user's trustworthiness or perceptions of risks.

3.4 Risk-Adaptable Access Control

Recently, researchers have proposed the notion of risk-adaptable access control [34, 40] as a way to facilitate dynamic adaptation of access control policies based on operational needs and situational awareness.

Cheng et al. [16] are among the earliest to argue that there are some degrees of uncertainty and risk involved in access control decisions. Ideally, access control models need to be able to express such risks explicitly and be adaptable to such risks. The authors propose the Fuzzy Multi-Level Security (Fuzzy MLS) model that allows risky information flow across security classes provided the risk can be accounted for and controlled. They also introduce methods that can be used to quantify risk associated with information access, and discuss a case study for implementing the Fuzzy MLS model.

McGraw [40] discusses the major reasons why risk-adaptable access control is preferred to other static models such as RBAC and ABAC and sketches the broad functionality of such models. The author argues that such models have the ability to encode more closely real world situations where access needs to be provided under less than ideal security settings. However, the author does not define any specific model. Kandala et al. [34] propose using the notion of attributes to explicitly encode the risk that is exposed when access decisions need to be made in less than ideal situations. This work introduces a formal framework for a risk-adaptable access control model in terms of a set of model elements and the interactions between them. However, the authors do not discuss how the risk values needed for the access transactions can be computed.

The notion of risk-adaptable access control looks promising for cloud environments. However, existing models are still in their infancies. They do not incorporate a notion of *history of access* by the user that plays a significant part in a measured service framework used by clouds. They fail to specify how delegation can be expressed within the model. In addition, there are no well-established risk models for computer security. Unless one has such a model, it remains to be seen whether risk-adaptable access control can scale well in elastic clouds.

3.5 Trust-Based Access Control

Researchers have started looking into incorporating notions of multiple levels of trust to access control models in order to address some of the challenges of dynamic systems. Multiple trust levels relate very easily to multiple risk levels. Thus, this approach seems to be the most appropriate to address the challenges of dynamic, elastic and highly scalable cloud systems.

In [4], the authors present X-RBAC, an XML-based RBAC policy specification framework to deal with access control issues in dynamic XML-based web services. They extend X-RBAC to a trust-enhanced version in [5] where the role assignment is based on trust. The authors define 'trust' as "the level of confidence associated with a user based on certain certified attributes". If attributes are certified and validated, the user is trusted. In this sense, X-RBAC is essentially an attribute-based/credential-based model. The authors use the trust management approach of trusted third parties (e.g., public key certification authority) and use certificates provided by the third party to assign roles to the users. The authors argue that traditional access control schemes following identity or capability-based approach for authorization do not scale well to the distributed web services architecture. To overcome this limitation, they describe a mechanism to configure X-GTRBAC to provide context-aware trust-based access control in Web services. X-GTRBAC is a framework based on Generalized Temporal RBAC (GTRBAC) [33]. It extends GTRBAC with XML to allow policy enforcement in heterogeneous and distributed environment.

Ya-Jun et al. [56] proposed the Trust Based Access Control (TBAC) model as an extension of the traditional RBAC for ubiquitous computing applications. In this model, the access privileges of a user depend on his trust level, which in turn depends on contextual information. Chakraborty and Ray [13] proposed another trust-based authorization model called TrustBAC. The model is the extension of the hierarchical RBAC model. In this model, a user can activate a role and invoke the permissions assigned to that role based on the user's trust level. However, the major problem with this work is that the model fails to capture the essential semantics of RBAC systems. The main idea in RBAC is that users are assigned to roles according to their organization's job requirements and responsibilities, not trust levels. Although TrustBAC claims to extend RBAC, its semantics are lost in the model. Moreover, the advantage of incorporating trust models into access control is to allow *operational needs* to override security risk.

3.6 History-Based Access Control

In [22] the authors present a history-based access control mechanism that is suitable for controlling access from mobile code. The scheme maintains a selective history of access requests made by individual programs and uses this history to measure the degree of safeness of a request. Another history-based access control for codes is presented in [1]. In this scheme, the access privileges of a code are determined in runtime by examining the attributes of the pieces of code that have previously run. The pieces of code that have run include the codes on stack as well as the codes that have been called and returned. However, neither of these two models are general enough to address a broad range of issues in clouds although the concept of history-based access control appears to be useful.

3.7 Break-Glass Access Control

In many real-world scenarios situations often arise that require violation of access control policies – users who do not have certain permissions are placed in situations that require these permissions nonetheless. While risk-adaptable access control models explicitly provision for such violations based on an evaluation of the risks involved, they still allow granting permissions from a previously well-defined set possible for the user. Researchers have recently proposed the notion of *Break-glass access control* to deal with situations where unanticipated interactions between entities take place that do not necessarily allow a careful evaluation of the risks. Brucker and Petritsch propose using the break-glass approach in standard access control models to provide more flexible and fine-grained policies [11]. They formalize a general break-glass model and show how it can be supported within SecureUML. They also develop transformation techniques for converting break-glass policies in SecureUML to XACML. They develop a security architecture to support the enforcement of such policies.

The Rumpole model [39] formalizes the concept of break-glass access control. It introduces two notions such as competence and empowerment for this purpose. Competence captures a user's ability to access a resource without security breach. Empowerment captures different constraints such as whether the desired access will compromise critical system policies. Competence and empowerment can each take one of the four values "true", "false", "conflict" or "unknown", to provide evidence to support the access control decision.

Break-glass is effectively a policy enforcement mechanism and a good way to achieve adaptability in access control.

3.8 Rule-Based Access Control

Role engineering [20, 27] is a process of defining, collecting and analyzing roles and RBAC structure that conforms to the practices and policies of an organization. In recent years some researchers have argued that RBAC models (e.g., any extension to RBAC) are challenging to implement because of limitations in role engineering. Rule-based access control has been suggested as an alternative.

Rule-based access control, although not formalized until recently, has often been used for access control in many systems. In this model, access decisions are made in real time by scripted policy rules. Rules are attached to each resource that governs the access that will be allowed when a user tries to use it. In this model, access decisions are made in real time by scripted policy rules. Rules are attached to each resource that governs the access that will be allowed when a user tries to use it.

One of the major concerns with rule-based access control is that there is no formal model, structure or language for expressing rules. The term "rule" is a generic term applied to encompass a broad range of organizational policies that are expressed in many different manners. For example, network routers often employ firewall rule sets to filter traffic. These "rules" use network address, domain and protocol information to determine whether users can be granted access or not. As a system becomes more complex, the number of rules that needs to be defined increases. Without a formal framework it is extremely difficult to define these rules in a rigorous manner and analyze the same to determine if there are conflicts among the rules. Moreover, history-based constraints and domain constraints are very difficult to express in logic as required by rule-based systems. This makes the applicability of rule-based access control very limited. However, rule-based access control appears to be a useful model to implement dynamic policies. Thus, combining rule-based system with other models may be useful for cloud environments.

4 Trust-Based Access Control Model for Secure Cloud Computing

We have discussed a number of novel access control models that have been proposed to address the requirements of different environments. However, there is not one single model that satisfies all the requirements of clouds. In addition, most models other than the trust-based models and the risk-based models do not explicitly support a method for expressing the basis for access control decision-making, which is one of the important features needed for facilitating collaborations in an ad hoc manner or dynamically adapting access control decisions in order to support elasticity and measured service. To support collaborations in the real world, decisions are frequently made to share sensitive information under less than ideal security conditions. Such decisions are made when the benefit of sharing the

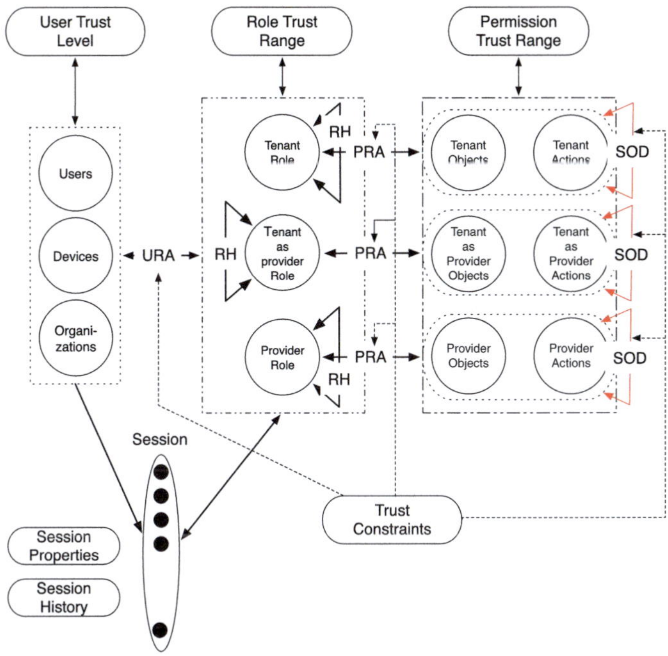

Fig. 4 Schematic overview of proposed trust-based access control model for clouds

information outweighs the risk of sharing the same. The basis for making these decisions is an understanding of the requirements of the mission, the potential security risks, and a knowledge of the effects of similar decisions in the past. These considerations motivate us to propose a trust-based approach to access control. Our proposed trust-based access control model is defined in terms of a set of elements and relations among those elements with trust-based constraints defined on these relations. Our approach is similar to an approach of Chen and Crampton [15].

Figure 4 provides a schematic overview of the proposed model. In the figure, there are entities termed *users*, *roles*, *objects*, and *actions*, which are described as follows.

- *Users*: The users are depicted in Fig. 4 as falling into three categories: (1) human users, (2) devices, and (3) organizations.
- *Roles*: The concept of role is same as in the RBAC model. A role is a job function with some associated semantics regarding the responsibilities conferred to the user. A user assigned to a role specifies the operational needs of the cloud. As shown in Fig. 4, the set of roles is further subdivided into three disjoint subsets as follows:

  1. TENANT-ROLES: The set of roles that are relevant for receiving services.
  2. TENANT-AS-PROVIDER-ROLES: The set of roles that are relevant for receiving services as well as offering other services.
  3. PROVIDER-ROLES: The set of roles that are relevant for providing services.

- *Objects*: An object is a data resource as well as a system resource. As indicated in Fig. 4, the set of objects is partitioned into three subsets:
  1. TENANT-OBJECTS: The set of objects that are accessed when some services are needed.
  2. TENANT-AS-PROVIDER-OBJECTS: The set of objects that are accessed for receiving as well as providing services.
  3. PROVIDER-OBJECTS: The set of objects that are accessed purely for providing services.
- *Action*: An action is an executable image of a program that operates on some object. As indicated in Fig. 4, the set of actions is partitioned into three subsets:
  1. TENANT-ACTIONS: The set of actions that act on objects from the set TENANT-OBJECTS.
  2. TENANT-AS-PROVIDER-ACTIONS: The set of actions that act on objects from the set TENANT-AS-PROVIDER-OBJECTS.
  3. PROVIDER-ACTIONS: The set of actions that act on objects from the set PROVIDER-OBJECTS.

A *permission* is an authorization to perform a certain task within the system. A permission is assigned to a role and consists of an (object, action) pair. The set of permissions is partitioned into three subsets as follows:

1. TENANT-PERMISSIONS: These permissions are assigned to the roles in TENANT-ROLES and are (tenant-object,tenant-action) pairs.
2. TENANT-AS-PROVIDER-PERMISSIONS: These permissions are assigned to the roles in TENANT-AS-PROVIDER-ROLES and are (tenant-as-provider-object, tenant-as-provider-action) pairs.
3. PROVIDER-PERMISSIONS: These permissions are assigned to the roles in PROVIDER–ROLES and are (provider-object, provider-action) pairs.

Figure 4 refers to the concepts of "user trust level", "trust range", and "trust constraints". These are explained as follows:

- *User trust level*: A user u associated with a role r will have a trust level $t(u,r)$, which is a real number between 0 and 1. The larger the value of the trust level, the less risky it is to grant access to the user.
- *Trust range*: A trust range is a subinterval of $[0,1]$ whose right endpoint is 1. Each role r is assigned a trust range $[t_r,1]$, and each permission p is assigned a trust range $[t_p,1]$.
- *Trust constraints*: For a user u in role r to acquire permission p, we impose the trust constraints that trust level $t(u,r)$ belongs to each of the trust ranges $[t_r,1]$ and $[t_p,1]$.

The remaining aspects of the Fig. 4 trust-based access control model that need to be explained are URA, PRA, RH and SoD:

- *User-role assignment (URA)*: URL is a set of user-role pairs that serves to identify what roles different users can potentially play in the cloud environment.

- *Permission-role assignment (PRA)*: PRL is a set of permission-role pairs that serves to identify the permissions for which the roles are authorized.
- *Role hierarchy (RH)*: RH is a partial order on the set of roles allowing for easy administration of roles by reducing the number of explicit assignments in the URA and PRA relations.
- *Separation of Duty (SoD)*: SoD is a set of SoD constraints for the model that serves either to guarantee that no user can be assigned to two conflicting roles, or to guarantee that no role can be assigned two conflicting permissions.

Once the users, roles, objects, actions, permissions, and trust levels/ranges are specified for the trust-based access control model of Fig. 4, there are still many choices for the final trust-based access control model that is designed, due to the flexibility in selecting the four entities URA, PRA, RH, SoD. In a trust-based access control model, a decision to allow an access request is made if all of the following three conditions are satisfied: (1) the relevant role is authorized for the requested permission; (2) the user can activate the role; and (3) the user is authorized for the permission; The particular selection of URA, PRA, RH, SoD that is made by the designer would depend upon what level of control the designer wishes the designed trust-based access control model to exert, that is, it would depend upon how restrictively one wishes to control the circumstances under which (1)–(3) are jointly valid.

For the purposes of this expository chapter, we have given here only a high-level description of our proposed trust-based access control model. We shall give a more detailed description in subsequent work, where we employ an *authorization graph* whose vertices are the users, roles, and permissions, and whose edges are certain pairs of vertices. Semantics can be given based on the authorization graph via which the constraints inherent in our proposed trust-based access control model can be managed efficiently. There are the trust constraints and also the constraints inherent in the choices of URA, PRA, RH, SoD.

5 Conclusion

In this chapter, we discussed issues related to access control in cloud computing. Cloud computing provides on-demand access to a shared pool of infrastructural assets across multiple clients, thereby reducing the need to maintain reserve capacity by each individual client. Access control in these environments, although not fundamentally different from other computing environments, poses unique challenges.

We began by identifying the various authorization and enforcement requirements for cloud systems. We observed that access control models need to strongly support delegation, SoD, and an ability to dynamically adapt access control decisions at run-time because of dynamic cloud environments such as the multi-tenancy, elasticity, massive scalability, and dynamicity properties of cloud systems. Moreover, proper access control cannot rely on an ability to authenticate a user because many situations arise in cloud systems that involve interacting with unknown users. Thus, traditional access control models such as DAC, MAC or RBAC should be precluded or enhanced for cloud computing environments. Next, we looked into some of the newer, more well-known access control models to determine suitability for clouds. We reviewed extensions to the RBAC model, the ABAC model, Risk-adaptable access control models, Trust-based models, History-based models, and the Break-glass models. We observed that although these newer models offer many novel features, none of them are completely appropriate for the dynamic cloud environments.

There is a need to develop trust-based access control models that are adaptable to clouds. As a step in this direction, we propose a new trust-based access control model that extends the RBAC model so as to make use of the most powerful features of RBAC that have made the model the de-facto standard for access control. Further development of trust-based models is needed to support delegation, to incorporate contextual information, and to incorporate more types of constraints that are suitable for secure cloud computing.

Acknowledgements This material is based upon work partially supported by the Air Force Office of Scientific Research (AFOSR)/the Air Force Research Laboratory (AFRL) Visiting Faculty Research Program (VFRP) extension grant LRIR 11RI01COR. The authors would like to thank Mr. John Graniero, AFRL Information Institute Director, for support for this research and the CyberBAT team members for their suggestions and comments. The views and conclusions contained in this document are those of the authors and should not be automatically interpreted as representing official policies, either expressed or implied, of the Air Force Research Laboratory or other federal government agencies.

References

1. Abadi, M., Fournet, C.: Access control based on execution history. In: Proceedings of the 10th Annual Network and Distributed System Security Symposium, NDSS'03, San Diego. The Internet Society (2003)
2. Bauer, L., Schneider, M.A., Felten, E.W.: A general and flexible access-control system for the web. In: Proceedings of the 11th USENIX Security Symposium, San Francisco, pp. 93–108. USENIX Association, Berkeley (2002)
3. Bertino, E., Bonatti, P.A., Ferrari, E.: TRBAC: a temporal role-based access control model. ACM Trans. Inf. Syst. Secur. **4**(3), 191–233 (2001). doi:10.1145/501978.501979
4. Bhatti, R., Joshi, J., Bertino, E., Ghafoor, A.: Access control in dynamic XM-based web-services with X-RBAC. In: Proceedings of the 1st International Conference on Web Services, San Diego, pp. 243–249. CSREA Press (2003)
5. Bhatti, R., Bertino, E., Ghafoor, A.: A trust-based context-aware access control model for Web-services. Distrib. Parallel Databases **18**(1), 83–105 (2005). doi:10.1007/s10619-005-1075-7
6. Blaze, M., Feigenbaum, J., Lacy, J.: Decentralized trust management. In: Proceedings of the 17th IEEE Symposium on Security and Privacy, SP'96, Oakland, pp. 164–173. IEEE Computer Society, Washington, DC (1996)
7. Blaze, M., Feigenbaum, J., Ioannidis, J., Keromytis, A.: ietf.org, the KeyNote trust management system (version 2). http://goo.gl/Bpfn0 (1999)

8. Bobba, R., Fatemieh, O., Khan, F., Gunter, C.A., Khurana, H.: Using attribute-based access control to enable attribute-based messaging. In: Proceedings of the 22nd Annual Computer Security Applications Conference, ACSAC'06, Miami Beach, pp. 403–413. IEEE Computer Society, Washington, DC (2006). doi:10.1109/ACSAC.2006.53
9. Bonatti, P.A., Samarati, P.: A uniform framework for regulating service access and information release on the web. J. Comput. Secur. **10**(3), 241–271 (2002)
10. Braynov, S., Sandholm, T.: Trust revelation in multiagent interaction. In: CHI 2002 Workshop on the Philosophy and Design of Socially Adept Technologies, Minneapolis, pp. 57–60 (2002)
11. Brucker, A.D., Petritsch, H.: Extending access control models with break-glass. In: Proceedings of the 14th ACM Symposium on Access Control Models and Technologies, SACMAT'09, Stresa, pp. 197–206. ACM, New York (2009). doi:10.1145/1542207.1542239
12. Chadwick, D.W., Otenko, A., Ball, E.: Role-based access control with X.509 attribute certificates. IEEE Internet Comput. **7**(2), 62–69 (2003). doi:10.1109/MIC.2003.1189190
13. Chakraborty, S., Ray, I.: TrustBAC: integrating trust relationships into the RBAC model for access control in open systems. In: Proceedings of the 11th ACM Symposium on Access Control Models and Technologies, SACMAT'06, Lake Tahoe, pp. 49–58. ACM, New York (2006). doi:10.1145/1133058.1133067
14. Chandran, S.M., Joshi, J.B.D.: LoT-RBAC: a location and time-based RBAC model. In: Proceedings of the 6th International Conference on Web Information Systems Engineering, WISE'05, New York, pp. 361–375. Springer, Berlin/Heidelberg (2005). doi:10.1007/11581062_27
15. Chen, L., Crampton, J.: On spatio-temporal constraints and inheritance in role-based access control. In: Proceedings of the 2008 ACM Symposium on Information, Computer and Communications Security, ASIACCS'08, Tokyo, pp. 205–216. ACM, New York (2008). doi:10.1145/1368310.1368341
16. Cheng, P.C., Rohatgi, P., Keser, C., Karger, P.A., Wagner, G.M., Reninger, A.S.: Fuzzy multilevel security: an experiment on quantified risk-adaptive access control. In: Proceedings of the 2007 IEEE Symposium on Security and Privacy, SP'07, Oakland, pp. 222–230. IEEE Computer Society, Washington, DC (2007). doi:10.1109/SP.2007.21
17. Cohen, E., Thomas, R.K., Winsborough, W., Shands, D.: Models for coalition-based access control (CBAC). In: Proceedings of the 7th ACM Symposium on Access Control Models and Technologies, SACMAT'02, Monterey, pp. 97–106. ACM, New York (2002). doi:10.1145/507711.507727
18. Covington, M.J., Long, W., Srinivasan, S., Dev, A.K., Ahamad, M., Abowd, G.D.: Securing context-aware applications using environment roles. In: Proceedings of the 6th ACM Symposium on Access Control Models and Technologies, SACMAT'01, Chantilly, pp. 10–20. ACM, New York (2001). doi:10.1145/373256.373258
19. Covington, M.J., Fogla, P., Zhan, Z., Ahamad, M.: A context-aware security architecture for emerging applications. In: Proceedings of the 18th Annual Computer Security Applications Conference, ACSAC'02, Las Vegas. IEEE Computer Society, Washington, DC (2002)
20. Coyne, E.J.: Role engineering. In: Proceedings of the 1st ACM Workshop on Role-Based Access Control, RBAC'95, Gaithersburg. ACM, New York (1996). doi:10.1145/270152.270159
21. Damiani, M.L., Bertino, E., Catania, B., Perlasca, P.: GEO-RBAC: a spatially aware RBAC. ACM Trans. Inf. Syst. Secur. **10**(1) (2007). doi:10.1145/1210263.1210265
22. Edjlali, G., Acharya, A., Chaudhary, V.: History-based access control for mobile code. In: Proceedings of the 5th ACM Conference on Computer and Communications Security, CCS'98, San Francisco, pp. 38–48. ACM, New York (1998). doi:10.1145/288090.288102
23. fas.org: DOD 5200-28-STD: trusted computer system evaluation criteria. http://goo.gl/L0fUw (1985)
24. Ferraiolo, D., Kuhn, R.: Role-based access controls. In: Proceedings of the 15th NIST-NCSC National Computer Security Conference, Baltimore, pp. 554–563 (1992)
25. Ferraiolo, D.F., Sandhu, R., Gavrila, S., Kuhn, D.R., Chandramouli, R.: Proposed NIST standard for role-based access control. ACM Trans. Inf. Syst. Secur. **4**(3), 224–274 (2001). doi:10.1145/501978.501980

26. Franco, L., Sahama, T., Croll, P.: Security enhanced Linux to enforce mandatory access control in health information systems. In: Proceedings of the 2nd Australasian Workshop on Health Data and Knowledge Management, HDKM'08, Wollongong, pp. 27–33. Australian Computer Society, Inc., Darlinghurst (2008)
27. Frank, M., Basin, D., Buhmann, J.M.: A class of probabilistic models for role engineering. In: Proceedings of the 15th ACM Conference on Computer and Communications Security, CCS'08, Alexandria, pp. 299–310. ACM, New York (2008). doi:10.1145/1455770.1455809
28. Georgiadis, C.K., Mavridis, I., Pangalos, G., Thomas, R.K.: Flexible team-based access control using contexts. In: Proceedings of the 6th ACM Symposium on Access Control Models and Technologies, SACMAT'01, Chantilly, pp. 21–27. ACM, New York (2001). doi:10.1145/373256.373259
29. Harrington, A., Jensen, C.: Cryptographic access control in a distributed file system. In: Proceedings of the 8th ACM Symposium on Access Control Models and Technologies, SACMAT'03, Como, pp. 158–165. ACM, New York (2003). doi:10.1145/775412.775432
30. Hu, V., Ferraiolo, D.F., Kuhn, D.R.: Assessment of access control systems. Interagency report 7316, National Institute of Standards and Technology (NIST) (2006)
31. Jin, S., Ahn, J., Cha, S., Huh, J.: Architectural support for secure virtualization under a vulnerable hypervisor. In: Proceedings of the 44th Annual IEEE/ACM International Symposium on Microarchitecture, MICRO'11, Porto Alegre, pp. 272–283. ACM, New York (2011). doi:10.1145/2155620.2155652
32. Joshi, J.B.D., Shafiq, B., Ghafoor, A., Bertino, E.: Dependencies and separation of duty constraints in GTRBAC. In: Proceedings of the Eighth ACM Symposium on Access Control Models and Technologies, SACMAT'03, Como, pp. 51–64. ACM, New York (2003). doi:10.1145/775412.775420
33. Joshi, J.B.D., Bertino, E., Latif, U., Ghafoor, A.: A generalized temporal role-based access control model. IEEE Trans. Knowl. Data Eng. **17**(1), 4–23 (2005). doi:10.1109/TKDE.2005.1
34. Kandala, S., Sandhu, R., Bhamidipati, V.: An attribute based framework for risk-adaptive access control models. In: Proceedings of the 6th International Conference on Availability, Reliability and Security, ARES'11, Vienna, pp. 236–241. IEEE Computer Society, Washington, DC (2011). doi:10.1109/ARES.2011.41
35. Kumaraswamy, S., Lakshminarayanan, S., Reiter, M., Stein, J., Wilson, Y.: cloudsecurityalliance.org, domain 12: guidance for identity & access management v2.1. http://goo.gl/Nnjg1 (2010)
36. Li, N., Mitchell, J.C.: DATALOG with constraints: a foundation for trust management languages. In: Proceedings of the 5th International Symposium on Practical Aspects of Declarative Languages, PADL'03, New Orleans, pp. 58–73. Springer, London (2003)
37. Li, N., Mitchell, J.C.: RT: a role-based trust management framework. In: Proceedings of the 3rd DARPA Information Survivability Conference and Exposition, Washington DC, vol. 1, pp. 201–212 (2003)
38. Li, N., Mitchell, J.C., Winsborough, W.H.: Design of a role-based trust-management framework. In: Abadi, M., Bellovin, S. (eds.) Proceedings of the 23rd IEEE Symposium on Security and Privacy, SP'02, Oakland, pp. 114–130. IEEE Computer Society, Washington, DC (2002)
39. Marinovic, S., Craven, R., Ma, J., Dulay, N.: Rumpole: a flexible break-glass access control model. In: Proceedings of the 16th ACM Symposium on Access Control Models and Technologies, SACMAT'11, Innsbruck, pp. 73–82. ACM, New York (2011). doi:10.1145/1998441.1998453
40. McGraw, R.W.: Risk-adaptable access control. In: Proceedings of the 1st NIST Privilege Management Workshop, Gaithersburg (2009)
41. Mell, P., Grance, T.: nist.gov, NIST special publication 800-145: the NIST definition of cloud computing. http://goo.gl/eBGBk (2011)
42. oasis-open.org: XACML language proposal, version 0.8. http://goo.gl/CXnLq (2002)
43. Ray, I., Toahchoodee, M.: A spatio-temporal role-based access control model. In: The 21st Annual IFIP TC-11 WG 11.3 Working Conference on Data and Applications Security, Redondo Beach, pp. 211–226. Springer, Berlin/Heidelberg (2007)

44. Ray, I., Toahchoodee, M.: A spatio-temporal access control model supporting delegation for pervasive computing applications. In: Proceedings of the 5th International Conference on Trust, Privacy and Security in Digital Business, TrustBus'08, Turin, pp. 48–58. Springer, Berlin/Heidelberg (2008). doi:10.1007/978-3-540-85735-8_6
45. Ray, I., Kumar, M., Yu, L.: LRBAC: a location-aware role-based access control model. In: Proceedings of the 2nd International Conference on Information Systems Security, ICISS'06, Kolkata, pp. 147–161. Springer, Berlin/Heidelberg (2006). doi:10.1007/11961635_10
46. Ristenpart, T., Tromer, E., Shacham, H., Savage, S.: Hey, you, get off of my cloud: exploring information leakage in third-party compute clouds. In: Proceedings of the 16th ACM Conference on Computer and Communications Security, CCS'09, Chicago, pp. 199–212. ACM, New York (2009). doi:10.1145/1653662.1653687
47. Ruj, S., Stojmenovic, M., Nayak, A.: Privacy preserving access control with authentication for securing data in clouds. In: Proceedings of the 12th IEEE/ACM International Symposium on Cluster, Cloud and Grid Computing, CCGRID'12, Ottawa, pp. 556–563. IEEE Computer Society, Washington, DC (2012). doi:10.1109/CCGrid.2012.92
48. Sampemane, G., Naldurg, P., Campbell, R.H.: Access control for active spaces. In: Proceedings of the 18th Annual Computer Security Applications Conference, ACSAC'02, Las Vegas, pp. 343–352. IEEE Computer Society, Washington, DC (2002)
49. Samuel, A., Ghafoor, A., Bertino, E.: A framework for specification and verification of generalized spatio-temporal role-based access control model. Technical report CERIAS TR 2007–08, Purdue University (2007)
50. Sandhu, R.S., Coyne, E.J., Feinstein, H.L., Youman, C.E.: Role-based access control models. IEEE Comput. **29**(2), 38–47 (1996). doi:10.1109/2.485845
51. Tassanaviboon, A., Gong, G.: OAuth and ABE based authorization in semi-trusted cloud computing: aauth. In: Proceedings of the 2nd International Workshop on Data Intensive Computing in the Clouds, DataCloud-SC'11, Seattle, pp. 41–50. ACM, New York (2011). doi:10.1145/2087522.2087531
52. Thomas, R.K.: Team-based access control (TMAC): a primitive for applying role-based access controls in collaborative environments. In: Proceedings of the 2nd ACM Workshop on Role-Based Access Control, RBAC'97, Fairfax, pp. 13–19. ACM, New York (1997). doi:10.1145/266741.266748
53. Toahchoodee, M., Ray, I.: On the formal analysis of a spatio-temporal role-based access control model. In: Proceeedings of the 22nd Annual IFIP WG 11.3 Working Conference on Data and Applications Security, London, pp. 17–32. Springer, Berlin/Heidelberg (2008). doi:10.1007/978-3-540-70567-3_2
54. van den Akker, T., Snell, Q.O., Clement, M.J.: The YGuard access control model: set-based access control. In: Proceedings of the 6th ACM Symposium on Access Control Models and Technologies, SACMAT'01, Chantilly, pp. 75–84. ACM, New York (2001). doi:10.1145/373256.373268
55. Wang, Q., Jin, H.: Data leakage mitigation for discretionary access control in collaboration clouds. In: Proceedings of the 16th ACM Symposium on Access Control Models and Technologies, SACMAT'11, Innsbruck, pp. 103–112. ACM, New York (2011). doi:10.1145/1998441.1998457
56. Ya-Jun, G., Fan, H., Qing-Guo, Z., Rong, L.: An access control model for ubiquitous computing application. In: Proceedings of the 2nd International Conference on Mobile Technology, Applications and Systems, Guangzhou, pp. 1–6 (2005)

Assured Information Sharing (AIS) Using Private Clouds

Vaibhav Khadilkar, Tyrone Cadenhead, Murat Kantarcioglu, and Bhavani Thuraisingham

Abstract Assured Information Sharing (AIS) is a framework that allows cooperating organizations to share information in a manner that respects the privacy, confidentiality and security of the data of each individual in each organization. In this chapter, we present an overview of AIS by detailing the motivations behind AIS, a goal-oriented architecture for AIS and challenges that must be overcome before the adoption of AIS. In addition, we present historical as well as recent research advances that have been made towards addressing the challenges that lie within an AIS framework. Finally, we describe the details of two novel cloud-based AIS implementations that support the high availability, scalability, agility and efficiency required for realizing the vision of AIS.

1 Introduction

One of the keys to effectively combat terror lies in the ability to share critical information that has been collected by multiple cooperating agencies (national and international). The process of data sharing, in this context, has several real-world implications, the foremost of which is the ability to make proactive decisions that could help save thousands of lives. However, the process must ensure the confidentiality of sensitive information and preserve the privacy of individuals. Additionally, the process must be able to integrate unanticipated variable data sets that arise from the ever-changing conditions in which agencies operate.

The Assured Information Sharing (AIS) framework is envisioned as a solution that addresses various issues that are integral to the process of data sharing. Our objectives in this chapter are to present the basics of an AIS framework as well

V. Khadilkar (✉) • T. Cadenhead • M. Kantarcioglu • B. Thuraisingham
University of Texas at Dallas, Richardson, TX, USA
e-mail: vvk072000@utdallas.edu; thc071000@utdallas.edu;
muratk@utdallas.edu; bxt043000@utdallas.edu

K.J. Han et al. (eds.), *High Performance Cloud Auditing and Applications*,
DOI 10.1007/978-1-4614-3296-8_9,

as to give details about contemporary cloud-based AIS implementations. First, we highlight the necessity of an AIS framework and subsequently, we outline the challenges associated with the adoption of an AIS framework. Next, we present the state of the art in the realm of Assured Information Sharing. Finally, we present the details of two pioneering, cloud-based AIS implementations that have extended the boundaries of the field of assured information sharing.

2 Assured Information Sharing (AIS)

In this section, we first present the requirements and architecture of an AIS framework that meets the DoD AIS requirements. Secondly, we present some of the motivating factors that drive the design of an AIS framework. Finally, we present some of the challenges that must be overcome before reaching our goal of having an effective process for assured information sharing.

2.1 Requirements and Architecture

There are several components that make up an AIS framework. However, the core of an AIS framework is made up of the following two themes that cut across all components:

- *Agile environment*: The ability to rapidly adapt to ever-changing ground conditions, thus ensuring that an implementation can readily integrate data created as a result of unexpected partnerships or events, which can then be quickly converted into actionable intelligence.
- *Security*: The capacity to ensure confidentiality of sensitive information, preserve the privacy of individuals, verify the integrity of data, enforce access control mechanisms when retrieving data as well as controlling how data is used upon release by a user.

We now present two DoD requirements that capture the intuitive concepts involved in the process of assured information sharing:

> Provide the ability to dynamically and securely share information at multiple classification levels among U.S., allied and coalition forces [5, 107].

> Deliver the power of information to ensure mission success through an agile enterprise with freedom of maneuverability across the information environment [38, 107].

As should be evident from the above requirements, the key to an AIS framework is to allow high quality information to be shared, leading to timely decision making. Towards this end, Fig. 1 presents a goal-oriented architectural overview of an AIS framework based on the approaches taken to achieve the information sharing goals outlined in the DOD Information Sharing Strategy [38]. The main component of

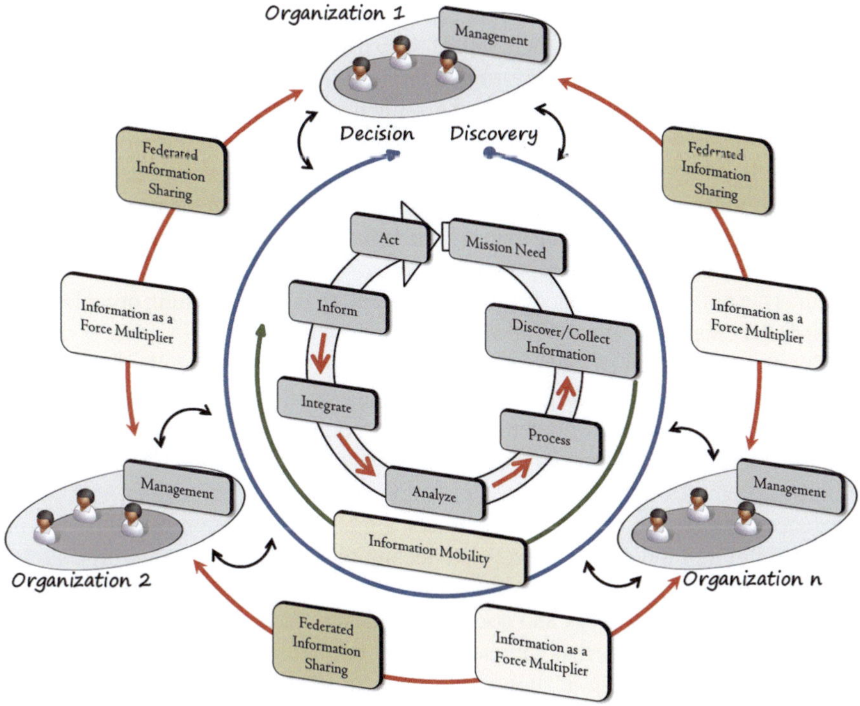

Fig. 1 AIS – Goal-oriented architectural overview

Fig. 1 is the ongoing cycle that defines the set of activities required to make a collaborative mission successful. The cycle itself is made up of several phases, beginning with *Mission Need*, in which a set of requirements for a particular mission is identified collaboratively by the organizations participating in the process of information sharing. This is followed by the *Discover/Collect Information* phase in which co-operating organizations use their ongoing data collection processes to capture data essential to the mission at hand. Then, the *Process* phase performs the task of information assimilation by checking the trustworthiness of data, resolving possible ambiguities in data and structuring data for the next phase. In the *Analyze* phase, the structured data is mined for possible trends or patterns associated with the current mission. After that, the *Integrate* phase links the knowledge extracted from various information sources that is relevant to the objectives of the ongoing mission. The *Inform* phase ensures that the knowledge derived from the integrated sources of information is shared with relevant personnel as well as with organizations that are presently a part of the information sharing process. Finally, the *Act* phase allows the participating organizations to operate on the actionable intelligence produced by the previous phases. A point to note is that the phases from *Process* to *Inform* are provided with a feedback mechanism through which they can advise their respective previous phases of improvements that can be immediately made so that the overall process is better aligned with the goals of the current mission.

The cyclic structure of the information sharing process allows the results of an iteration of the cycle to be used for the refinement of mission objectives, which are then cascaded over to the remaining phases of the next iteration of the cycle. Additionally, this ensures that information sharing becomes an ongoing activity in which various organizations can join/leave the process based on the requirements of a specific mission that they want to carry out. Finally, the phases defined in the cyclic information sharing process represent a *"Discovery to Decision"* continuum that captures the notion of continued collaboration among the participating organizations towards the creation of timely and actionable intelligence [38].

The phases between *Discover/Collect Information* and *Inform* collectively create the concept of *Information Mobility*. This idea captures the readiness required within an organization to address predefined mission requirements as well as unforeseen requirements of unanticipated partners and events. Consequently, information mobility promotes a higher degree of intra- and inter-organizational situational awareness among the cooperating partners. Finally, information mobility can be achieved through the use of appropriate technological tools (viz. enterprise architectures, services and communications infrastructure), trust building/maintenance tools (viz. training exercises and social networking) and security tools (viz. intra- and inter-domain policies that address integrity, confidentiality, etc.).

The outer layer of Fig. 1 shows *n* different organizations participating in the information sharing process. Each of these organizations is an autonomous entity and as such defines its own *Management* structure. This existing structure allows an organization to effectively streamline currently operational budgets towards the task of information sharing. Additionally, an organization should also be provided with incentives in future budgets (through federal funding, for example) to encourage information sharing since this leads to higher quality data that can then be used to make informed decisions. The collaboration between participating organizations also leads to the creation of a *Federated Information Sharing* environment. The federated approach strengthens the relationships between the collaborating organizations and creates an information sharing framework that easily incorporates trust mechanisms, best-of-breed technological tools and compliance procedures. In addition, federation also allows all organizations to support an agile environment that is conducive to rapid integration of unanticipated partners and events. Finally, the federated approach also creates the notion of *Information as a force multiplier*, since when a large amount of information is shared, it can become a force if it is used appropriately while verifying its accuracy, consistency, authority and completeness.

As stated earlier, the two themes of agile environment and security cut across all phases and goals described in Fig. 1. The agility aspect refers to the ability of an organization to dynamically handle the data and requirements of unanticipated partners and events to subsequently produce timely decisions. The security aspect refers to the capability of an AIS framework to ensure that information is protected at all levels (viz. intra- and inter-organization) through the use of appropriate policies and procedures.

2.2 Motivation

In this subsection, we present some of the motivating factors that necessitate the development of an AIS framework. First, we present a motivating scenario in which co-operating agencies would greatly benefit from the use of an AIS framework. This is followed by a general discussion of the basic reasons that compel the immediate development of various implementations of AIS frameworks.

Motivating Scenario As stated earlier, an AIS framework provides a mechanism through which multiple, co-operating agencies are able to dynamically and securely share information.

A relevant use-case where the features of an AIS framework would provide significant value-added benefits pertains to the Distributed Common Ground System (DCGS) [87]. The DCGS is a global, internet-like network where both military and national agencies have access to time sensitive intelligence, surveillance and reconnaissance (ISR) data. The main motivation behind the DCGS project is to provide real-time access to actionable intelligence to the various U.S. combat forces that are deployed throughout the world. Each of the three military departments, namely the Army, the Navy and the Air Force have developed their own versions of the DCGS [1] which are interfaced through the DCGS Integration Backbone (DIB) [42]. The DIB provides the architecture, standards, tools, and documentation for connecting current and future systems to the Global Information Grid (GIG) [39].

An AIS implementation could be built on top of the DIB software stack that encourages co-operating agencies (military and national) to proactively share information through the use of flexible security mechanisms. Additionally, the implementation would support rapid integration of evolving data sources that arise due to the volatile environment in which agencies function. Finally, the implementation would foster trusted relationships between various stakeholders while allowing each of them to operate in an agile enterprise environment that accommodates unforeseen circumstances.

Motivating Factors As illustrated by the use-case we just presented, there are several motivating factors that encourage the immediate development of various AIS implementations:

- *Encourage/Incentivize information sharing*: The various coalition partners are encouraged/incentivized to participate in the process of information sharing, which creates a win-win situation for them, since it provides them with higher quality data that can then be used to build finer data analytics tools (e.g., data mining models, trend analysis and pattern recognition).
- *"Need to Share" over "Need to Know"*: The ability to proactively share information between co-operating agencies through the use of security mechanisms that support the "need to share" philosophy rather than the traditional "need to know" paradigm.
- *Rapid integration of evolving data sources*: The capability of quickly identifying and integrating data from sources that are continuously evolving entities.

- *Promoting trusted relationships between stakeholders*: The promotion of trusted relationships between the different stakeholders ensured through the use of appropriate trust mechanisms (e.g., cryptographic mechanisms, access control and redaction).
- *Agile enterprise environment*: The creation of an agile enterprise environment for each stakeholder thus enabling them to adapt to unexpected partners (e.g., international intelligence agency) and events such as an imminent terrorist threat or activity.

2.3 Challenges

In this subsection, we outline the different challenges associated with the adoption of an AIS framework within an organization. The challenges are based on the touchstones presented in the *DoD Information Sharing Strategy* [38] and are therefore broadly classified into the following five categories:

- *Cultural*: Since the process of information sharing always involves a tradeoff between resource allocation/consumption and security risks, many agencies would be averse to participating in this process. Therefore, it is necessary to change the mindset of an organization from one of information "ownership" to that of information "stewardship". Consequently, a solution that addresses this category would involve the following open-ended list of activities:
  - Inculcate the information sharing methodology into training curricula across the organization.
  - Promote an environment for information sharing throughout the organization through the development/maintenance of a risk-based management approach.
- *Policy*: The existing laws, policies, regulations, and rules within an organization may not readily support the process of information sharing. Therefore, it may be necessary to restructure an organization's internal policies towards the common vision of assured information sharing. As a result, a solution that addresses this category would involve the following open-ended list of activities:
  - Evaluate existing policies within an organization and promote the development/modification of rules that encourage information sharing.
  - Develop techniques that allow sharing information while preserving the confidentiality of sensitive information and respecting the privacy of individuals.
- *Governance*: The management framework currently in place within an organization may not be equipped for creating/sustaining an information sharing environment. Hence, it may be necessary to modify this framework so that various processes/policies associated with information sharing can be monitored effectively. Additionally, a solution that addresses this touchstone must involve the following, not necessarily complete, list of activities:

- Develop a management structure within an organization to govern the various activities associated with the process of information sharing.
- Create/maintain practices within an organization that are aligned with the technical, legal and performance aspects of information sharing.

• *Economics and Resources*: The economics of moving towards an assured infor mation sharing framework may not be feasible for an organization. Consequently, an organization needs to put in place budgetary planning, management and performance methodologies to ensure that the information sharing process is cost-effective while providing a significant return-on-investment. Some of the activities that are a part of a solution that addresses this challenge can be summarized as follows:
  - Inculcate the vision of information sharing into the financial processes within an organization.
  - Develop/maintain practices as a part of the budgeting process to ensure a cost-effective information sharing process.

• *Technology Infrastructure*: The current technological infrastructure within an organization may not be completely equipped to handle the various components of an AIS framework. Therefore, an organization may require the integration of newly developed technologies with their existing IT environment to support an information sharing framework. A solution that addresses this challenge would be composed of the following list of activities:
  - Ensure that newly developed information sharing technologies are compliant with existing federal IT design and technology standards.
  - Enforce the implementation of federally approved security mechanisms within a technology/product/service.

3 State of the Art in AIS

There is a significant body of research work that addresses the challenges associated with AIS that we outlined in the previous section. In this section, we present a summarization of this work categorized on the basis of different research themes, that cut across the challenge areas we identified earlier. Note that the summarization we present is in no way complete; we have simply tried to focus on research work that can provide concrete solutions to challenges faced in the adoption of AIS.

3.1 Access Control Models

The phenomenon of access control is pervasive, in that it can be found in everyday tasks such as gaining access to a room/building using a keycard or using a personal identification number (PIN) when withdrawing money from an automatic teller machine (ATM) as well as in more complex computer-based systems such as allowing only authorized users to access a particular file within a file system. Moreover, it is critical for organizations to put into place appropriate access control mechanisms without which the disclosure of sensitive data could lead to loss of competitive advantage or damaging lawsuits. We first present a historical perspective on access control models surveyed in [17]. In addition, we briefly mention some of the languages that have been developed for the construction and enforcement of access control policies. However, existing access control models and languages are not directly equipped to model the complex policies that arise when collaborating organizations want to share information securely. Therefore, in the second part of this subsection, we present recent research efforts that address this discrepancy.

3.1.1 A Brief History

The initial efforts of researchers in the area of access control models focused around the development of classes of models based on *discretionary access control policies* and *mandatory access control policies* for relational database systems. Discretionary access control models, mandatory access control models and access control languages will be discussed below:

Discretionary Access Control Models In models based on discretionary access control (DAC) policies, the access of a user to a particular resource is based on their identity as well as the authorization rules implemented in the system. Furthermore, these models are discretionary since they allow a user to grant authorizations on data items that they own to other users. One of the first implementations of discretionary access control models was made in the context of relational databases for the System R DBMS [11, 47, 56]. In this model, the tables and views in the database form the objects that need to be protected while the SQL functions that can be executed on the tables form the set of permissible operations for users. The enforcement of policies in System R is achieved by looking at the ownership of a table as well as the possible chain of privileges that may be delegated to allow other users to access that table. This basic access control model was further enhanced with various features that allows it to better handle modern application requirements through the use of more expressive authorization languages. These enhancements include:

- *Negative authorizations*: In this extension, a user can explicitly "deny" other users the right to perform certain operations on a table owned by that user [18].
- *Temporal authorizations*: This extension allows users to access tables/views only for a specified period of time after which their authorizations are revoked [19].
- *Context-aware authorizations*: In this model, user queries against tables/views are rewritten to incorporate contextual information (e.g., time, location and user's organizational position) that is part of the access control policies [13, 30].
- *Content-Based Access Control*: This extension makes use of *views*, which can be defined as queries on tables, for the purpose of defining access control policies

for users [16]. Then, a query on a view is rewritten by appending to the original `WHERE` clause, the `WHERE` clause of the view definition query. The effect of this mechanism is that tuples that do not match the predicates in the view definition query are filtered out, thus providing an efficient way to limit access to data.

- *RBAC Model*: This model uses the notion of a "role" as a bridge between users and authorizations. A role incorporates all the necessary authorizations needed to perform a specific function within an organization. In addition, users are assigned to various roles thus allowing them to gain authorizations associated with those roles and thereby performing the required operations on protected resources [48, 105]. The RBAC model has been extended further along the following dimensions: the development of temporal constraints in RBAC, called the TRBAC model [20, 60], the development of administration models [35, 69, 73], the construction of security analysis techniques [83], and more recently, the development of RBAC in a semantically rich language [50].

Mandatory Access Control Models In models based on mandatory access control (MAC) policies, a user's access to resources is determined based on a predefined classification of users and resources in the system. In particular, the classification is made on the basis of a set of *access classes*, that are defined over all users and resources in the system. Then, a user can access a specific resource only if an ordered relationship exists between the access classes for the user and the resource. A well known example of this mechanism is based on the Bell and LaPadula model [15] wherein a user can only read resources whose access classes are dominated by the user's access class (*No read-up*) and a user can only write resources whose access classes dominate the access class of the user (*No write-down*). There has also been a lot of research work devoted to addressing some of the issues posed by MAC policies in the context of multilevel secure DBMS's [21, 92, 94].

Access Control Languages The general goals for an access control language are to enable the creation and enforcement of access control policies. At a general level, these languages make assertions about users and their permissible operations [103]. Additionally, languages have been defined for making authorization decisions for policies that have been combined from various sources [23] as well as for supporting trust management [22, 108, 124]. Furthermore, since the advent of Extensible Markup Language (XML) and its acceptance as a de facto standard for information exchange, a number of access control languages based on XML have been proposed [36, 78]. A universally accepted XML-based access control language is XACML [96] that allows the definition of policies in XML for resources that may also be expressed in XML. There have also been access control languages that are designed in a logic-based framework [82, 84, 85]. The additional expressive power and formal, verifiable methodology offered by logic-based languages are particularly useful in the context of access control. Finally, access control languages have also been defined in the context of Semantic Web languages (e.g., "Rei" [61] and "KAoS" [115]). The Semantic Web languages are based on description logics, which are a decidable subset of first order logic, and hence provide benefits that are similar to logic-based languages.

3.1.2 Recent Results for AIS

As stated earlier, existing access control models cannot be directly adapted for the purposes of assured information sharing. Therefore, new models, implementations and mechanisms need to be developed to address this shortcoming.

An example of a family of access control models that is tailored to suit the requirements of information sharing is Group-Centric Secure Information Sharing (g-SIS) [74,76,77]. The family of models in g-SIS is based on the notion of bringing users and resources together into a group for the purposes of information sharing. In other words, this means that users and resources must be present in the system simultaneously for the users to be able to access resources. In addition, the family of g-SIS models is based on the following two principles:

- *Share but differentiate*: The sharing aspect is achieved through users joining a group and then adding information to that group. However, the access that a user is granted to resources is based on several factors such as, the time at which a user is added to the group relative to the time at which the information was posted to the group as well as other such user-configurable attributes.
- *Multiple groups*: This principle refers to the notion that various types of relationships can hold between different groups (e.g., hierarchical, mutual exclusion, etc.) with a possibly overlapping set of users.

Research on g-SIS has employed linear temporal logic (LTL) [90] to develop a formal model for g-SIS and its core properties. In addition, researchers [75] developed g-SIS model extensions that include a "stateful" authorization policy specification and enforcement of g-SIS for group-centric secure information sharing in order to be more secure and agile than traditional access control techniques [106].

A separate line of research work has focused on the development of a comprehensive "**E**nvironment for **X**ACML policy **A**nalysis and **M**anagement (EXAM)" [86, 100] in the context of XACML policies. This work is motivated by the requirements of policy integration and interoperability that arise when organizations with widely varying access control policies want to collaborate. EXAM provides a variety of policy analysis techniques where each approach is implemented through *analysis queries*. An analysis query is a function that allows a user who has designed or deployed a policy to validate the accuracy of the properties of that policy. Additionally, EXAM provides a number of analysis queries for the single as well as multiple policy scenarios. Furthermore, EXAM also allows users to combine analysis queries to conduct arbitrarily complex analyses. EXAM also includes a policy similarity analyzer that implements the strategies necessary for the execution of analysis queries, and is based on the Multi-Terminal Binary Decision Diagram (MTBDD) [52] and SAT-solver techniques [4, 12, 52].

A separate direction of research has focused on the development of access control languages based on semantically rich languages such as Resource Description Framework (RDF) [120] and OWL [121] that are more expressive than XML-based languages such as XACML. In [50], the authors present R*OWL*BAC, a language that uses OWL ontologies to model the various components of RBAC and furthermore, they also show how these ontologies can be used to capture and enforce access control policies. In [49], the authors define XACML+OWL, a framework that integrates OWL ontologies with XACML policies for supporting RBAC. The rationale is to model role hierarchies and constraints using an OWL ontology and to design authorization policies using XACML. To achieve this goal, XACML+OWL defines new functions that extend traditional XACML policies with reasoning mechanisms. Furthermore, XACML's reference architecture and data-flow, which is used for making authorization decisions, is modified to include the newly defined functions.

A language that is tailored to suit the requirements (e.g., policy enforcement based on an aggregation of applicable policies and redaction policies) of access control in provenance was defined in [93]. The language uses an XML syntax and specifies a variety of tags (e.g., policy, target, condition and effect) to capture various access control use-cases that arise in the domain of provenance. However, this language is not able to capture resources with arbitrary path lengths that occur within a provenance graph. Therefore, a resource to be protected must be identified beforehand, rather than being passed as a parameter at runtime. The task of identifying resources beforehand will be infeasible if there is an extremely large number of resources in a provenance graph. An extension of the original language that addresses this drawback is given in [26], where the authors make use of regular expressions to define resources that need to be protected. The use of regular expressions allows resources with arbitrary path lengths to be defined and used at runtime rather than having to create resources beforehand. Finally, in [25], the authors extend the notion of redaction to provenance graphs. They use a graph grammar approach [34,45,102] that rewrites a high-level specification of a redaction policy into a graph grammar rule that transforms the original provenance graph into a redacted provenance graph.

3.2 Privacy-Preserving Data Management

In today's digital age, data has become a priceless commodity. There are a multitude of organizations that collect information about people that is then analyzed for widely varying purposes such as targeted advertising, medical research, and personalized recommendations. Additionally, organizations may share data with other organizations or they may allow access to their data. Thus, a large number of data sources exist, which could be analyzed to infer information that is private to a person or an organization. In the context of AIS, privacy is an even bigger concern since information sharing forms the backbone of an AIS framework. Therefore, it is necessary to have in place appropriate safeguards that preserve the privacy of individuals when their data is shared between cooperating organizations. In this subsection, we first present a brief overview of existing privacy-preserving data management techniques available in the literature [89]. After that, we describe some recent research efforts that aim to address privacy issues in the context of AIS.

3.2.1 A Brief History

We discuss several techniques relevant to AIS that have been introduced in the literature to address the issue of preserving data privacy.

One line of research focuses on preserving privacy when data is released to third parties. If the data is released as is, the organization that owns the data cannot enforce any control on how the data will be used by the receiving organization, since they lose control of the data once it has been released. A solution to this problem is to modify the data before it is released so that no private information can be inferred by the receiving organization. This solution can be implemented through the process of *data anonymization*. There are several techniques to perform data anonymization [88,104,111,122] that overcome re-identification attacks [111]. Although personally identifiable information (e.g., names and social security numbers) may be removed from a database, an attacker might be able to infer private information by using other attributes from the released data (e.g., zip code, birth date and sex) in conjunction with publicly available data sources (e.g., voter registration lists and online social networks). An anonymity-based technique was developed in [104,111] for thwarting such a re-identification attack. Subsequently, a second technique was developed that provides a higher degree of privacy protection by paying attention to the diversity of data attributes [88, 122].

A second line of research work focuses on preserving privacy in the context of data mining. This work is important because even though private information may be removed from a dataset using some type of data anonymization technique, sophisticated data mining techniques could still be used to get back the private information. All approaches that have been proposed to perform privacy-preserving data mining generally use some form of data perturbation [2, 68]. The challenge for all approaches is to avoid a considerable modification of the data, since otherwise the dataset could become useless. An approach that takes into account an estimate of the data modification error when perturbing the data is given in [97]. A separate method for preserving privacy, based on the notion of data sampling is presented in [32]. In this method, a subset of data is released where the subset is selected in such a way that any inference that is made from the released dataset is highly unlikely to be true. Techniques based on commutative encryption methods [62, 116] have also been developed that allow multiple sites to share information for computing some pre-determined function such that no private information is released from any site.

A separate direction of research focuses on the concept of Hippocratic databases [3, 72]. The idea behind this concept is to have privacy policies and authorizations directly built into a relational database. The policy defines the specific characteristics of attributes in a table while the authorization defines the users authorized to query the table. The architecture also maintains a separate "purpose" column for every table that records the different purposes for which users will use the data. Then, during query processing, the architecture ensures the following: (1) The querying user is authorized to execute the query for the purpose given with the query. (2) The query only accesses those parts of relations that match the intended purpose of the query. (3) During query processing, only tuples for which the purpose attribute matches the purpose of the query are allowed to be accessed by the query. Research on Hippocratic databases was largely theoretical, and the authors proposed several technical challenges that need to be overcome before a complete, concrete implementation is possible.

3.2.2 Recent Results for AIS

In the context of AIS, several of the existing privacy-preserving techniques can be reused based on the specific requirements of the domain in which information is being shared. Additionally, several new techniques have also been developed to address issues that arise with reference to AIS.

An example illustrating if and when organizations should invest in privacy-preserving technologies is provided in [66, 67]. This is an important consideration, since consumers would always want to know the privacy guarantees that an organization is able to provide before using that organization's product/service. In [66], the authors propose a set of guidelines using a novel static game-theoretic framework, based on which an organization can decide if they want to invest in privacy-preserving technologies.

This framework is extended in [67], by computing an organizational project's value under uncertainty because of consumers' nuisance decision-making, and considering the effect that market competition may have on a firm's investment in privacy-preserving technologies.

A separate line of work has focused on applying privacy-preserving techniques, specifically for performing record linkage within the healthcare domain. For example, in [44], the authors conduct a comprehensive evaluation of different privacy-preserving string comparators that are used for performing privacy-preserving record linkage, taking into account correctness of record linkage, computational complexity and security. Their results show that string comparators that use a partition, encode and compare strategy lead to a higher accuracy of record linkage, but are less secure than comparators that use a map and compare strategy. In [46], the authors develop a secure multi-party computation protocol using the Paillier cryptosystem that allows healthcare providers to share data in a way that meets the confidentiality requirements of providers while generating utility from the data being shared. Another research work [80] aims at presenting a formal security analysis of the use of Bloom filter encoding (BFE) techniques for the task of performing record linkage [80]. The authors first introduce an attack on BFE procedures for private record linkage that uses the constraint satisfaction problem (CSP) [24] as its primitive building block. Using this attack as a foundation, the authors propose a rigorous analysis of BFE and furthermore, they provide guidelines on how to mitigate the risks of facing such an attack. Finally, in [43], the authors propose a modified version of an existing privacy-preserving string comparator that is tailored for comparing fields that capture demographics in a medical dataset. Then, they use an existing private record linkage algorithm that uses this modified similarity measure to perform record linkage in a real-world medical dataset.

Finally, there has also been significant new research work in the area of privacy-preserving data mining. In [65], the authors contend that the existing secure protocols used to compute the dot product (used as a sub-protocol in many data mining tasks) are computationally intensive, and therefore, infeasible. They introduce an approximation technique for computing the dot product based on Bloom filters and furthermore, also show how this approximation method can be used as a sub-protocol in a protocol used to mine association rules while preserving privacy. In [64], the authors analyze the tradeoffs between performance and security for existing privacy-preserving distributed data mining techniques in models that assume semi-honest vs. malicious adversaries. To perform their analysis, the authors extend secure sub-protocols from the semi-honest model with threshold decryption and zero knowledge proofs to make them secure in the malicious model. Using these techniques, they then develop specialized algorithms (secure dot product) for protocols in the malicious model. Finally, they discuss how their extensions could be used as a part of larger protocols that perform privacy-preserving k-means clustering and association rule mining. In [117], the authors present a generalized privacy-preserving variant of the well-known ID3 algorithm for vertically partitioned data that is distributed over multiple sites. Their algorithm reveals the basic structure of the decision tree and the site that is responsible for making a decision for a node; they further present the necessary means to make their protocol completely secure along with an analysis of the computational and communications complexity. Finally, in [118], the authors present a privacy-preserving version of the Naïve Bayes classifier [91], a popular classification algorithm that is used as a baseline to measure the performance of other classifiers. They assume a semi-honest model and provide privacy-preserving versions of Naïve Bayes for both horizontally and vertically partitioned data.

3.3 Incentive-Based Information Sharing

In many organizations, there is an urgent need to form coalitions for sharing critical information. However, members of a coalition will share information only in accordance with their own objectives. We survey recent research that addresses the challenge of incentivizing organizations to honestly share information, which is a major challenge that needs to be overcome in order for AIS to be widely adopted. Organizations may need to decide on criteria for joint implementation of an AIS framework for managing, sharing and protecting *critical and sensitive information* within the organizations.

An approach based on an evolutionary game theoretic framework [81] studies the effect of trust and incentive issues existing among cooperating organizations. The paper also explicitly considers the cost of determining whether the information shared among cooperating organizations is accurate, which is a relatively difficult task. Furthermore, the paper tries to establish the potential of enforcing punishments on organizations by isolating them from other organizations in the context of trust computations that measure the level of trust between collaborating organizations.

A separate line of work focuses on applying game-theoretic techniques for performing privacy-preserving data analysis tasks. In [95], the authors propose game-theoretic frameworks to encourage cooperating parties to truthfully share data for the purposes of distributed data mining. Their approaches encourage parties to tell the truth only on the basis of the data mining result without resorting to actually validating the data that the parties have provided. Thus, their frameworks are applicable in cases when data validation is not possible due to privacy concerns. Finally, the authors also prove that their mechanisms are incentive compatible under certain assumptions. In a separate but related research effort, the authors of [63] contend that unless proper incentives are set, privacy-preserving data analysis techniques cannot verify that participating parties have shared information truthfully. Furthermore, they explore what analysis tasks can be implemented such that the participating entities are incentivized to provide the right information. They also provide a theoretical foundation for checking if inputs could be modified for the computation of a given analysis task, and if they can be, then complex protocols need not be designed for such tasks.

3.4 Technological Infrastructure

In the context of AIS, the term "technological infrastructure" explicitly refers to the software ecosystem that allows cooperating agencies to efficiently, economically, flexibly and scalably share information. There are several major components to a concrete implementation of an AIS framework: (1) *User Interface*: This component determines how users will interact with an AIS implementation. (2) *Access Control Interface*: This component specifies how cooperating organizations will be allowed to gain access to data. (3) *Data Interface*: This component provides an interface to the actual information that is shared between organizations. Based on the decision made by the access control component, only authorized users will be able to access the shared data.

In this subsection, we first present a short overview of the available storage system technology that can be used as the backbone of an AIS implementation. This is followed by a description of recent research implementations of AIS frameworks.

3.4.1 A Brief History

The last few decades have seen an extraordinary development of data storage systems. These systems have evolved from simple file-based systems that store data organized as files to relational databases that provide storage of and access to structured data and have now further evolved to the cloud computing paradigm that neither enforces that data be structured nor that data be accessed in a particular fashion. We now present a brief historical perspective of these systems.

File System A file system is a method by which a computer's operating system is able to store, organize and retrieve files. A file system also stores metadata (e.g., modification date and owner) about every file that it manages. A good survey on the history of file systems is given in [101]. One of the earliest file systems was tape-based (DECTape) and stored a very small amount of data per table (184 KB). The next stage of file system development evolved to systems using a file allocation table (FAT) to manage files. A separate file system was developed for the Macintosh called Hierarchical File System (HFS) that provided improvements on earlier file systems by introducing features specific to hard drives. The Unix File System (UFS) introduced concepts such as *inodes* (a data structure that holds file information) and *fsck* (File System Check). The design of UFS inspired many subsequent file systems such as ext2 and ext3. The NTFS (Windows NT File System) provided several new ideas such as the use of B+ trees and fine-grained security using Access Control Lists (ACL). Finally, the MAC OS X brought Apple's old file system MAC OS up to the latest standards by introducing various features such as support for journaling and fine-grained file permissions. Although file systems represent an important first step in the design of storage systems, one would have to develop a multitude of technologies on top of file systems to make them suitable for use in AIS implementations.

Relational Database A relational database is a mechanism that provides an organized collection of data such that the organization conforms to the relational model [33]. A relational database consists of a set of relations where each relation is a set of tuples that have the same attributes. An application can access the data in a relational database by specifying queries that perform operations such as *selecting* a subset of tuples in a relation, *projecting* a subset of all attributes in a relation, *joining* two relations and so on. A relational database provides an efficient way to structure data and obtain access to this data using a query language. Additionally, relational databases also provide *atomicity, consistency, isolation, durability (ACID)* properties that guarantee transactional reliability [55]. A relational database was traditionally stored on a single computer; this barrier was broken down with the advent of distributed databases that allowed data to be stored on physically separate computers that are connected through a network. In addition to advantages provided by standard relational databases, distributed databases offered an increased reliability, scalability, availability and monetary savings. Relational databases provide an effective storage medium for AIS implementations due to their numerous benefits. However, the relational model may not be the best fit for capturing the ever-changing environment in which organizational mission requirements are rapidly changing. Moreover, it will be fairly hard to scale up or down in terms of relational databases as the peak performance demand of AIS implementations fluctuates with changing ground conditions.

Cloud Computing The Cloud Computing paradigm emerged after the seminal work done in the implementation of the Google File System [53] as well as the MapReduce framework [37]. The development of Hadoop and a subsequent list of open source tools opened up the cloud computing ecosystem to a host of new users at virtually no cost. The cloud computing paradigm considerably overlaps with the Software as a Service (SaaS) movement whereby software services are provided to an end user through a web browser, a desktop application or a mobile app. In addition to providing all the advantages of relational databases, cloud computing also allows agility in terms of restructuring an organization's technological infrastructure as peak demand fluctuates with changing conditions. These advantages make cloud computing ideal for being used as the storage backbone of an AIS implementation.

3.4.2 Recent Results for AIS

There are several recent developments in AIS frameworks which are now briefly discussed.

In [112], the authors design and develop an AIS framework for information dissemination during emergencies based on the idea of a Trusted Computing Base (TCB) that also supports the "need to share" paradigm. The idea is to use the TCB to construct a hierarchy of users in the system based on the level of trust for each user. Then, when an emergency occurs, the system disseminates information based on the severity of the emergency and the level of trust in a user.

In [51], the authors propose an information sharing lifecycle (AISL) to realize the information sharing value chain of the DoD [38]. AISL consists of three major phases: (1) *Information discovery and advertising*. (2) *Information acquisition, release and integration*. (3) *Information usage and control*. Each of these phases entail a number of challenges such as, finding out what information can be shared and with whom, in phase (1), verifying if a participating entity really needs the information that is being requested, in phase (2), and ensuring that information is used appropriately once it is released to a party, in phase (3). The authors provide a general set of ideas that span various research themes and lead to a number of challenges within each theme.

In a separate line of research, the authors of [123] propose a Trustworthiness-centric Identity, Usage, Provenance, and Attack Management (TIUPAM) framework. The authors contend that with the shift towards the "need to share" paradigm, it is imperative to build a framework that can manage the trustworthiness of information as well as the risks associated with using or not using the information. Towards this end, they propose a framework that includes the following components: (1) *Identity management*: Allow participating entities to verify the trustworthiness of shared information. (2) *Usage management*: Check whether users are authorized to perform activities. (3) *Provenance management*: Verify the accuracy of information, data sources, people, and so forth. (4) *Attack management*: Handle procedures that need to be followed when the system is attacked or unauthorized activities are detected.

The authors of [113] present a cloud-based framework to achieve assured information sharing. In particular, they use a Hadoop and Hive implementation as the storage subsystem, and XACML to construct access control policies. Additionally, they provide a web-based application that allows authorized users to interact with the system and perform various operations such as data loading and querying as well as policy construction. On the other hand, the authors of [28] present a general-purpose RDF policy engine that provides support for various types of security policies such as access control, redaction and policies for information sharing. In addition, they provide a cloud-based framework that implements the RDF policy engine, thus allowing cooperating agencies to share information. Additional details of both implementations are given in subsequent Sects. 4.3 and 4.4.

4 Cloud-Based AIS

In this section, we focus on "technological infrastructure", which is a critical requirement for the adoption of AIS. Although a number of AIS implementations have been developed, none of these are cloud-based and therefore, they lack the scalability, flexibility and efficiency that is needed for supporting a large user-base that utilizes vast quantities of data. In this section, we present the details of two recent prototype systems that provide cloud-based AIS capabilities, thus addressing the issues that were just outlined. We begin by outlining the need for cloud-based AIS systems. After that, we present an overview of cloud computing and semantic web tools and technologies that are leveraged by both prototypes. Finally, we give a detailed description of the architectures employed by the two prototypes.

4.1 Motivation for Cloud-Based AIS

There is a tremendous need for the development of efficient and scalable AIS tools and technologies for various organizations. The exigence of this need was recently expressed in an NSA document [57] outlining the agency's focus on using a "cloud-centric" approach in the agency's effort towards improving its information sharing capabilities with other intelligence agencies. Similarly, the DoD has endorsed the use of the cloud computing paradigm to meet its vision of "delivering the power of information to ensure mission success through an agile enterprise with freedom of maneuverability across the information environment" [38, 40, 41]. Furthermore, the adoption of cloud computing by various big businesses has validated the possible advantages of efficiency, flexibility, scalability and cost-effectiveness that can be made by embracing this paradigm. Additionally, the SaaS model of cloud computing has made it feasible to conveniently provide AIS as a cloud-based service. Finally, even though a number of tools have been developed for policy-based information sharing [51, 100, 112], none of these tools operate in a cloud environment, thus being unable to address the core requirement of large scale data analysis by a significant number of users. This deficiency necessitates the urgent development of various cloud-based AIS tools and technologies; preliminary steps in that direction have been made through the implementation of two prototype systems that are described in subsequent subsections.

4.2 Overview of Tools and Technologies Used

In this subsection, we provide high-level descriptions of various tools and technologies that have been leveraged in the development of two cloud-based AIS implementations that are presented in subsequent subsections. First, we highlight the various open source cloud computing tools, namely Hadoop, HBase and Hive, that form the basis of both cloud-based AIS implementations. After that, we give high-level details about the different Semantic Web technologies (RDF, Jena, Jena-HBase) that have been used by one of the cloud-based AIS implementations.

4.2.1 Hadoop

Apache Hadoop [6] is an open source software framework that allows batch processing tasks to be performed on vast quantities of data. Hadoop uses the Hadoop Distributed File System (HDFS), a Java-based open source distributed file system based on the Google File System [53], as its underlying storage mechanism. HDFS provides several advantages such as data replication and fault tolerance.

HDFS uses a master/slave architecture that consists of a single namenode process (running on the master node) and several datanode processes (usually one per slave node). The namenode process manages the entire file system by tracking how blocks of individual files are distributed over all datanodes in a cluster. A datanode process runs on every node and is used to keep track of file blocks stored on that node. Hadoop uses the MapReduce paradigm to perform batch processing tasks in a parallelized fashion over data stored in HDFS.

The MapReduce framework consists of a single JobTracker process (running on the master node) and several TaskTracker processes (usually one per slave node). The JobTracker process is tasked with the responsibility of scheduling a MapReduce job among the slave nodes while a TaskTracker process is responsible for controlling the execution of the sub-task given to its hosting slave. A MapReduce job consists of three phases: (1) A *Map* phase in which each slave node performs some computation on the data blocks of the input that it has stored. The output of this phase is a key-value pair based on the computation that is performed. (2) An intermediate *Sort* phase in which the output of the Map phase is sorted based on keys. (3) A *Reduce* phase in which a reducer aggregates various values for a shared key and then further processes them before producing the desired result.

4.2.2 HBase

Apache HBase [7] is a column-oriented store that is modeled after Google's BigTable [31]. HBase allows the creation and storage of very large tables – billions of rows × millions of columns, on clusters made up of commodity hardware. Additionally, HBase provides several advantages such as query predicate push down and optimizations for real time queries. HBase makes use of HDFS as the underlying storage system for its tables. Due to this reason, HBase inherits Hadoop's master/slave configuration whereby its architecture consists of a single HBase master node that coordinates the distribution of data among the HBase region servers (usually one per slave).

A HBase table is composed of column families and rows; these may be physically separated over region servers. A row in a table is defined by a row key and all rows in a table are stored based on a lexicographic sorting of row keys. A column family contains a collection of columns where all columns share the same prefix. For example, the columns person:name and person:age are both columns of the column family person. The colon (:) character delimits a column family from a column. A column family is defined at the time of schema creation whereas columns can be declared on the fly as needed. A cell in a HBase table is identified as a three tuple (row, column, version). The version dimension of a tuple allows a user to store multiple versions of a column value for the same row key value. Finally, HBase requires the periodic use of a "commit" operation to reflect any changes (viz. inserts, deletes or updates) to a table.

4.2.3 Hive

Apache Hive [8] is a data warehousing framework that provides the ability to manage, query and analyze large datasets stored in HDFS or HBase [114]. Hive provides basic tools to perform Extract-Transfer-Load (ETL) operations over data, project structure onto the extracted data and query the structured data using a SQL-like language called HiveQL. HiveQL performs query execution using the MapReduce paradigm while allowing advanced Hadoop programmers to plug in their custom-built MapReduce programs to perform advanced analytics not supported by the language. Some of the design goals of Hive include dynamic scale-out, user defined analytics, fault-tolerance and loose coupling with input formats.

The essence of the Hive framework is the ability to structure data as well as to conduct sophisticated analyses of data using HiveQL. Hive reuses the concepts of databases, tables, partitions and data types defined by relational databases to impart structure to data stored in HDFS. In particular, Hive uses the Data Definition Language (DDL) component of HiveQL for managing these tasks. Additionally, Hive uses the Data Manipulation Language (DML) component of HiveQL to perform different types of operations on data stored in tables or partitions. A few of these operations include: selecting only particular columns in a table, selecting rows in a table that match given filters, and performing an equi-join operation between two tables.

4.2.4 RDF

The Resource Description Framework (RDF) [119] is a language that is used to represent metadata about resources on the World Wide Web (WWW). In addition, RDF can be used to represent information about the resources themselves. The RDF language defines a simple graph data model to capture relationships between resources using the concept of a triple. Each triple is of the form (S,P,O), where S is the *subject*, P is the *predicate*, and O is the *object*; a set of triples forms an RDF graph. A RDF triple asserts that the relationship represented by the *predicate* is held between the *subject* and *object* of the triple.

The RDF data model is extremely flexible since a resource is not constrained based on its type or the properties that can be associated with it. Moreover, RDF uses the open-world assumption that allows anyone to make statements about any resource. The RDF data model is different from other data models in the sense that it does not depend on fixed schemas or type hierarchies.

The main challenge for the RDF data model is to provide a scalable and efficient persistent storage mechanism. A simple approach could be to convert the RDF data into XML using the work available on optimized storage of XML. However, there are several drawbacks to this approach. The first is that several XML serializations are available for the same RDF graph that is ineffective to retrieve complex features. The second is that this conversion process is unable to support advanced features of RDF such as reification and the ability to treat properties as resources. The use of relational or object databases for RDF storage and retrieval may also not be suitable in all cases. This is because the semantics of these data models is different from the open and flexible RDF data model. Towards this end, we present below a storage system for RDF (Jena-HBase) that makes use of the HBase platform that was introduced earlier.

4.2.5 Jena

The Jena [9] framework provides a rich set of features for building and querying RDF graphs. An overview of the Jena architecture is given in [59]. At its core, Jena uses the RDF API to allow users to create and manipulate RDF graphs. At this level, users interact with constructs called `Models` that allow an English-like representation of triples. In addition, the RDF API allows users to read RDF data from external sources (files or URLs) as well as to de-serialize/serialize RDF graphs in the most common RDF syntaxes.

While the `Model` interface provides a high-level abstraction for users, internally an RDF graph is stored in a simpler construct called `Graph`. This allows Jena to use various storage mechanisms such as in-memory, relational database (RDB and SDB) or a custom-built disk-based tuple index (TDB) equivalently. Additionally, the `Graph` interface provides a plug-in point for users to connect their own customized triple stores with Jena.

The Inference API provides the ability to infer additional information that is not stated as a part of a graph through the use of different rulesets (built-in or custom-defined). In addition, the inference API also allows external reasoners to be connected with Jena for performing the inference task.

The SPARQL API allows users to query their underlying RDF graphs through the SPARQL query language. Jena supports both the query and update functionalities of SPARQL and also tracks the revisions and updates in the areas of SPARQL that are currently in development.

The Ontology API provides different methods for interacting with RDF Schema (RDFS) and OWL, the two ontology languages that are richer than RDF. These languages capture the real-world domain that an application tries to model in a logical description.

Finally, Fuseki acts as a data publishing server, thus allowing RDF graphs to be accessed and updated using SPARQL and HTTP.

4.2.6 Jena-HBase

Jena-HBase [54] provides a HBase-backed storage subsystem for use with the Jena framework [70]. Jena-HBase leverages the design architecture of Jena SDB [10] by using the concept of a store to provide data loading and querying capabilities on underlying HBase tables. A store represents a single RDF dataset and can be composed of several RDF graphs, each with its own storage layout. A layout uses several HBase tables with different schemas to store RDF triples; each layout provides a tradeoff in terms of query performance/storage. A distinct advantage of using such a pluggable approach is that it is easy for application programmers to plug in their own storage layouts into Jena-HBase. Additionally, Jena-HBase provides support for other operations such as reification, inference, and SPARQL query processing.

Figure 2 presents an overview of the Jena-HBase architecture. Since Jena-HBase uses the concept of a store, composed of several RDF graphs where each graph can have its own storage layout, all operations on a RDF graph are implicitly converted into operations on the underlying layout. These operations include the following: (1) Formatting a layout (i.e., deleting all triples while preserving the tables in a layout) as represented by the *Formatter* block. (2) Loading-unloading triples into a layout (*Loader* block). (3) Querying a layout for triples that match a set of given triple patterns (S, P, O) using the *Query Runner* block. (4) Additional operations include the following: (i) Maintaining a HBase connection (*Connection* block). (ii) Maintaining configuration information for each RDF graph (*Config* block).

We now give a brief summary of the storage schemas used by the various layouts in Table 1. A more detailed description of each layout detailing the ways in which they handle query processing operations is given in [70].

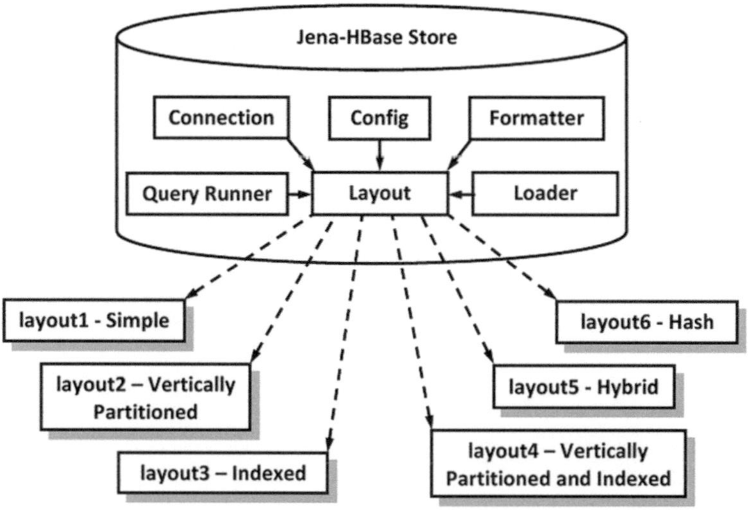

Fig. 2 Jena-HBase architectural overview (Reproduced from Khadilkar et al. [70, 71])

Table 1 Storage schemas for Jena-HBase layouts

Layout type	Storage schema
Simple	Three tables each indexed by subjects, predicates and objects
Vertically partitioned (VP)	For every unique *predicate*, two tables, each indexed by subjects and objects
Indexed	Six tables representing the six possible combinations of a triple namely, SPO, SOP, PSO, POS, OSP and OPS
VP and indexed	VP layout with additional tables for SPO, OSP and OS
Hybrid	Simple + VP layouts
Hash	Hybrid layout with hash values for nodes and a separate table for hash-to-node mappings

4.3 Cloud-Centric Assured Information Sharing System with XACML Policies (CAISS-X)

In this subsection, we present the details of a recent prototype, called Cloud-centric Assured Information Sharing System with XACML policies (CAISS-X), that employs a cloud-centric framework to store and query large amounts of data via a non-cloud policy engine that enforces XACML-based policies [113]. This prototype presents a significant improvement over prior non-cloud AIS efforts, and is based in part upon the recommendations of the Cloud Security Alliance (CSA) for Identity and Access Management [79] as well as previously developed work involving XACML policies [99, 112].

Figure 3 presents an overview of the architecture employed by CAISS-X. The architecture is broadly divided into the Web Application, Access Control, Hive

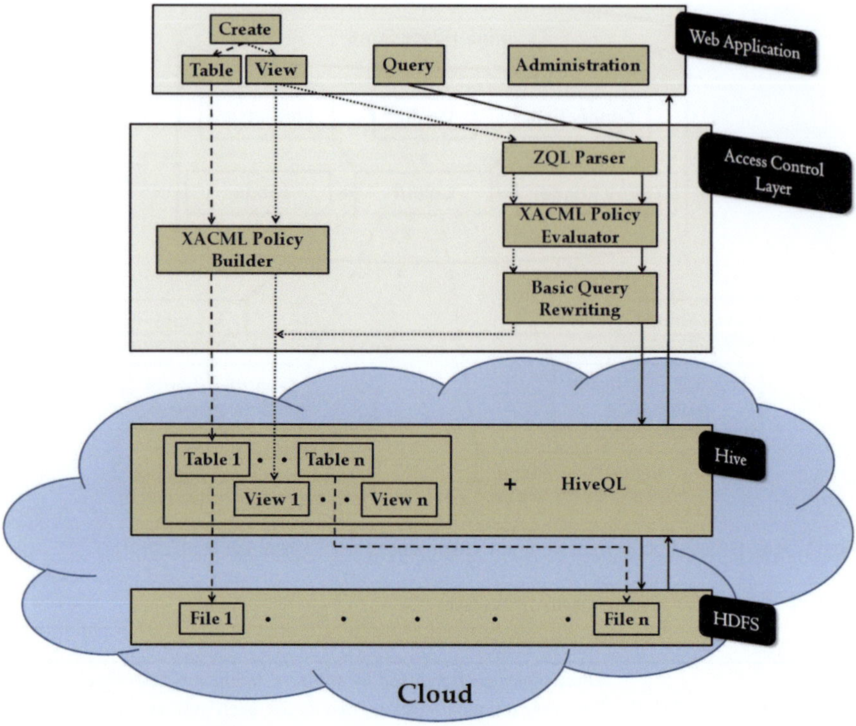

Fig. 3 CAISS-X architecture

and HDFS layers that are further described in detail below. A point to note is that the different line styles used in Fig. 3 denote the flow of control through CAISS-X for various tasks supported by the system. In particular, the *solid line* denotes the flow of control in CAISS-X for the task of querying data stored in existing Hive tables/views, the *dashed line* shows the sequence of steps in CAISS-X for creating/loading a new table in Hive and finally, the *dotted line* represents the procedure followed by CAISS-X when a new view needs to be defined in Hive. We now present the details of each of the layers of CAISS-X that we identified earlier.

Web Application Layer This layer of CAISS-X provides the only point of entry for users to interact with the cloud infrastructure. CAISS-X currently uses a tight coupling between user privileges (viz. read-only, read/write and administrative) and operations (viz. query-only, create/load/query and create/load/query/user-assignment) to define the interaction of a user with the system. In particular, the interaction takes on one of the following three user roles:

- *Read-only, query-only*: In this role, a user is only allowed to query pre-existing Hive tables/views.

- *Read/write, create/load/query*: In this role, users can create/load tables/views as well as define XACML policies on them, in addition to being able to query all tables/views.
- *Administrative, create/load/query/user assignment*: This is a specially reserved role for an "admin" user, who in addition to the previously defined functions (create/load/query), can also assign new users to either of the above roles.

Additionally, the web application provides a basic authentication mechanism in the form of a login page to identify legitimate users. The usernames and passwords of all users are securely stored in a file using the Java simplified encryption (JASYPT) library's [58] salted hash technique. Furthermore, this sensitive file is stored in a secure location that is inaccessible to any user.

Access Control Layer This layer forms a core component of CAISS-X and is itself composed of several parts, each of which is described below.

- *ZQL Parser*: ZQL is a compiler for processing SQL queries. The ZQL parser [110] is used to validate the accuracy of queries that are submitted by a user to the Web Application layer. To accomplish this task, ZQL parses a query submitted by a user and if it is unsuccessfully parsed, then an appropriate error message is passed back to the user through the Web Application layer. On the other hand, if the process of parsing is successful, ZQL fills in various Java Vector [98] objects with results of the parsing process. The Vector objects that are needed are then distributed among the different layers of CAISS-X. For example, given the query

```
Q = SELECT Student.id, Student.name
    FROM Student
    WHERE Student.Grade = 'A';
```

  ZQL would, in parsing `Q`, construct a separate Java Vector object for each of the distinct parts of `Q`, namely `SELECT`, `FROM` and `WHERE`. Furthermore, each of these objects would be filled in with values found in `Q`:

```
SELECT = {Student.id, Student.name}
FROM   = {Student}
WHERE  = {Student.Grade = 'A'}
```

  The `SELECT` Vector is then passed back to the Web Application layer and is used to display column headers for a table containing the results of query `Q`. The `FROM` Vector is passed to the XACML Policy Evaluator to ensure that the user submitting query `Q` has the necessary permissions to access all tables/views specified in the query. If the user has the appropriate permissions, the query (`Q` in this case) is evaluated further, else an error message is returned to the user through the Web Application layer. Finally, ZQL currently provides parsing support only for queries involving the `SELECT`, `INSERT`, `DELETE` and `UPDATE` keywords.

- *XACML Policy Layer*: The XACML Policy Builder and Evaluator together form the XACML Policy layer that is used to construct and enforce XACML policies. We now describe each of these components in additional detail:
  - *XACML Policy Builder*: In CAISS-X, the tables/views defined in Hive are treated as resources that need to be protected through the use of XACML policies. A point to observe here is that since a user can define policies on both tables and views, CAISS-X is automatically able to support a finer level of access control. Furthermore, CAISS-X uses the Role Based Access Control mechanism [48, 105] for allowing only authorized users to access tables/views defined in Hive. Towards this end, CAISS-X defines a mapping between the permissible types of queries and the users allowed to run those types of queries. We point out that

    ```
    SELECT root userA userB
    INSERT root userA userC
    ```

    is one example of such a mapping. As should be evident in this example, the `root` (i.e., root user or super user) and the `userA` have the permission to execute both `SELECT` and `INSERT` queries, while the `userB` can only execute `SELECT` queries and the `userC` can only execute `INSERT` queries. Additionally, when a user wants to create a new table/view in Hive, CAISS-X allows the user to upload their own predefined XACML policy for that table/view or have the framework build a new policy for them. In the latter case, the user also needs to specify the types of queries that can be executed on the newly constructed table/view. CAISS-X also maintains a separate mapping between the table/view name and the types of queries that can be executed on that table/view. For example, consider the `userA` wants to create a new table `table1` and allow `SELECT`, `INSERT` and `DELETE` queries to be executed on `table1`. The mapping file would then be appropriately updated with this new information:

    ```
    table2 SELECT INSERT
    view1  SELECT INSERT
    table1 SELECT INSERT DELETE
    ```

    Once the mapping file has been updated, CAISS-X uses Sun's XACML implementation [109] to construct a policy for the newly created table/view with the associated query types specified by the user.
  - *XACML Policy Evaluator*: This component makes use of the two mappings defined during policy construction as well as the entire set of XACML policies to ensure that a user is authorized to execute a particular query (such as `SELECT` and `INSERT`) over the tables/views defined in that query. In practice, CAISS-X uses Sun's XACML implementation to enforce XACML policies. If a user is authorized to access all tables/views defined in the submitted query, then the query is evaluated further, else an appropriate error message is passed back to the Web Application layer. Furthermore, the XACML

Policy Evaluator is used both during query execution as well as during view construction. A view in Hive can only be constructed by executing a `SELECT` query on existing tables/views. Therefore, in order to create a view, a user must have access to all tables/views that are a part of the view definition's `SELECT` query. The policy evaluator is therefore required to ensure that the querying user is authorized to access all the necessary tables/views.

- *Basic Query Rewriting*: This component allows users to submit traditional SQL queries which are then rewritten to suit HiveQL's syntax, thereby adding another layer of abstraction between the users and Hive. CAISS-X currently supports the following simple rewrite rules that can be used in conjunction with user submitted SQL queries:
  – HiveQL does not permit the use of multiple tables/views in the `FROM` clause of an input query, but rather expects this type of query to be defined as a sequence of `JOIN` statements. CAISS-X allows a user to submit a query with multiple tables in the `FROM` clause thereby abstracting a user from HiveQL's syntax. The rewriting component then transforms the submitted query into a new query in which the tables in the `FROM` clause are replaced with a sequence of corresponding `JOIN` statements. For example, the original SQL query

```
Q = SELECT Employee.name, Department.name
    FROM Employee, Department;
```

  is transformed into the following HiveQL query

```
Q' = SELECT Employee.name, Department.name
     FROM Employee JOIN Department;
```

  – HiveQL supports the following version of SQL's `INSERT-INTO-SELECT` statement,

```
INSERT OVERWRITE TABLE <tablename>
SELECT <statement>;
```

  rather than the traditionally defined,

```
INSERT INTO <tablename> SELECT <statement>;
```

  In this case, CAISS-X allows users to use the traditional definition of the `INSERT-INTO-SELECT` statement, which is then rewritten to conform to HiveQL's syntax by simply replacing the keyword `INTO` with the keywords `OVERWRITE TABLE`.

Hive Layer As stated earlier, Hive [114] is an open source framework that provides data warehousing capabilities on top of Hadoop. In addition, Hive allows the definition of tables/views over structured data that is stored in the underlying HDFS. Finally, Hive also provides an SQL-like language, HiveQL, to conduct complex analyses over the stored data. In Fig. 3, the arrows between tables defined in this layer and files stored in the HDFS layer indicate a one-to-one correspondence between tables (e.g. Table1) and files (e.g. File1). Furthermore, the lack of arrows between views defined in this layer and files stored in the HDFS layer suggests that a view is simply a logical concept in Hive that is defined as such using a `SELECT` query. CAISS-X uses Hive to allow cooperating organizations to share structured data as well as to perform complex analyses on the shared data using a familiar SQL-like syntax.

HDFS Layer As was presented earlier, the HDFS is a distributed file system that is designed to run on commodity hardware while providing typical features such as fault tolerance, data replication, and so forth. The HDFS layer in CAISS-X is used to store structured data files that correspond to tables created in the Hive layer. The security assumption made in CAISS-X is that users can only access HDFS files through the Web Application Layer of CAISS-X; thus, users cannot access these files through Hadoop's command line or web interface.

4.4 Cloud-Centric Assured Information Sharing System (CAISS)

In the previous subsection, we presented CAISS-X, an AIS implementation that uses a cloud-based framework for data storage and retrieval and a non-cloud policy engine to enforce XACML policies. In this subsection, we outline a separate prototype, called CAISS, which uses a cloud-based framework for both data storage and retrieval as well as policy enforcement. Although CAISS-X represents a discernible enhancement over prior AIS implementations, it still has the following constraints that are addressed in CAISS:

- *XACML-based policies*: The types of inter-organizational use-cases that can be captured by XACML-based policies is restricted by the limited expressive power of XACML. On the other hand, CAISS uses the power of successively more expressive languages such as RDF and OWL to capture complex use-cases that may arise when multiple organizations want to share information.
- *Cloud-based policy engine*: Since CAISS-X uses a non-cloud policy engine, it may not be able to provide the scalability required for some of the necessary information sharing tasks that involve a large number of users. Conversely, CAISS uses a cloud-resident policy manager to store and enforce policies defined in RDF. Therefore, CAISS achieves the desired goals of scalability, flexibility and efficiency that are integral to mission-critical information sharing tasks.
- *Support for unstructured data*: CAISS-X uses the relational model for data representation and therefore requires an a priori definition of a schema, thus making it infeasible for many critical information sharing tasks that operate under rapidly changing mission requirements. Moreover, the relational model is also incapable of representing unstructured data such as text, images, audio and video

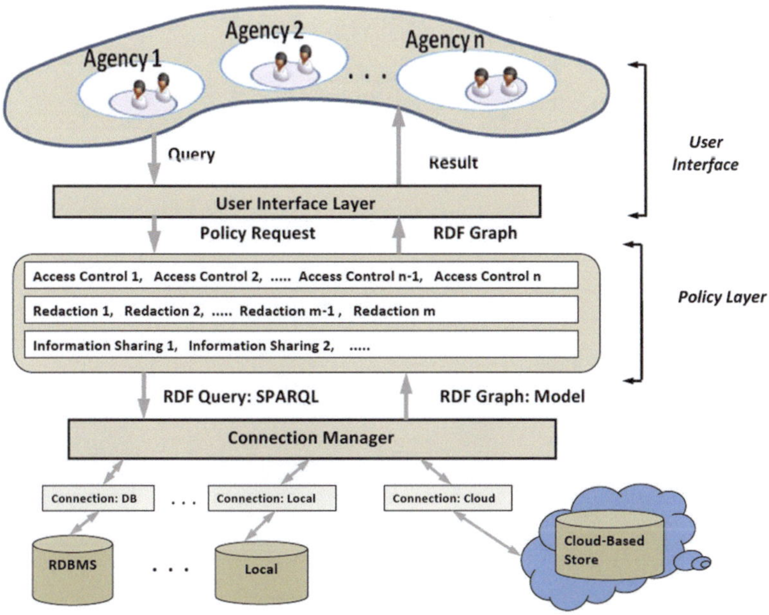

Fig. 4 Cloud-centric assured information sharing system (CAISS) architecture (Modified from Cadenhead et al. [27])

files. Since CAISS uses the RDF model for data representation, it can easily adapt to changing mission requirements since RDF is schemaless. Furthermore, RDF can also easily represent and process unstructured data.

We now present the architectural details of CAISS followed by a description of the various novel features supported by CAISS [27].

4.4.1 Architecture

We used a layered approach in designing the architecture of CAISS (see Fig. 4). The architecture of CAISS consists of three layers: *User Interface Layer*, *Policy Engine Layer*, and *Data Layer*. We discuss each of these layers below.

User Interface Layer This layer enables interaction between a user and the policy framework and is the entry point for the user to interact with CAISS. A user must first be registered with CAISS in order to access the resources stored in the data layer. In addition, an agency must also register its resources and configure security policies for its resources. A detailed overview of the configuration of the framework, which includes the sequence of steps followed during user and agency registration can be found in [29].

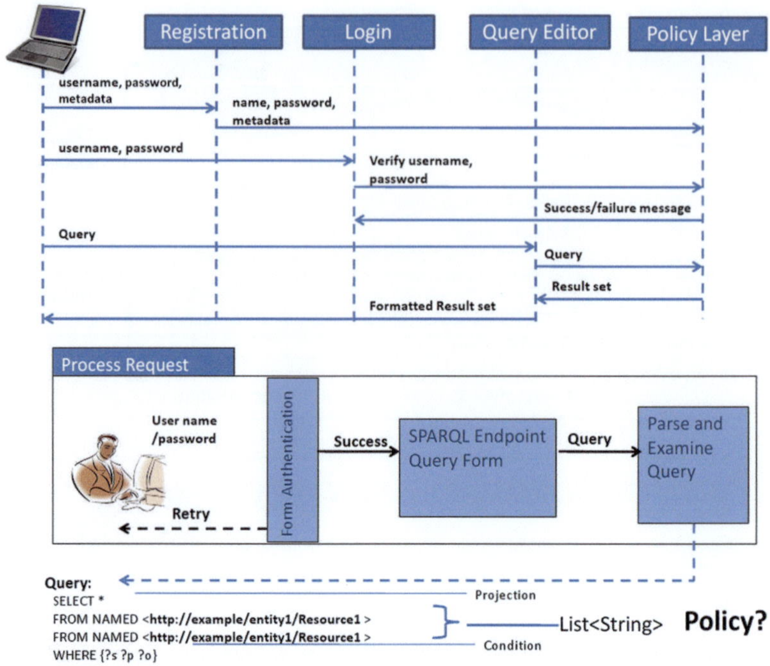

Fig. 5 User registration sequence

A user is first authenticated in CAISS before being able to access any resource stored in the data layer (see Fig. 5). A user is presented with the opportunity to register with the system using a web registration page. The registration page captures the user's name, password and metadata about the user. The metadata is additional information about the user such as the agency that the user belongs to, or the data used for mapping the user's credentials to a role that has been assigned to the user.

Query Editor A query editor is provided as a web form that allows users to write a SPARQL query; the form ensures that a query is answerable before passing it on to the underlying services, which process the query against an RDF document. The web form is used as a SPARQL endpoint that enables users to query an agency's resources using the SPARQL language. The results for the query are returned in one or more machine-processable formats such as the RDF/XML, TURTLE, N-TRIPLE or JSON formats. Therefore, the SPARQL endpoint can be conceived as a machine-friendly interface towards an RDF knowledge base.

A user formulates valid SPARQL queries and submits them via the query editor, which eventually passes them to the policy engine layer. An input query must first be examined before it is passed to the policy engine layer and then on to the data layer. The lower half of Fig. 5 provides a breakdown of a typical SPARQL SELECT query, which is composed of three parts: the projection, the set of named graphs, and the where clause. Each of these parts are targets for enforcing policies via a policy engine. The projection can be examined so that sensitive information is removed or blocked out before returning it to the querying user. The list of named graphs can be examined to determine which agency's policies to apply, or whether more than one agency's resources are involved, or whether the policies allow users to combine resources from different agencies in a single query. The condition on the where clause can be examined to determine whether the RDF graph triples can be modified via a query rewriting or match-and-apply pattern technique to yield modified triples. The decision as to where to apply a policy for a given query is normally encoded in a policy engine.

Policy Engine Layer This layer is designed to allow agencies to evaluate their pre-defined security policies over their respective resources. These resources are RDF documents, which contain information (also referred to as objects or resources in the policy community) that can be shared with other agencies. A RDF document may have different levels of classification, for example, a document could be classified as sensitive (or private in case it contains personal data about individuals). In addition, the classification may vary based on a querying user's credentials. Therefore, all policy engines are designed to accept as input two standard parameters, an agency's *resource* and a user's *credentials*. In response to these inputs, a policy engine evaluates the underlying logic of a policy over the inputs, and then generates as output a new RDF graph. (We refer to both, the default and named graphs of a SPARQL query, as *resource* input parameter; we also refer to the result of a SPARQL query as an output graph or model. This is normal, since a RDF graph is a set of triples, which may have one or more machine-processable formats such as RDF/XML, TURTLE or N-TRIPLE. It is sometimes customary to display an output graph in a tabular format, where each column stores the binding for a variable of a given SPARQL query.)

We now discuss the details of the kinds of policy engines that are currently supported by our policy engine layer. These are the access control engine, the redaction engine, the information sharing engine, and the provenance engines.

- *Access Control Policy Engine*: An access control policy authorizes a set of *users* to perform a set of *actions* on a set of *resources* within an *environment*. Unless authorized through one or more access control policies, users have no access to any resource in the system. We currently support three major classes of access control policies, namely role-based access control (RBAC), which restricts access based on roles; discretionary access control (DAC), which controls access based on the identity of the user; and mandatory access control (MAC), which controls access based on mandated regulations determined by a central authority. There are two major concerns with these policies. The first is the number of user to object assignments and the second is that these policies are defined over a single resource. In some cases, the resource may require extending these existing policies as discussed in [25, 26, 93].
- *Redaction Policy Engine*: A redaction policy blocks out sensitive information from a document. Unlike access control policies, which restrict access, redaction

policies encourage sharing of information, by ensuring that sensitive or proprietary information is removed (or obscured) before providing the final RDF graph (referred to as a redacted graph) that corresponds to the result of the user's query back to the user. Our redaction policy engine relies on a graph transformation technique. This technique is based on a graph grammar method (given in [34,45,102] and implemented for provenance graphs in [25]). There are two basic steps to applying a redaction policy over a directed labeled RDF graph: (1) Identify a resource (or subgraph) in the original RDF graph that we want to protect. This can be done with a graph query (i.e., a query equipped with regular expressions). (2) Apply a redaction policy to this identified resource in the form of a graph transformation rule.

- *Information Sharing Policy Engine*: The policy engines discussed so far implement logic for executing a policy's rules over a single RDF graph whenever the rules are applied. An information sharing policy engine has logic for processing a query requesting information on two or more RDF graphs simultaneously. For example, suppose there are two RDF graphs G_1 and G_2. The information policy engine could then generate RDF graph $G_1 \odot G_2$, where the operation $\odot$ could be one of the following three operations: (1) form the RDF graph $G_1 \cup G_2$ whose triples are the union of the triples in G_1 and G_2; (2) form the RDF graph $G_1 \cap G_2$ whose triples are the intersection of the triples in G_1 and G_2; (3) form the RDF graph $G_1 - G_2$ whose triples consist of the triples that belong to G_1 but do not belong to G_2. An information sharing policy engine allows agencies to determine the context in which their resources are shared or combined with resources from other agencies.
- *Provenance Policy Engines*: Sometimes, the relationships among the data items in a resource need to be taken into consideration when defining policies. The three policy types discussed so far fail to address the cases where sensitive information is implicit in the various paths within an RDF graph resource. The provenance policy engines focus on definitions of policies that are tailored to the execution of access control and redaction policies over a provenance graph. The theory behind the logic of these policy engines is based on [26], which discusses an access control policy language for provenance and [25], which discusses how to perform redaction over provenance. These policies can complement the information sharing policies we discussed earlier for sharing resources between agencies.

Data Layer At the core of the implementation of this layer is a connection factory, which acts as a facade for creating connection objects. These connection objects expose a common set of properties in the form of publicly accessible methods to the policy designer. This makes it easier for the policy designer to concentrate on the policy engine design. The policy designer makes a call to a RDF Policy Factory, which returns a RDF model object. This RDF model object is backed by a connection store, which can be a local connection or a relational database connection or a cloud connection to the Jena-HBase framework.

During registration, an agency is given an opportunity to decide where it wants to store its resources and configuration files. It is recommended that smaller configuration files be stored locally on disk (or in a local database) to enable quick access to them. However, an agency may decide to store them on a private cloud (or on a remote database server) to take advantage of the added protection there.

4.4.2 Features of Our Policy Engine Framework

We now present some novel features of our policy engine framework that are tailored to suit the requirements of an AIS framework.

Policy Reciprocity Policy reciprocity enables agencies to specify policies when knowledge of other agencies, their resources or policy specifications is available. This is made possible via the registration process, where agencies make metadata available about themselves, their resources and associated policies. We provide two scenarios for policy reciprocity among two agencies, *Agency*1 and *Agency*2:

1. *Policy Symmetry*: A consequence of policy reciprocity is symmetry in the sharing of policies. For example, *Agency*1 shares its resources with *Agency*2 using operator $\odot$ if *Agency*2 also shares its resources with *Agency*1 using the same operator $\odot$. That is, we have the following information sharing policy:

   $\forall r1 \in Agency1, \forall r2 \in Agency2, Agency1$ uses r1 $\cup$ r2 if $Agency2$ uses r2 $\cup$ r1.

2. *Conditional Policies*: Another consequence of policy reciprocity is allowing the use of conditional sharing policies. For example, *Agency*1 shares its resources with *Agency*2 if *Agency*2 does not share *Agency*1's resources with *Agency*3. That is, we have the following information sharing policy:

   $\forall r1 \in Agency1, \forall r2 \in Agency2$, $Agency1$ defines r1 $\cap$ r2. Given $\forall r3 \in Agency3$, then

   - *Agency*2 does not define any sharing policy of the form r1 $\cap$ r3,
   - or *Agency*2 does not define any sharing policy of the form r1 $\subseteq$ r2 $\odot$ r3, where $\odot \in \{\cup, \cap\}$.

Develop and Scale Policies To enable freedom of maneuverability across the information environment and to deliver the power of information to ensure mission success, an agency should be able to rapidly develop policies and deploy them as needed. We next discuss the features that are available to an agency during and after development of its policies.

1. *Policy Development*: *Agency*1 wishes to simulate a live environment and create test scenarios to visualize the results of each policy configuration. Our policy framework provides three configurations: (1) a standalone version for development and testing; (2) a version backed by a relational database; and (3) a cloud-based version that achieves high availability and scalability while maintaining low setup and operation costs.

2. *Rapid Elasticity*: *Agency*1 identifies security vulnerabilities in its existing policy configurations and wishes to extend (or grow) its existing policy set with support for policies at a finer granularity. Our policy framework provides a policy interface that should be implemented by all policies; therefore, we can add newer types of policies as needed. In addition, our policy framework gives an agency rapid elasticity, whereby the capabilities provided by our policy framework appear unlimited.

Justification of Resources Provenance [14] makes available an explanation about why information was manipulated and provides a trace to the source that manipulated the information. This establishes trust amongst agencies, thus facilitating partnerships for common goals.

For example, *Agency*1 asks *Agency*2 for a justification of resource R2. The current commercial access control policies are mainly designed to protect single data items, while current redaction policies are designed for redacting text and images. Our policy framework allows agents to define policies over provenance; therefore, *Agency*2 can provide the provenance of R2 to *Agency*1, while protecting sensitive portions of R2 by using access control or redaction policies.

5 Conclusion

In this chapter, we highlighted the basic characteristics of an Assured Information Sharing (AIS) framework, we presented the state of the art in the realm of AIS, and finally, we detailed two contemporary, cloud-based AIS implementations (CAISS-X and CAISS). We began this chapter by presenting the key requirements of AIS as well as a description of a goal-oriented architecture of AIS, which could serve as the foundation of an AIS implementation. After that, we presented some of the motivating factors that could encourage organizations to adopt an AIS framework, and we also outlined several challenges that must be overcome before AIS frameworks can be widely adopted. The next part of the chapter discussed recent advances in the realm of AIS. In particular, we discussed how AIS access control models allow an organization to more effectively control the disclosure of sensitive information to authorized users than do traditional access control models. We also reviewed how AIS research has focused on protecting the privacy of individuals when their data is being shared by cooperating organizations. Finally, we elaborated on how AIS research has focused on incentivizing participating organizations to honestly share information. In the last part of the chapter, we discussed the following two cloud-based AIS implementations: (1) CAISS-X, which employs a cloud-centric framework to store and query large amounts of data via a non-cloud policy engine that enforces XACML-based policies. (2) CAISS, which uses a cloud-based framework for both data storage and retrieval as well as policy enforcement.

Although we have presented significant steps towards the vision of a unified solution that addresses various issues that are integral to the process of data sharing, there are additional issues that need to be addressed before this vision is complete. One issue is that given the multitude of cloud deployment models, there is a need for a detailed study that evaluates these models from the perspectives of performance, security, and financial costs. Another issue is that since participating organizations would necessarily want their sensitive data to be protected, an AIS implementation must therefore store data using various protection mechanisms; given the variety of data protection techniques currently available, a detailed analysis of these techniques in terms of their information disclosure risk, performance and associated overhead is required.

Acknowledgements This material is based upon work partially supported by the Air Force Office of Scientific Research (AFOSR) MURI-grant FA-9550-08-1-0265 and AFOSR grant FA9550-12-1-0082, the National Institutes of Health (NIH) grant 1R01LM009989, the National Science Foundation (NSF) CAREER-grant CNS-0845803, NSF grants CNS-0964350, CNS-1016343, CNS-1111529 and CNS-1228198, the Army Research Office (ARO) grant 58345-CS, and the Air Force Research Laboratory (AFRL) Visiting Faculty Research Program (VFRP) extension grant LRIR 11RI01COR. We thank Dr. Robert Herklotz for his support.

References

1. af.mil, Air Force Distributted Common Ground System (DCGS). http://goo.gl/qLkFc
2. Agrawal, R., Srikant, R.: Privacy-preserving data mining. In: Proceedings of the 2000 ACM SIGMOD International Conference on Management of Data, SIGMOD'00, Dallas, pp. 439–450. ACM, New York (2000). doi:10.1145/342009.335438
3. Agrawal, R., Kiernan, J., Srikant, R., Xu, Y.: Hippocratic databases. In: Proceedings of the 28th International Conference on Very Large Data Bases, VLDB'02, Hong Kong. VLDB Endowment, pp. 143–154 (2002)
4. Agrawal, D., Giles, J., Lee, K.W., Lobo, J.: Policy ratification. In: Proceedings of the 6th IEEE International Workshop on Policies for Distributed Systems and Networks, POLICY'05, Stockholm, pp. 223–232. IEEE Computer Society, Washington, DC (2005). doi:10.1109/POLICY.2005.25
5. Anderson, T.E.: dtic.mil, Making GIG information assurance better through portfolio management. http://goo.gl/RTSS3 (2008)
6. apache.org, Apache Hadoop. http://goo.gl/tnkf
7. apache.org, Apache HBase. http://goo.gl/2kbIQ
8. apache.org, Apache Hive. http://goo.gl/Jxwgx
9. apache.org, Apache Jena. http://goo.gl/l5mYK
10. apache.org, Apache Jena, SDB – persistent triple stores using relational databases. http://goo.gl/QCO8L
11. Astrahan, M.M., Blasgen, M.W., Chamberlin, D.D., Eswaran, K.P., Gray, J.N., Griffiths, P.P., King, W.F., Lorie, R.A., McJones, P.R., Mehl, J.W., Putzolu, G.R., Traiger, I.L., Wade, B.W., Watson, V.: System R: relational approach to database management. ACM Trans. Database Syst. **1**(2), 97–137 (1976). doi:10.1145/320455.320457
12. Backes, M., Karjoth, G., Bagga, W., Schunter, M.: Efficient comparison of enterprise privacy policies. In: Proceedings of the 2004 ACM Symposium on Applied Computing, SAC'04, Nicosia, pp. 375–382. ACM, New York (2004). doi:10.1145/967900.967983

13. Bacon, J., Moody, K., Yao, W.: A model of OASIS role-based access control and its support for active security. ACM Trans. Inf. Syst. Secur. **5**(4), 492–540 (2002). doi:10.1145/581271.581276
14. Bates, A., Mood, B., Valafar, M., Butler, K.: Towards secure provenance-based access control in cloud environments. In: Proceedings of the 3rd ACM Conference on Data and Application Security and Privacy, CODASPY'13, San Antonio, pp. 277–284. ACM, New York (2013). doi:10.1145/2435349.2435389
15. Bell, D.E., LaPadula, L.J.: nist.gov, Secure computer systems: unified exposition and multics interpretation. http://goo.gl/uVC7c (1976)
16. Bertino, E., Haas, L.M.: Views and security in distributed database management systems. In: Proceedings of the 1988 International Conference on Extending Database Technology: Advances in Database Technology, EDBT'88, Venice, pp. 155–169. Springer, London (1988)
17. Bertino, E., Sandhu, R.: Database security-concepts, approaches, and challenges. IEEE Trans. Dependable Secure Comput. **2**(1), 2–19 (2005). doi:10.1109/TDSC.2005.9
18. Bertino, E., Samarati, P., Jajodia, S.: An extended authorization model for relational databases. IEEE Trans. Knowl. Data Eng. **9**(1), 85–101 (1997). doi:10.1109/69.567051
19. Bertino, E., Bettini, C., Ferrari, E., Samarati, P.: An access control model supporting periodicity constraints and temporal reasoning. ACM Trans. Database Syst. **23**(3), 231–285 (1998). doi:10.1145/293910.293151
20. Bertino, E., Bonatti, P.A., Ferrari, E.: TRBAC: a temporal role-based access control model. ACM Trans. Inf. Syst. Secur. **4**(3), 191–233 (2001). doi:10.1145/501978.501979
21. Bertino, E., Castano, S., Ferrari, E.: Securing XML documents with Author-X. IEEE Internet Comput. **5**(3), 21–31 (2001). doi:10.1109/4236.935172
22. Bonatti, P.A., Samarati, P.: A uniform framework for regulating service access and information release on the web. J. Comput. Secur. **10**(3), 241–271 (2002)
23. Bonatti, P., De Capitani di Vimercati, S., Samarati, P.: An algebra for composing access control policies. ACM Trans. Inf. Syst. Secur. **5**(1), 1–35 (2002). doi:10.1145/504909.504910
24. Bulatov, A.A.: Complexity of conservative constraint satisfaction problems. ACM Trans. Comput. Logic **12**(4), 24:1–24:66 (2011). doi:10.1145/1970398.1970400
25. Cadenhead, T., Khadilkar, V., Kantarcioglu, M., Thuraisingham, B.: Transforming provenance using redaction. In: Proceedings of the 16th ACM Symposium on Access Control Models and Technologies, SACMAT'11, Innsbruck, pp. 93–102. ACM, New York (2011). doi:10.1145/1998441.1998456
26. Cadenhead, T., Khadilkar, V., Kantarcioglu, M., Thuraisingham, B.: A language for provenance access control. In: Proceedings of the 1st ACM Conference on Data and Application Security and Privacy, CODASPY'11, San Antonio, pp. 133–144. ACM, New York (2011). doi:10.1145/1943513.1943532
27. Cadenhead, T., Kantarcioglu, M., Khadilkar, V., Thuraisingham, B.: Design and implementation of a cloud-based assured information sharing system. In: Proceedings of the 6th International Conference on Mathematical Methods, Models and Architectures for Computer Network Security, MMM-ACNS'12, St. Petersburg, pp. 36–50. Springer, Berlin/Heidelberg (2012). doi:10.1007/978-3-642-33704-8_4
28. Cadenhead, T., Khadilkar, V., Kantarcioglu, M., Thuraisingham, B.: A cloud-based RDF policy engine for assured information sharing. In: Proceedings of the 17th ACM Symposium on Access Control Models and Technologies, SACMAT'12, Newark, pp. 113–116. ACM, New York (2012). doi:10.1145/2295136.2295157
29. Cadenhead, T., Khadilkar, V., Kantarcioglu, M., Thuraisingham, B.: A general purpose RDF policy engine for assured information sharing. Technical report, University of Texas at Dallas (2012)
30. cgisecurity.com, The virtual private database in Oracle9iR2. http://goo.gl/Lijl1 (2002)
31. Chang, F., Dean, J., Ghemawat, S., Hsieh, W.C., Wallach, D.A., Burrows, M., Chandra, T., Fikes, A., Gruber, R.E.: Bigtable: a distributed storage system for structured data. ACM Trans. Comput. Syst. **26**(2), 4:1–4:26 (2008). doi:10.1145/1365815.1365816

32. Clifton, C.: Using sample size to limit exposure to data mining. J. Comput. Secur. **8**(4), 281–307 (2000)
33. Codd, E.F.: A relational model of data for large shared data banks. Commun. ACM **26**(1), 64–69 (1983). doi:10.1145/357980.358007
34. Corradini, A., Montanari, U., Rossi, F., Ehrig, H., Heckel, R., Löwe, M.: Algebraic approaches to graph transformation – part i: basic concepts and double pushout approach. In: Rozenberg, G. (ed.) Handbook of Graph Grammars, pp. 163–246. World Scientific, Singapore (1997)
35. Crampton, J., Loizou, G.: Administrative scope: a foundation for role-based administrative models. ACM Trans. Inf. Syst. Secur. **6**(2), 201–231 (2003). doi:10.1145/762476.762478
36. Damiani, E., De Capitani di Vimercati, S., Paraboschi, S., Samarati, P.: A fine-grained access control system for XML documents. ACM Trans. Inf. Syst. Secur. **5**(2), 169–202 (2002). doi:10.1145/505586.505590
37. Dean, J., Ghemawat, S.: MapReduce: simplified data processing on large clusters. Commun. ACM **51**(1), 107–113 (2008). doi:10.1145/1327452.1327492
38. defense.gov, Department of Defense Information sharing strategy. http://goo.gl/3IcDm (2007)
39. defense.gov, The Department of Defense's Global Information Grid (GIG) architectural vision. http://goo.gl/Ph6Kd (2007)
40. defense.gov, Department of Defense Information enterprise strategic plan. http://goo.gl/3eA65 (2010–2012)
41. defensemarket.com, Department of Defense Embraces cloud computing. http://goo.gl/FrPuF (2009)
42. dtic.mil, DCGS Integration Backbone (DIB) v4.0 overview. http://goo.gl/6KYKr (2012)
43. Durham, E., Xu, Y., Kantarcioglu, M., Malin, B.: Private medical record linkage with approximate linkage. In: American Medical Informatics Association Annual Symposium, Washington, DC (2010)
44. Durham, E., Xue, Y., Kantarcioglu, M., Malin, B.: Quantifying the correctness, computational complexity, and security of privacy-preserving string comparators for record linkage. J. Inf. Fusion **13**(4), 245–259 (2012). doi:10.1016/j.inffus.2011.04.004
45. Ehrig, H., Ehrig, K., Prange, U., Taentzer, G.: Fundamentals of Algebraic Graph Transformation. Monographs in Theoretical Computer Science. An EATCS Series. Springer, Secaucus (2006)
46. Emam, K.E., Hu, J., Mercer, J., Peyton, L., Kantarcioglu, M., Malin, B., Buckeridge, D., Samet, S., Earle, C.: A secure protocol for protecting the identity of providers when disclosing data for disease surveillance. J. Am. Med. Inf. Assoc. **18**(3), 212–217 (2011). doi:10.1136/amiajnl-2011-000100
47. Fagin, R.: On an authorization mechanism. ACM Trans. Database Syst. **3**(3), 310–319 (1978). doi:10.1145/320263.320288
48. Ferraiolo, D., Kuhn, R.: Role-based access control. In: Proceedings of the 15th NIST-NCSC National Computer Security Conference, Baltimore, pp. 554–563 (1992)
49. Ferrini, R., Bertino, E.: Supporting RBAC with XACML+OWL. In: Proceedings of the 14th ACM Symposium on Access Control Models and Technologies, SACMAT'09, pp. 145–154. ACM, New York (2009). doi:10.1145/1542207.1542231
50. Finin, T., Joshi, A., Kagal, L., Niu, J., Sandhu, R., Winsborough, W., Thuraisingham, B.: ROWLBAC: representing role based access control in OWL. In: Proceedings of the 13th ACM Symposium on Access Control Models and Technologies, SACMAT'08, Estes Park, pp. 73–82. ACM, New York (2008). doi:10.1145/1377836.1377849
51. Finin, T., Joshi, A., Kargupta, H., Yesha, Y., Sachs, J., Bertino, E., Li, N., Clifton, C., Spafford, G., Thuraisingham, B., Kantarcioglu, M., Bensoussan, A., Berg, N., Khan, L., Han, J., Zhai, C., Sandhu, R., Xu, S., Massaro, J., Adamic, L.: Assured information sharing life cycle. In: Proceedings of the 2009 IEEE International Conference on Intelligence and Security Informatics, ISI'09, Richardson, pp. 307–309. IEEE, Piscataway (2009)

52. Fisler, K., Krishnamurthi, S., Meyerovich, L.A., Tschantz, M.C.: Verification and change-impact analysis of access-control policies. In: Proceedings of the 27th International Conference on Software Engineering, ICSE'05, St. Louis, pp. 196–205. ACM, New York (2005). doi:10.1145/1062455.1062502
53. Ghemawat, S., Gobioff, H., Leung, S.T.: The Google file system. In: Proceedings of the 9th ACM Symposium on Operating Systems Principles, SOSP'03, Bolton Landing, pp. 29–43. ACM, New York (2003). doi:10.1145/945445.945450
54. github.com, Hbase-rdf. http://goo.gl/4DxN5
55. Gray, J.: The transaction concept: virtues and limitations (invited paper). In: Proceedings of the 7th International Conference on Very Large Data Bases, VLDB'81, Cannes. VLDB Endowment, vol. 7, pp. 144–154 (1981)
56. Griffiths, P.P., Wade, B.W.: An authorization mechanism for a relational database system. ACM Trans. Database Syst. **1**(3), 242–255 (1976). doi:10.1145/320473.320482
57. informationweek.com, NSA pursues intelligence-sharing architecture. http://goo.gl/LYrRX (2011)
58. jasypt.org, Java simplified encryption. http://goo.gl/htlXQ
59. jena.apache.org, Jena architecture overview. http://goo.gl/cbfnZ
60. Joshi, J., Bertino, E., Latif, U., Ghafoor, A.: A generalized temporal role-based access control model. IEEE Trans. Knowl. Data Eng. **17**(1), 4–23 (2005). doi:10.1109/TKDE.2005.1
61. Kagal, L., Finin, T., Joshi, A.: A policy language for a pervasive computing environment. In: Proceedings of the 4th IEEE International Workshop on Policies for Distributed Systems and Networks, POLICY'03, Lake Como, pp. 63–74. IEEE Computer Society, Washington, DC (2003)
62. Kantarcioglu, M., Clifton, C.: Privacy-preserving distributed mining of association rules on horizontally partitioned data. IEEE Trans. Knowl. Data Eng. **16**(9), 1026–1037 (2004). doi: 10.1109/TKDE.2004.45
63. Kantarcioglu, M., Jiang, W.: Incentive compatible privacy-preserving data analysis. IEEE Trans. Knowl. Data Eng. (99) (2012). doi:10.1109/TKDE.2012.61
64. Kantarcioglu, M., Kardes, O.: Privacy-preserving data mining in the malicious model. ACM Trans. Inf. Syst. Secur. **2**(4), 353–375 (2008). doi:10.1504/IJICS.2008.022488
65. Kantarcioglu, M., Nix, R., Vaidya, J.: An efficient approximate protocol for privacy-preserving association rule mining. In: Proceedings of the 13th Pacific-Asia Conference on Advances in Knowledge Discovery and Data Mining, PAKDD'09, Bangkok, pp. 515–524. Springer, Berlin/Heidelberg (2009). doi:10.1007/978-3-642-01307-2_48
66. Kantarcioglu, M., Bensoussan, A., Hoe, S.: When do firms invest in privacy-preserving technologies? In: Proceedings of the 1st International Conference on Decision and Game Theory for Security, GameSec'10, Berlin, pp. 72–86. Springer, Berlin/Heidelberg (2010)
67. Kantarcioglu, M., Bensoussan, A., Hoe, S.: Investment in privacy-preserving technologies under uncertainty. In: Proceedings of the 2nd International Conference on Decision and Game Theory for Security, GameSec'11, College Park, pp. 219–238. Springer, Berlin/Heidelberg (2011). doi:10.1007/978-3-642-25280-8_17
68. Kargupta, H., Datta, S., Wang, Q., Sivakumar, K.: On the privacy preserving properties of random data perturbation techniques. In: Proceedings of the 3rd IEEE International Conference on Data Mining, ICDM'03, Melbourne, pp. 99–106. IEEE Computer Society, Washington, DC (2003)
69. Kern, A., Kuhlmann, M., Kuropka, R., Ruthert, A.: A meta model for authorisations in application security systems and their integration into RBAC administration. In: Proceedings of the 9th ACM Symposium on Access Control Models and Technologies, SACMAT'04, Yorktown Heights, pp. 87–96. ACM, New York (2004). doi:10.1145/990036.990050
70. Khadilkar, V., Kantarcioglu, M., Castagna, P., Thuraisingham, B.: utdallas.edu, Jena-HBase: a distributed, scalable and efficient RDF triple store. Technical report, University of Texas at Dallas (2012)

71. Khadilkar, V., Kantarcioglu, M., Thuraisingham, B.M., Castagna, P.: Jena-HBase: a distributed, scalable and efficient RDF triple store. In: Proceedings of the 11th International Semantic Web Conference Posters & Demonstrations Track, ISWC-PD'12, Boston, pp. 85–88. CEUR-WS.org (2012)
72. Kirchberg, M., Link, S.: Hippocratic databases: extending current transaction processing approaches to satisfy the limited retention principle. In: Proceedings of the 43rd Hawaii International Conference on System Sciences, HICSS'10, Honolulu, pp. 1–10. IEEE Computer Society, Washington, DC (2010). doi:10.1109/HICSS.2010.220
73. Koch, M., Mancini, L.V., Parisi-Presicce, F.: Administrative scope in the graph-based framework. In: Proceedings of the 9th ACM Symposium on Access Control Models and Technologies, SACMAT'04, Yorktown Heights, pp. 97–104. ACM, New York (2004). doi: 10.1145/990036.990051
74. Krishnan, R., Sandhu, R.: A hybrid enforcement model for group-centric secure information sharing. In: Proceedings of the 12th IEEE International Conference on Computational Science and Engineering, CSE'09, Vancouver, vol. 03, pp. 189–194. IEEE Computer Society, Washington, DC (2009). doi:10.1109/CSE.2009.397
75. Krishnan, R., Sandhu, R.: Authorization policy specification and enforcement for group-centric secure information sharing. In: Proceedings of the 7th International Conference on Information Systems Security, ICISS'11, Kolkata, pp. 102–115. Springer, Berlin/Heidelberg (2011). doi:10.1007/978-3-642-25560-1_7
76. Krishnan, R., Sandhu, R., Niu, J., Winsborough, W.H.: A conceptual framework for group-centric secure information sharing. In: Proceedings of the 4th International Symposium on Information, Computer, and Communications Security, ASIACCS'09, Sydney, pp. 384–387. ACM, New York (2009). doi:10.1145/1533057.1533111
77. Krishnan, R., Sandhu, R., Niu, J., Winsborough, W.H.: Foundations for group-centric secure information sharing models. In: Proceedings of the 14th ACM Symposium on Access Control Models and Technologies, SACMAT'09, Stresa, pp. 115–124. ACM, New York (2009). doi: 10.1145/1542207.1542227
78. Kudoh, M., Hirayama, Y., Hada, S., Vollschwitz, A.: Access control specification based on policy evaluation and enforcement model and specification language. In: SCIS, Okinawa (2002)
79. Kumaraswamy, S., Lakshminarayanan, S., Stein, M.R.J., Wilson, Y.: cloudsecurityalliance.org, Domain 12: Guidance for Identity & Access Management V2.1. http://goo.gl/Iwhaq (2010)
80. Kuzu, M., Kantarcioglu, M., Durham, E., Malin, B.: A constraint satisfaction cryptanalysis of bloom filters in private record linkage. In: Proceedings of the 11th International Conference on Privacy Enhancing Technologies, PETS'11, Waterloo, pp. 226–245. Springer, Berlin/Heidelberg (2011)
81. Layfield, R., Kantarcioglu, M., Thuraisingham, B.M.: Incentive and trust issues in assured information sharing. In: Proceedings of the 4th International Conference on Collaborative Computing: Networking, Applications and Worksharing, CollaborateCom'08, Orlando, pp. 113–125. Springer (2008). doi:10.1007/978-3-642-03354-4_10
82. Li, N., Mitchell, J.C.: DATALOG with constraints: a foundation for trust management languages. In: Proceedings of the 5th International Symposium on Practical Aspects of Declarative Languages, PADL'03, New Orleans, pp. 58–73. Springer, London (2003)
83. Li, N., Tripunitara, M.V.: Security analysis in role-based access control. ACM Trans. Inf. Syst. Secur. **9**(4), 391–420 (2006). doi:10.1145/1187441.1187442
84. Li, N., Mitchell, J.C., Winsborough, W.H.: Design of a role-based trust-management framework. In: Proceedings of the 2002 IEEE Symposium on Security and Privacy, SP'02, Oakland, pp. 114–130. IEEE Computer Society, Washington, DC (2002)
85. Li, N., Grosof, B.N., Feigenbaum, J.: Delegation logic: a logic-based approach to distributed authorization. ACM Trans. Inf. Syst. Secur. **6**(1), 128–171 (2003). doi:10.1145/605434.605438

86. Lin, D., Rao, P., Bertino, E., Li, N., Lobo, J.: EXAM: a comprehensive environment for the analysis of access control policies. Int. J. Inf. Secur. **9**(4), 253–273 (2010). doi:10.1007/s10207-010-0106-1
87. lockheedmartin.com, distributed common ground system (dcgs). http://goo.gl/1dYox
88. Machanavajjhala, A., Gehrke, J., Kifer, D., Venkitasubramaniam, M.: l-Diversity: privacy beyond k-anonymity. In: ICDE, Atlanta, p. 24. IEEE (2006). doi:10.1109/ICDE.2006.1
89. Malone, P., McLaughlin, M., Leenes, R., Ferronato, P., Lockett, N., Guillen, P.B., Heistracher, T., Russello, G.: Endorse: a legal technical framework for privacy preserving data management. In: Proceedings of the 2010 Workshop on Governance of Technology, Information and Policies, GTIP'10, Austin, pp. 27–34. ACM, New York (2010). doi:10.1145/1920320.1920325
90. Manna, Z., Pnueli, A.: The Temporal Logic of Reactive and Concurrent Systems – Specification. Springer, New York (1992)
91. Mitchell, T.M.: Machine Learning. McGraw Hill Series in Computer Science. McGraw-Hill, New York (1997)
92. Moss, J.E.: Nested Transactions: An Approach to Reliable Distributed Computing. MIT, Cambridge (1985)
93. Ni, Q., Xu, S., Bertino, E., Sandhu, R., Han, W.: An access control language for a general provenance model. In: Proceedings of the 6th VLDB Workshop on Secure Data Management, SDM'09, Lyon, pp. 68–88. Springer, Berlin/Heidelberg (2009). doi:10.1007/978-3-642-04219-5_5
94. nist.gov, Department of Defense Trusted computer system evaluation criteria, DOD5200.28-STD. http://goo.gl/wt4in (1975)
95. Nix, R., Kantarcioglu, M.: Incentive compatible privacy-preserving distributed classification. IEEE Trans. Dependable Secure Comput. **9**(4) (2011). doi:10.1109/TDSC.2011.52
96. oasis-open.org, eXtensible Access Control Markup Language (XACML) version 3.0. http://goo.gl/m9hqv
97. Oliveira, S.R.M., Zaïane, O.R.: Privacy preserving frequent itemset mining. In: Proceedings of the IEEE International Conference on Privacy, Security and Data Mining, CRPIT'14, Maebashi City, vol. 14, pp. 43–54. Australian Computer Society, Darlinghurst (2002)
98. oracle.com, Java Platform, standard edition 7, API specification. http://goo.gl/lYUSS
99. Parikh, P.: Secured information integration with a semantic Web-based framework. Technical report, The University of Texas at Dallas (2009)
100. Rao, P., Lin, D., Bertino, E., Li, N., Lobo, J.: EXAM: an environment for access control policy analysis and management. In: Proceedings of the 9th IEEE Workshop on Policies for Distributed Systems and Networks, POLICY'08, Palisades, pp. 238–240. IEEE Computer Society, Washington, DC (2008). doi:10.1109/POLICY.2008.30
101. Reimer, J.: arstechnica.com, From BFS to ZFS: past, present, and future of file systems. http://goo.gl/44KTi (2008)
102. Rozenberg, G. (ed.): Handbook of Graph Grammars and Computing by Graph Transformations, Volume 1: Foundations. World Scientific, Singapore (1997)
103. Samarati, P., de Capitani di Vimercati, S.: Access control: policies, models, and mechanisms. In: Revised versions of lectures given during the IFIP WG 1.7 International School on Foundations of Security Analysis and Design on Foundations of Security Analysis and Design: Tutorial Lectures, FOSAD'00, Bertinoro, pp. 137–196. Springer, London (2001)
104. Samarati, P., Sweeney, L.: Generalizing data to provide anonymity when disclosing information (abstract). In: Proceedings of the 17th ACM SIGACT-SIGMOD-SIGART Symposium on Principles of Database Systems, PODS'98, Seattle, p. 188. ACM, New York (1998). doi:10.1145/275487.275508
105. Sandhu, R.S., Coyne, E.J., Feinstein, H.L., Youman, C.E.: Role-based access control models. Computer **29**(2), 38–47 (1996). doi:10.1109/2.485845

106. Sandhu, R., Krishnan, R., Niu, J., Winsborough, W.H.: Group-centric models for secure and agile information sharing. In: Proceedings of the 5th International Conference on Mathematical Methods, Models and Architectures for Computer Network Security, MMM-ACNS'10, St. Petersburg, pp. 55–69. Springer, Berlin/Heidelberg (2010)
107. sbir.gov, Assured information sharing in clouds. http://goo.gl/FxSbp (2011)
108. Seamons, K., Winslett, M., Yu, T., Smith, B., Child, E., Jacobson, J., Mills, H., Yu, L.: Requirements for policy languages for trust negotiation. In: Proceedings of the 3rd International Workshop on Policies for Distributed Systems and Networks, POLICY'02, Monterey, p. 68. IEEE Computer Society, Washington, DC (2002)
109. sourceforge.net, Sun XACML implementation. http://goo.gl/huKM
110. sourceforge.net, Zql: a Java SQL parser. http://goo.gl/ruUUX
111. Sweeney, L.: k-Anonymity: a model for protecting privacy. Int. J. Uncertainty Fuzziness Knowl. Based Syst. **10**(5), 557–570 (2002)
112. Thuraisingham, B.M., Kumar, Y.H., Khan, L.: Design and implementation of a framework for assured information sharing across organizational boundaries. Int. J. Inf. Secur. Privacy **2**(4), 67–90 (2008)
113. Thuraisingham, B.M., Khadilkar, V., Gupta, A., Kantarcioglu, M., Khan, L.: Secure data storage and retrieval in the cloud. In: Proceedings of the 6th International Conference on Collaborative Computing: Networking, Applications and Worksharing, CollaborateCom'10, Chicago, pp. 1–8. IEEE (2010)
114. Thusoo, A., Sarma, J.S., Jain, N., Shao, Z., Chakka, P., Anthony, S., Liu, H., Wyckoff, P., Murthy, R.: Hive: a warehousing solution over a map-reduce framework. Proc. VLDB Endow. **2**(2), 1626–1629 (2009)
115. Tonti, G., Bradshaw, J.M., Jeffers, R., Montanari, R., Suri, N., Uszok, A.: Semantic Web languages for policy representation and reasoning: a comparison of KAoS, Rei, and Ponder. In: Proceedings of the 2nd International Semantic Web Conference, ISWC'03, Sanibel Island, vol. 2870, pp. 419–437. Springer (2003). doi:10.1007/978-3-540-39718-2_27
116. Vaidya, J., Clifton, C.: Privacy preserving association rule mining in vertically partitioned data. In: Proceedings of the 8th ACM SIGKDD International Conference on Knowledge Discovery and Data Mining, KDD'02, Edmonton, pp. 639–644. ACM, New York (2002). doi:10.1145/775047.775142
117. Vaidya, J., Clifton, C., Kantarcioglu, M., Patterson, A.S.: Privacy-preserving decision trees over vertically partitioned data. ACM Trans. Knowl. Discov Data (TKDD) **2**(3), 14:1–14:27 (2008). doi:10.1145/1409620.1409624
118. Vaidya, J., Kantarciouglu, M., Clifton, C.: Privacy-preserving naive bayes classification. VLDB J. **17**(4) (2008). doi:10.1007/s00778-006-0041-y
119. w3.org, Resource Description Framework (RDF). http://goo.gl/A3eq
120. w3.org, RDF current status. http://goo.gl/zPes
121. w3.org, OWL Web ontology language guide. http://goo.gl/sQrDQ
122. Xiao, X., Tao, Y.: Personalized privacy preservation. In: Proceedings of the 2006 ACM SIGMOD International Conference on Management of Data, SIGMOD'06, Chicago, pp. 229–240. ACM, New York (2006). doi:10.1145/1142473.1142500
123. Xu, S., Sandhu, R.S., Bertino, E.: TIUPAM: a framework for trustworthiness-centric information sharing. In: Trust Management III: IFIP Advances in Information and Communication Technology, vol. 300, West Lafayette, pp. 164–175. Springer, Berlin/Heidelberg (2009). doi: 10.1007/978-3-642-02056-8_11
124. Yu, T., Winslett, M., Seamons, K.E.: Supporting structured credentials and sensitive policies through interoperable strategies for automated trust negotiation. ACM Trans. Inf. Syst. Secur. **6**(1), 1–42 (2003). doi:10.1145/605434.605435

Part III
High Performance Cloud Computing

GPGPU Computing for Cloud Auditing

Virginia W. Ross and Miriam E. Leeser

Abstract With the increasing computational complexity of cloud auditing and other data-intensive analysis applications, there is a growing need for computing platforms that can handle massive data sets and perform rapid analysis. These needs are met by systems with accelerators, such as Graphics Processing Units (GPUs), that can perform data analysis with a high level of parallelism employing tools like Hadoop MapReduce to handle massively parallel computing jobs. Applying GPUs to general purpose processing is known as GPGPU. This chapter uses an introductory approach to cover the basics of GPUs and GPGPU computing and their application to cloud computing and handling of large data sets. The main aim is to give the reader a broad background on how GPGPUs are used and their contribution to advances in cloud auditing.

1 Introduction

As information management applications grow in size and complexity, and as the amount of data transferred across the Internet and other networks continues to expand rapidly, the need for a technology to effectively and efficiently deal with the big data transfer issue and the associated analysis becomes critical. A cluster containing nodes with CPU-GPU combinations capitalizes on the GPU's many-

---

"Approved for Public Release; Distribution Unlimited: 88ABW-2013-0081, 09-Jan-2013"

V.W. Ross (✉)
Air Force Research Laboratory, WPAFB, OH, USA
e-mail: virginia.ross@us.af.mil

M.E. Leeser
Northeastern University, Boston, MA, USA
e-mail: mel@coe.neu.edu

K.J. Han et al. (eds.), *High Performance Cloud Auditing and Applications*,
DOI 10.1007/978-1-4614-3296-8_10,

core parallelism and high memory bandwidth. Clusters with GPUs provide the opportunity to speed up applications by one or more orders of magnitude, enabling more effective problem solution and data handling, whether for cloud auditing or for other data-intensive or compute-intensive applications. This speedup is most effective when the data is easily decomposable into many subsets. Modern GPUs can perform double precision floating point operations efficiently, enabling them to handle many of the most challenging analytical tasks.

The objectives of this chapter are to describe GPUs, explain their features and capabilities, and discuss GPGPU applications [55] and frequently used software tools, including programming languages such as Compute Unified Device Architecture (CUDA) and OpenCL. Various libraries and toolsets that can facilitate greater ease of programming are also discussed, such as using Hadoop MapReduce for data parallelization. In addition, methods to enhance application performance employing GPGPUs, including providing the computational speed and data storage availability for high performance cloud auditing are covered.

This chapter is organized as follows. Section 2 covers the background of GPGPU computing. Section 3 describes heterogenous CPU-GPU systems and performance optimization. Section 4 presents GPGPU programming, such as CUDA, OpenCL and conjugate gradient, and programming methods for running cloud applications with GPGPUs. In Section 5, high performance systems for cloud auditing are introduced. Section 6 discusses GPGPU security issues. Conclusions are presented in Section 7.

2 GPGPU Computing

With GPUs produced widely and inexpensively for graphical applications for the commercial gaming market, the computing industry has decided to capitalize on their computing capabilities. The low cost and high performance of commercial gaming technologies enable systems to efficiently handle data-intensive and compute-intensive applications, while capitalizing on the affordability of mass-produced commercial computing hardware. Because of the performance advantages, including significant computational speedup for certain applications, and reasonable cost, GPUs are being considered for an increasing number of computing applications.

2.1 History and Technical Discussion of GPU

General-purpose programming on graphics processing unit (GPGPU) has been an active area of research since the early 1980s. Graphics Processing Units (GPUs) are cards designed for computers that are optimized to process graphics. Graphics hardware is highly parallel and designed to handle compute intensive tasks. It also tends to be inexpensive due to high volume sales. Early on, scientists recognized that such hardware could be used as a cost effective platform for accelerating compute intensive problems. In the early years, graphics hardware was designed for specific tasks such as texture, shader and vertex processing. Languages such as DirectX (DX) and Open Graphics Library (OpenGL) were designed around this hardware. Scientists who wished to write code for such languages and hardware had to understand the GPU processing pipeline and write their code using graphics primitives. For this reason, until the mid 2000s, GPGPU was extremely onerous and its practitioners were rare.

Several things changed in the mid 2000s. Graphics hardware became increasingly more capable, more parallel and more general purpose. This was largely driven by computer games. Designers writing games wanted more and more video features such as lifelike skin and hair, realistic movement of armies and groups of people, as well as weather such as realistic wind and rain. Programming such features requires general purpose computer architectures; specialized graphics tasks could easily be mapped to fast general purpose architectures. As a result, the graphics hardware became more general purpose as did the programming languages to program the hardware. In 2006, NVIDIA introduced Compute Unified Device Architecture (CUDA). Note that CUDA refers to both the hardware architecture and the language, a C like language for programming NVIDIA GPUs [36].

Open Computing Language (OpenCL) 1.0 was introduced in December 2008 [26]. OpenCL is an open standard that can be used to program a wide variety of devices including multicore processors and GPUs. There are OpenCL compilers available from the two major GPU vendors: AMD [1] and NVIDIA [43]. CUDA and OpenCL are the two major frameworks for programming GPUs today.

Scientists rapidly adopted the new GPGPU paradigm. GPUs have dozens of "cores" each with multiple parallel units. Thousands of parallel operations can be launched as threads. GPUs are highly parallel, but lack the flexibility of general purpose processors. Threads are launched in lockstep in a Single Instruction Multiple Thread (SIMT) programming model. Interthread and inter-core communication is limited. If the ultimate goal is to generate pixels on a screen, such communication is rarely required. Despite these limitations, scientists continue to be attracted by the high compute intensity and low cost of GPU processors. Many scientific problems have been mapped to GPUs with a great deal of success. Since 2008, both NVIDIA and AMD have supported double precision floating point in hardware, further helping scientific applications.

2.2 Data-Intensive Cloud Computing

GPUs fill a specific niche in cloud computing since they are high compute intensity devices that can perform calculations involving large data sets rapidly and with high fidelity. The parallel computational capability of the multiple stream processors available on a single GPU has the ability to handle the demanding computing needs required for processing big data sets.

Computational tasks involving large data sets, as used in cloud auditing, must have adequate resources to perform in real-time. These include hardware, software, network, and infrastructure resources. For a hybrid solution consisting of a mix of CPUs and GPUs, the approach normally involves one or more powerful CPU host processors to handle control intensive code, input and output, memory access, and extensive data storage capabilities. The compute intensive, parallel operations are performed on the GPUs. These systems need to have all of the appropriate software for performing I/O, data manipulation, and storage functions. The CPUs and storage can be hosted on virtual machines (VMs) on the cloud [38]. To incorporate GPUs, a high bandwidth connection is required between CPUs and GPUs, necessitating that the GPUs are also supported in the cloud. These platforms will need a high throughput network to transport data from the source to the CPU-GPU system in the cloud. System administration and adequate infrastructure support are required, including system, power, air conditioning, network, and facility maintenance. System backup and contingency arrangements to handle failures or utility interruptions should also be available if needed.

To meet the needs of a wide variety of users, steps must be taken to provide adequate resources for the planned system users. In addition to taking adequate measures to provide user authentication and partitioning of files, steps to ensure resource availability and accessibility should also be made.

2.3 GPU Architecture Description

There are two main types of GPU devices; devices where the GPU is integrated with the main processor on a single chip, and ones where the GPU is a separate card, referred to as a device, with its own local memory. In this chapter we will focus on the second type as this is the most common type used for GPGPU. Single chip solutions, including AMD Fusion and Intel's Sandy Bridge architectures, are becoming more capable, but most scientific computing is done on a discrete GPU sold as a card that slots into a computer system.

A GPU device is a card with local memory that communicates with a CPU or host processor, typically over the Peripheral Component Interconnect (PCI) Express bus. The host loads programs onto the GPU and transfers data to and from the GPU. In some configurations, more than one GPU may be connected to a single host, but a one to one configuration is the most common. Direct GPU communication is still an area of research and has not been widely adopted.

The GPU device itself generates performance with a massive amount of low level parallelism. While a CPU core devotes considerable area to control, caches and memory management, a GPU minimizes this overhead and devotes more chip area to computation. A GPU accomplishes this goal through a number of techniques. One is using data parallelism for processing. NVIDIA uses a Single Instruction Multiple

Data (SIMD) style of computation where the same instruction is used to launch many threads of computation. Each thread uses its own identity (id) to determine its particular computation. The AMD model is similar; however, it uses a vector processing programming model to express data parallelism. With either style, the user can express a great deal of parallel execution with very few instructions. In hardware, a large number of cores that support floating point execute these parallel instructions. Thus, the compute density of GPUs is much higher than that of CPUs.

Another advantage of GPUs is high memory bandwidth. On board device memory is currently up to 6 GB for high end NVIDIA and AMD cards. While this is not a huge amount of memory, the bandwidth to the GPU is much higher than the memory bandwidth to the CPU. For example, high end NVIDIA cards deliver 150 GB/s of memory. Such high memory bandwidth is required to keep the large number of parallel cores busy. This memory bandwidth allows for high throughput processing at the expense of latency. The latency to device memory can be very long, on the order of hundreds of clock cycles. GPUs provide this high memory bandwidth by requesting many different pieces of input data in parallel, and only sending individual operations on to a compute unit once their data have arrived. With enough memory requests in parallel, memory latency is hidden by the interleaving of data transfer with computation. This process reduces the need for caches.

GPUs provide very little automatic support for the memory hierarchy. The large, multi-level caches found in CPUs are not present. Small caches for groups of parallel cores are provided, but much of the memory management has to be done manually by the programmer. While caches make sense for typical CPU workloads, many of the data access patterns found in large data parallel programs, such as streaming data, do not make effective use of the caching hierarchy.

There are some problems that do not lend themselves to GPGPU solutions. Problems that do not exhibit large amounts of data parallelism cannot take advantage of a GPU architecture. In addition, support for communications between threads and blocks on a GPU is very limited, so problems that require a great deal of communication will not run fast on a GPU. Similarly, problems with a great deal of control flow, where different choices are made at run time, do not run effectively on GPU architectures.

There has been an increasing interest in power efficiency and reducing energy usage in computation. GPUs are very power hungry; however, since they significantly speedup computations, the resulting energy usage may be significantly lower than running the same computation on a CPU. When looking at performance per watt, GPUs can efficiently deliver performance, especially for those applications that run at close to peak performance on a GPU. The Green 500 list [14], managed by Feng and Cameron at Virginia Tech, lists the most energy efficient supercomputers in the world. Several of the top ten systems employ GPUs. The primary means of developing an energy efficient computer are to either use low power CPUs or to add energy-efficient accelerators from the graphics or gaming markets, such as GPUs or Cell Broadband Engine (BE) processors, to complement the CPU processing.

2.4 GPGPU Computer Architectures

For efficiency and improved interfacing with users, GPU-based computing systems normally employ a CPU host and GPU-based processors, with the host interacting with users and the GPU interacting with the CPU. As mentioned previously, the host and GPU processors can be separate devices, though combining them on a single chip can boost performance.

The most common GPGPU architecture [19, 25] for scientific and technical computing employs NVIDIA GPUs [6, 16, 28], because of the relative maturity of the CUDA software, which only runs on NVIDIA GPUs. However, with the growing interest in and increased maturity of OpenCL, which is not platform specific, GPGPUs from AMD and other vendors are getting more attention for high performance computing.

2.5 GPGPU System and Programming Options

The basic options in designing a GPGPU system include the choice of hardware and the programming language. If CUDA is used, NVIDIA GPUs will be used. If OpenCL is used, GPUs from NVIDIA or other vendors, such as AMD, can be used. Also, the type of CPU used, the data storage and memory options will need to be determined. Many of these systems can be virtual systems hosted on a cloud. The connectivity between devices and with the outside network also needs to be determined. Any resource limitations need to be considered.

3 Heterogeneous CPU-GPU Systems

Heterogeneous CPU-GPU systems [64] are required for high performance applications since the GPU is programmed from the CPU; high performing applications capitalize on the strengths of both architectures. Many popular cloud computing languages and frameworks, such as MapReduce, are developed primarily for running on a CPU. CPUs also excel in sequential computations and I/O and memory latency limited applications, while GPUs, with their multi-threaded architecture can provide one or more orders of magnitude speedup on computations for highly parallel applications. Such capabilities are invaluable in solving many Big Data applications, including cloud auditing.

3.1 CPU-GPU Systems

Some of the most powerful computers in the world are heterogeneous supercomputers combining the benefits of high end CPUs and GPUs. For example, three of the top ten systems on the Top 500 list of supercomputers released in June 2012 are heterogeneous systems with high end CPUs – either Cray or Xeon and high end NVIDIA 20XX series GPUs [58]. These top ten systems each have demonstrated throughput in excess of 1 petaFLOPs or one quadrillion floating point operations per second.

While large heterogeneous CPU-GPU systems can be very powerful, smaller CPU-GPU systems can meet the needs of many other research and computational groups, including the high throughput needs of cloud auditing. For example, the HokieSpeed system at Virginia Tech, which uses 126 kW of power and employs NVIDIA GPUs, has an output of over 100 teraFLOPS, and is used in support of multiple research efforts for academic and defense applications.

3.2 Condor Supercomputer

Another example of a heterogeneous supercomputer employing GPUs is the Condor Cluster. One compute node contains dual Intel Nehalem x5650 processors and 2 NVIDIA Tesla GPUs, as well as 24 GB RAM, 2 TB HD, 1200 W power supply, 40 GB/s InfiniBand connection, and Dual 10 GB Ethernet connections. The complete Condor system combines these compute nodes with Cell Broadband Engine (BE) processors. Fig. 1 shows a picture of a single node and a schematic of the full system, which has a theoretical peak throughput of 500 teraFLOPS or 0.5 petaFLOPS. Adding Cell BE processors in the form of PlayStation3 (PS3) gaming consoles to the CPU and GPU mix adds increased computational power at low cost. This system was designed with an emphasis on affordable supercomputing, with an aim to demonstrate energy efficiency [37, 54]. This system supports a wide range of applications. One area is neuromorphic computing, where text recognition software processes around 16–18 pages of occluded text per second for pattern recognition [46]. Other applications include synthetic aperture radar image formation using GPGPUs, high definition video processing, and large matrix multiplication [49]. Another application of this system is a computational chemistry research on chemical compounds [22]. For all these applications, the high throughput system enables faster time to solution and also the ability to solve more computationally intense problems than can be solved on less powerful computers. GPUs contribute significantly to this capability.

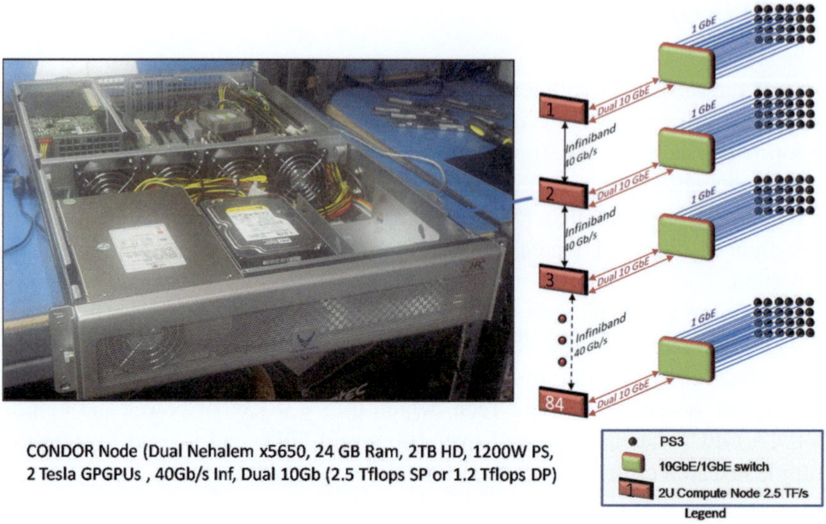

Fig. 1 A Condor system node (Reproduced from Ross [49])

3.3 Performance Optimization

There are multiple approaches to maximize the performance of a CPU-GPU heterogeneous system [5, 41, 56]. One primary approach is to capitalize on the advantages of both types of hardware. With the GPUs able to perform rapid computations on massive amounts of data with highly parallelized code [7], they are often the hardware of choice for such computations. However, control, input/output functions and data transfer are handled more effectively by CPUs.

Significant speedups by employing GPGPUs can be obtained by a series of steps. First, the code must be highly parallelizable to benefit from GPU acceleration. Also, by mapping the number of threads and GPU blocks used to the available GPU architecture, the computation can optimize its benefit from GPU acceleration [35]. Note that such optimization can be tricky since the best performance for GPU code does not necessarily occur when the architecture has the highest occupancy [62]. One study that highlighted the variable speedup on several types of architectures for different types of computation and communication patterns was performed at Virginia Tech using OpenCL and the 13 Dwarfs from Berkeley [13]. These 13 Dwarfs encompass frequent types of high performance computations encountered in parallel computing. No one type of GPU performed best on every application, with less powerful GPUs outperforming the more powerful ones on one type of Fast Fourier Transform (FFT) that relies heavily on a single GPU processing unit. Communication within a CPU was often found to be slower than between a CPU and a GPU. This study demonstrated the value of architecture aware optimization of computing applications. In addition to achieving high performance on the GPU, for overall system performance it is also important to take advantage of the CPU. Tasks that require extensive memory access or are less adaptable to parallelization are best performed on the CPU. Taking best advantage of both CPUs and GPUs available can maximize system throughput for a given computation.

3.3.1 Input/Output (I/O) Issues

Since most GPU-based systems are heterogeneous systems consisting of both CPUs and GPUs, and possibly other accelerators, any discussion of input and output issues will need to cover I/O between the network and the system and also I/O between different components within the system. For the sake of simplicity, this discussion will concentrate primarily on CPU-GPU based systems, though some systems contain additional accelerators, such as the Cell Broadband Engine (BE) processor or Field Programmable Gate Arrays (FPGAs).

A common approach in CPU-GPU heterogeneous systems is to have a CPU host that handles I/O between the system and the network along with some computing tasks, and GPU boards that communicate with the CPU and perform a major share of the computational tasks.

3.3.2 Performance Optimization of I/O Operations

There are various approaches to optimize performance between CPU and GPU components [23]. For example, the Tianhe-1 system uses a software pipelining [47, 48] technique when performing GPU computing to reduce the effect of the communication overhead between the memory on the CPU and GPU [65]. This technique separates each task into three components: input, computation, and output to host. Data transfer and computation tasks are overlapped for higher efficiency. More recently, researchers have introduced CUDA Direct Memory Access (DMA) [30, 50] to support overlapping computation and data transfer just on the GPU [4]. Code to support CUDA DMA is available as an open source library.

A method for optimizing the time to fetch memory from on-board GPU memory, called warp specialization, is available in recent GPUs. Warp specialization allows the user to launch kernels to fetch data into the GPU at the same time as other kernels are operating on data already fetched. The third stage of the pipeline stores results back in device memory.

4 GPGPU Programming

The most popular languages for programming GPGPU are CUDA and OpenCL. These are both relatively low level and based on the C programming language. CUDA is a language developed by NVIDIA for programming their GPUs. CUDA describes both their architecture and the programming language used. CUDA is the first GPGPU language, and is widely used by scientists. It has very well developed libraries and some advanced language features not available in OpenCL. OpenCL has the advantage of being an open standard, and is supported by both NVIDIA and AMD as well as by multicore manufacturers including Intel and IBM. At the time of writing, OpenCL libraries for GPGPU are not as well developed as CUDA libraries.

CUDA organizes a parallel computation using the abstractions of threads, blocks and grids when expressing parallelism in their programs. Groups of threads form blocks and groups of blocks are organized into grids. In December 2011, NVIDIA announced that they were going to make their CUDA compiler runtime source code and internal representation format public, opening up the technology for different programming languages and processor architectures. There are numerous tutorials and textbooks available for programming in CUDA [12,29,44]. Along with CUDA, a programmer gets the NVIDIA C compiler for CUDA (nvcc), a profiler, a version of the gdb debugger for the GPU, and a CUDA runtime driver for launching kernels. In addition, there are many libraries available for CUDA.

CUDA files use the naming convention *.cu. CUDA files include both the host and GPU (device) code. A simple CUDA program creates data structures on the host computer, transfers the data from the host to the device using the cudaMemcpy command, calls the computation (referred to as the kernel) on the device, and then copies the results back to the host. More complex programs may leave the data in device memory and launch several computations before the results are transferred to host memory. Since data transfers can take a long time compared to computation, this process of bringing data to the device increases computational efficiency.

4.1 CUDA Programming

All CUDA [15] kernels have a similar kernel structure. CUDA works with a three dimensional built in structure of grids, blocks and threads. CUDA uses the following built in variables: blockIdx contains the block index within the grid, threadIdx contains the thread index within the block, and blockDim contains the number of threads in a block. Rather than loop through elements of an array, in CUDA multiple threads are launched, each of which operates on its own element of the array. Thus, a CUDA kernel operating on array A, contains the line:

```
int idx = blockIdx.x*blockDim.x + threadIdx.x;
```

This line of code has each thread determining its own id for the element of the array it is working on. To reference its array element, the construct `A[idx]` is used in the CUDA code. To call a kernel, the number of blocks and block size must be determined before the kernel is invoked. The syntax `"<<<"` and `">>>"` tells the CUDA compiler that this kernel is going to be launched on the GPU device.

The reader is referred to an example presented in an article by Farber [12]. This article presents a simple CUDA kernel, written to run on the GPU, that increments each element of a floating point array.

It is not necessary for users to write all of their routines at such a low level. There are many libraries of CUDA routines available for frequently used programming constructs. NVIDIA libraries include The CUDA Basic Linear Algebra Subroutines (cuBLAS) library and the CUDA Fast Fourier Transform (CUFFT) library . Thrust is an open-source template library that is used for data parallel CUDA applications. Thrust has an interface that is similar to the C++ Standard Template Library (STL). These libraries provide a high level interface to ease the task of GPGPU code development.

4.2 OpenCL Programming

OpenCL [27,31,42], an open, industry-standard approach to programming GPUs, is growing in popularity. Like CUDA, OpenCL provides a general purpose approach to parallelizing data for computations on GPUs. OpenCL has the advantage of not being vendor-specific. As such, applications written in OpenCL are portable between GPUs and CPUs from different vendors. Similar to CUDA, OpenCL has the host CPU control operation of the GPU. OpenCL is currently supported by NVIDIA, AMD, Intel and IBM among other vendors. However, OpenCL does not have the rich set of libraries that have been developed for CUDA. In addition, while the same code will run on both NVIDIA and AMD GPUs, the code is not "performance portable." Since the underlying architectures are different, a programmer who wishes to get peak performance may need to tune OpenCL code to the specific platform. There are numerous sources for OpenCL tutorials.

A major difference between OpenCL and CUDA is that, since OpenCL is designed for a wide variety of platforms, an OpenCL programmer must define the context in which the program is intended to run. In an OpenCL context, the programmer defines the devices that the code will run on, which may be one or more CPUs (`CL_DEVICE_TYPE_CPU`) or GPUs (`CL_DEVICE_TYPE_GPU`) or a combination. From the OpenCL code, the programmer can execute a query to discover which devices are part of the current context using:
`context.getInfo<CL_CONTEXT_DEVICES>()`. Once the context is defined, a programmer will notice many similarities between CUDA and OpenCL code. Memory needs to be allocated to run a program. Kernels that run on devices are programmed similarly to CUDA programs. OpenCL has work items, which are similar to CUDA threads, and work groups, which are similar to CUDA blocks. Instead of grids, an OpenCL programmer refers to the ND-Range, an n-dimensional arrangement of work groups. Below are the details of the program that increments a floating point array. This performs the exact same computation as the example from

Farber [12], except in OpenCL rather than CUDA. Notice that OpenCL is more verbose than CUDA, as evident from this example.

In OpenCL, the kernel for incrementing an array is defined as a string. Note how similarly the kernel computation is described in the two languages.

```
const char * gpuKern = "__kernel void
    incrementArrayOnDevice(__global float *a, int N) {
    int idx = get_global_id(0); if (idx<N) a[idx] = a[
    idx]+1.f; }";
```

The calling code for this kernel involves the following steps:

1. Get platform and device IDs:

```
cl_platform_id plat;
clGetPlatformIDs(1,&plat,NULL);
cl_device_id dev;
clGetDeviceIDs(plat, CL_DEVICE_TYPE_GPU, 1, &dev,
    NULL);
```

2. Create context and command queue:

```
cl_context_properties cprops[3] = {
    CL_CONTEXT_PLATFORM, (cl_context_properties)plat
    , 0};
cl_context ctx = clCreateContext(cprops, 1, &dev,
    NULL, NULL, &err);
cl_command_queue que = clCreateCommandQueue(ctx,
    dev, 0, &err);
```

3. Create the buffer on the device:

```
cl_mem devBuff = clCreateBuffer(ctx,
    CL_MEM_READ_WRITE, size, NULL, &err);
```

4. Create and build the program:

```
cl_kernel krn;
cl_program prg = clCreateProgramWithSource(ctx, 1,
    &gpuKern, (size_t*)&krnLen, &err);
clBuildProgram(prg, 1, &dev, bldOpt, NULL, NULL);
clCreateKernelsInProgram(prg, 1, &krn, NULL);
```

5. Do calculation on device:

```
// Compute execution configuration
    size_t blockSize = 4;
    size_t nBlocks = N/blockSize + (N%blockSize ==
        0?0:1);
    size_t gSize = blockSize * nBlocks;
// Set up kernel arguments and launch kernel
```

```
clSetKernelArg(krn, 0, sizeof(float*), &devBuff)
   ;
clSetKernelArg(krn, 1, sizeof(int), &N);
clEnqueueNDRangeKernel(que, krn, 1, NULL, &gSize
   , &blockSize, 0, NULL, NULL);
clEnqueueBarrier(que);
```

6. Retrieve result from device and store in `b_h`:

```
clEnqueueReadBuffer(que, devBuff, CL_TRUE, 0, size,
    b_h, 0, NULL, NULL);
clFinish(que);
```

7. Cleanup (not shown)

4.3 Higher Level GPGPU Programming Tools

Higher level tools for GPGPU programming are also available, both commercial and open source. The Portland Group sells compilers for compiling Fortran and C++ to CUDA. Their OpenCL compiler targets 32-bit Reduced Instruction Set Computer (RISC) processors developed by ARM; GPUs are not currently supported. PyCUDA and PyOpenCL are freely available tools that generate either CUDA or OpenCL from Python source code. These tools use autotuning to obtain efficient solutions on different hardware devices.

The MathWorks Parallel Computing Toolbox (PCT) supports the use of NVIDIA GPUs and CUDA from MathWorks source code. There are two different models of computation. One is to call CUDA functions that were coded at MathWorks and ship with the PCT. The second is for a programmer to code his or her own CUDA functions and have them called from within MATLAB GPU computing. Both use the data type "GPUArray" in MATLAB to represent data that is stored on the GPU. The company AccelerEyes also sells GPU solutions for MATLAB code.

Microsoft has recently announced C++ Accelerated Massive Parallelism (C++ AMP), a language that promises to support CPU and GPU programming from C. NVIDIA, Cray and the Portland Group have announced OpenACC, a programming standard for parallel computing based on C++ or Fortran. It remains to be seen whether either of these standards will be popular with programmers.

4.4 GPGPU Programming Process

This chapter has outlined many programming options for GPGPU programming [33]. In this section we describe the typical programming process as outlined in Fig. 2. Typically a programmer starts with some code that he or she would

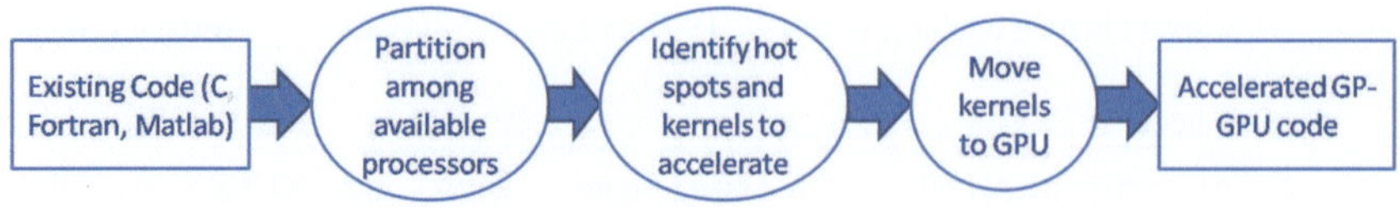

Fig. 2 Flowchart of programming process

like to accelerate. This code may be written in any of a variety of programming languages such as C, Fortran or MATLAB. A programmer may have several options of systems, including single processor, multiple processors, multiple threads with either a single processor or multiprocessor and using accelerators such as GPUs. Parallel programming patterns suitable for multiprocessors have been identified in an influential book by Mattson et al. [40]. Note that this book predates the widespread use of GPGPU. The most common parallel programming approaches and languages use OpenMP [45] for shared memory multi-processing and Message Passing Interface (MPI) [2] for networked processors.

Once the program has been distributed over the number of processors available, the next step is to partition the code on a single processing node between CPU and GPU. The programmer may know which portions of the code can require acceleration or may use a profiler to identify the portions of the code that take up most of the execution time. Once such kernels have been identified, a good first step is to look at available libraries to see what has already been implemented for GPU acceleration. [54] Many of these are mentioned above, including the CUDA Thrust and Basic Linear Algebra Subroutine (BLAS) libraries. If the algorithm to be accelerated is not available in a preexisting library, or the overall performance or features do not meet the programmer's needs, then the next step is to use one of the tools described in this chapter or to write one's own CUDA or OpenCL kernel to implement the necessary code.

It is important to note that a programmer may write code that moves parts of their algorithm to the GPU even if the resulting kernel runs slower on the GPU than on the CPU. The main reason for doing so is to avoid transferring data between GPU and CPU since such transfers can be very costly in terms of performance. When choosing which code to move to the GPU it is important to consider end-to-end system performance and not just the speedup of a particular kernel. A successful project results in end-to-end speedup by using an appropriate combination of multicore processing and GPU acceleration.

4.5 GPU-Based Conjugate Gradient Programming

Leeser and Yablonski developed a Conjugate Gradient (CG) solver for GPGPU. CG is an algorithm for the numerical solution of systems of linear equations whose matrix is symmetric and positive definite. The conjugate gradient method is iterative;

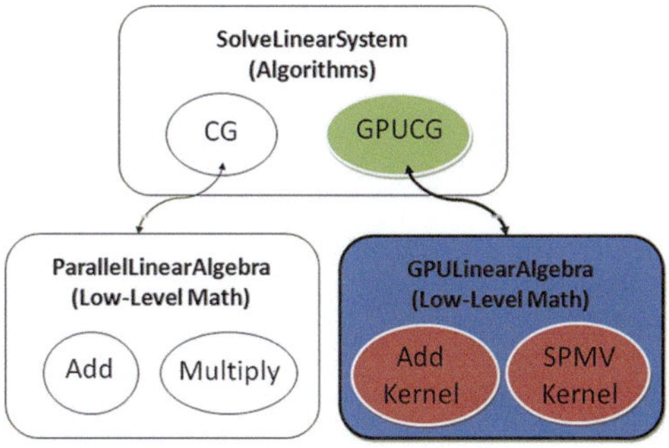

Fig. 3 Integrating GPU CG into SCIRun

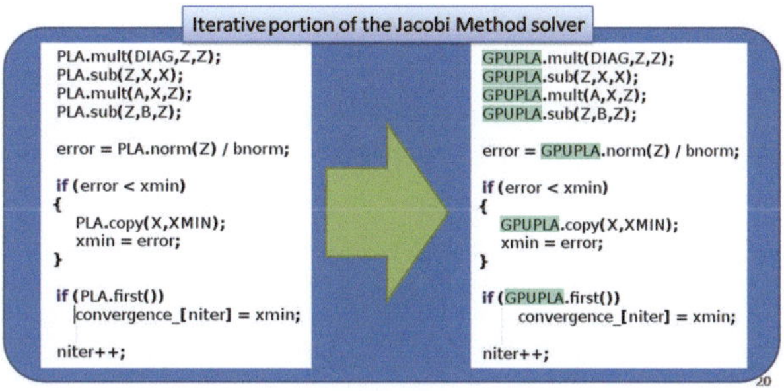

Fig. 4 Turning CPU code into GPU code

it can be applied to sparse systems that are too large to be handled by direct methods [52]. The solver was designed to be called from the `SolveLinearSystems Module`, part of the SCIRun Problem Solving Environment [51].

A software implementation of CG exists in SCIRun. To add a GPU version, low level components such as Sparse Matrix Vector multiply and matrix addition were added that were implemented to run on the GPU. The hierarchy of software modules is shown in Fig. 3. This is similar to the approach used by Mathworks in their Parallel Computing Toolbox [39]. An alternative is to tune the entire CG algorithm to the GPU and call that GPU version. This alternative was not chosen for several reasons. First, it is more difficult to interact with the visualization tools in SCIRun if CG is a monolithic implementation. Second, a third party CG implementation was tried and the overall system speedup was found to be only a moderate improvement over this approach. Finally, the approach used can be applied to many other modules in the SCIRun system. This approach is illustrated in Fig. 4 for another solver, the Jacobi method. PLA refers to the Parallel Linear Algebra module; GPUPLA calls the GPU versions instead.

Conjugate Gradient was run on several examples, including the heart ischemia module distributed with SCIRun as well as several large matrices from the University of Florida's Sparse Matrix Collection [59]. Note that these are large sparse matrices that vary in size from 6,300 to 3.5 million rows and have between 42,000

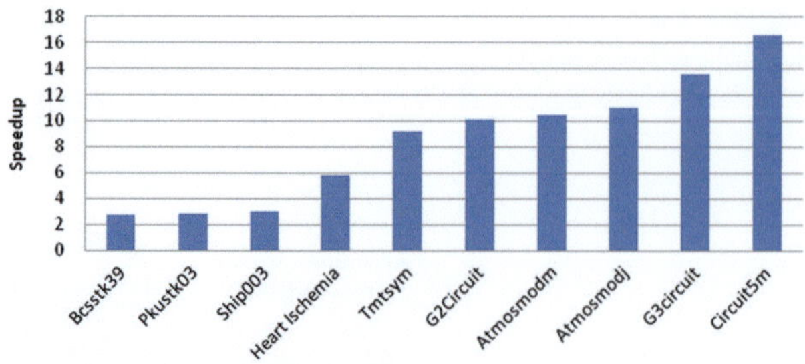

Fig. 5 Speedup of CG on a GPU over a CPU

and 14.8 million non-zero elements. The speedup for several of these examples is shown in Fig. 5. The CPU code was run on an Intel 2 1.86 GHz processor and compared to an NVIDIA GeForce GTX 280 GPU. Double precision floating point arithmetic was used for all examples. Other results from the same study, including CG with a preconditioner, have been published [32].

4.6 Programming Methods for Cloud Computing

Any cloud computing application that can benefit from the application acceleration possible with GPGPUs can be a candidate for running on a GPU. Intrusion detection and other compute intensive cloud auditing work is run on GPUs. Encryption is another cloud computing application that runs efficiently on GPUs. Other highly parallelizable applications also run on GPUs.

The book by Mattson et al. [40] identifies patterns for parallel programming. It describes commonly used patterns for expressing concurrency including task and data parallelism, patterns for expressing algorithm structures such as pipelining and divide and conquer approaches, and supporting patterns including Single Program Multiple Data (SPMD) and loop parallelism. The book was published before GPGPU was widely adopted; however, many of the concepts, especially those relating to SIMD and data level parallelism, directly apply.

A parallel programming pattern that is particularly applicable to GPU computing is MapReduce. MapReduce grew out of the concepts of mapping and reduction in functional programming models. Its advantage as a parallel programming model was first published by researchers at Google [10]. MapReduce operates by a combination of a "map" step that breaks up a computation into smaller, parallel computations, and dispatches them to a large number of nodes for processing. The "reduce" step collects and compiles the results from the "map" step to create a consolidated solution to the computation.

While MapReduce traditionally has run on CPU clusters, the multi-threaded parallelism and massive number of compute units per GPU offer the potential for significant speedup by running applications on GPUs versus on CPUs. Several frameworks have been created to enable users to run MapReduce applications on

GPUs [8,18,57]. Each of these frameworks has its own emphasis. These applications can be designed for different scenarios: running on a single GPU or on multiple GPUs, or running on NVIDIA GPUs or on GPUs from other vendors, such as AMD. Since MapReduce also runs on CPUs, some of these frameworks have the advantage of running on both CPUs and GPUs, which provides more code portability than traditional GPU languages, such as CUDA or OpenCL.

One of the most widely used MapReduce frameworks is Hadoop from Apache [3]. Hadoop is open source software for the distributed processing of large data sets on clusters of computers. Hadoop is designed to scale to thousands of machines, and includes support for high availability, even when individual nodes fail. Hadoop replicates data on multiple racks to lessen the impact of power outages or switch failures. There are several projects that combine Hadoop and CUDA to make the Hadoop framework available on clusters with GPGPU. One challenge is that Hadoop users tend to be Java programmers, while CUDA programs are called from C++. This issue can be addressed by writing the code in C and using Hadoop Streaming to enable the executable to be used as a Hadoop mapper or reducer. Hadoop will then run the application on each node.

Several other MapReduce frameworks for GPUs are also available. Mars is a MapReduce framework for GPUs that was jointly developed a number of years ago by researchers at the Hong Kong University of Science and Technology, Sina Corporation, and Microsoft Corporation [17]. Mars simplifies programming on a GPU, by use of MapReduce, and, using the NVIDIA G80 GPU, it offers up to 16 times speedup compared to a quad-core CPU on 6 common applications.

A more recent development is MapCG, a MapReduce framework that writes code that is portable between CPUs and GPUs [18]. An implementation of MapCG by Tsinghua National Laboratory and the China Research Lab of IBM, employing both multi-core CPUs and NVIDIA GPUs, demonstrated an average of 1.6–2.5 times speedup over previous MapReduce implementations on eight common applications.

StreamMR is a MapReduce framework optimized to capitalize on the architectural advantages of the AMD GPUs [11]. This OpenCL MapReduce framework capitalizes on atomic free algorithms, to reduce the memory access delays that occur with executing atomic operations on different threads of a GPU. StreamMR, optimized for AMD GPUs, can outperform atomic-based MapReduce implementations and has demonstrated over five times performance improvement over prior atomic-free MapReduce applications on the AMD Radeon HD 5870.

All of these MapReduce frameworks reduce the programming overhead and make GPU programming easier for the average user.

5 High Performance Systems for Cloud Auditing

Cloud auditing has two primary areas of emphasis: reducing cloud security risks and increasing availability and performance of cloud computers in a contested network environment. The Cloud Security Alliance (CSA) [9] is an interagency group promoting security assurance for cloud computing. They promote methods to streamline the cloud audit process for evaluating cloud services, including Infrastructure as a Service (IaaS), Platform as a Service (PaaS), and Software as a Service (SaaS). According to Viswanathan [61], as cloud computing permeates more enterprises, cloud auditing is needed to ensure that clouds work as expected, becoming more vital. Areas to audit that he mentions include regulatory compliance, disaster recovery and business continuity plans, cloud security, performance and reliability, and the return on investment for using a cloud. All of these audit techniques can capitalize on the computing capabilities of GPUs [61].

GPUs can add significant performance benefits to CPU clusters, with the ability to accelerate highly parallel computations on GPU clusters. An increasing number of systems containing application accelerators, including GPUs, are appearing on the TOP 500 list, the list of the world's most powerful supercomputers, as evaluated using a specific Linpack benchmark suite. One example is the Tianhe-1 system, located at China's National University of Defense Technology, that employs Xeon Quad-core CPUs and AMD 4870 GPUs, which was ranked number 5 in November 2009 [65].

Also, with the need to save money and be more environmentally aware, organizations are becoming more conscious of the power usage of their computers. The Green500 Supercomputer list [14], compiled by Feng and Cameron at Virginia Tech, ranks supercomputers by performance in millions of floating point operations (MFLOPS) per watt (W). This list includes systems with a variety of architectures, including many high performance systems that capitalize on graphics accelerators and that can be energy efficient in calculating large numbers of highly parallel computations.

To evaluate energy efficiency, Huang, Xiao, and Feng [21] compared the operating efficiency of running a common biological analysis code, GEM, in three manners; on a serial CPU, on a multi-threaded CPU, and on a hybrid CPU-GPU. While the CPU-GPU version used a considerable amount of power at any given time, it performed GEM computations 46 times faster than the serial CPU and 25 times faster than the multi-threaded CPU. Also, the CPU-GPU hardware had a lower execution time and lower energy consumption for a given computation as the number of GPU blocks and the number of threads increased. This decreased computation time when employing GPUs contributed to the increased energy efficiency when employing GPUs. GPU acceleration also has the potential to contribute similar performance and energy efficiency benefits to some highly parallel applications when run on a cloud computer.

6 GPGPU Security Issues

Since cloud computing involves performing computational tasks over a network, security issues can arise from the users' devices, on the Internet, and on the cloud system. As a result, malware can be introduced at any of these places. Some of the primary security concerns on a cloud computer are regarding access control and data integrity. While GPU security research is less extensive than CPU-based security research, there is growing interest in GPU security, both in areas where GPUs can enhance security applications such as intrusion detection or data encryption, and in ways to ensure security on the GPUs. Currently, many GPGPU security issues are delegated to the end user.

One vital area of cloud auditing is intrusion detection. With the ability of GPUs to perform real-time, detailed rendering of 3-D animations, the same compute power makes them invaluable for the compute-intensive pattern matching that is needed for intrusion detection [20]. Huang, who did early work in this area, using OpenGL on NVIDIA GPUs, proposed a GPU-based solution for high performance pattern matching. In 2011, Vasiliadis et al. introduced the Multi-Parallel Intrusion Detection Architecture (MIDeA) [60]. The MIDeA has been used to accelerate network intrusion detection systems with a parallel implementation that uses both CPUs and GPUs. MIDeA is an example architecture that can take excellent advantage of cloud computing systems.

In addition, GPUs can capitalize on their ability to speed up computations to accelerate data encryption. Li et al. [34] demonstrated a 3–5 times speedup in employing the MD5-RC4 encryption algorithm using CUDA on an NVIDIA GPU, as compared to running it on an AMD Sempron CPU. Jang et al. [24] demonstrated a speedup of Secure Sockets Layer (SSL) encryption ranging from 22–33 times over the fastest CPU implementation. These research efforts demonstrated that a GPU could be a viable platform to efficiently encrypt data.

GPUs can be even more dramatic in speedup for other applications. For example, in 2010, Shimokawabe et al. [53] demonstrated an 80-fold speedup of non-hydrostatic weather code using single precision computations with a full GPU implementation running CUDA on NVIDIA GT200 Tesla GPUs, as compared to running them on a single CPU core.

As a counterpoint, Vuduc et al. [63] compared the performance on a CPU-GPU system to that on a CPU only system for applications with complex behavior, variable computational intensity, and irregular memory access, and found that with CPU tuning, the system performance running applications on the two types of systems was comparable. Since these applications were less consistent than some of the highly parallelizable applications with limited memory access that run so well on GPUs, these findings point out the value of tailoring the choice of computing system to the type of application.

Techniques to ensure the security of data on GPUs and to protect against malware are less developed. As GPUs continue to handle an increasing amount of computational applications, the threat of malware and malicious activities against GPUs is expected to grow. To counter this threat, security measures to protect GPUs are being developed.

7 Conclusion

This chapter has provided the history of general purpose programming on graphics hardware, has introduced programming languages and frameworks that ease the burden on application programmers, and has discussed several applications that are important for security in cloud computing and that benefit from acceleration on GPUs. The current state of affairs is that GPU hardware is changing rapidly, and in order to achieve performance, programmers need to understand the details of the underlying hardware architectures. Nevertheless, many researchers have found that the benefits of GPGPU programming outweigh the problems. Future languages and frameworks should make programming GPUs more accessible to researchers in a variety of areas, including cloud computing and auditing.

Acknowledgements This material is based upon work partially supported by the National Science Foundation (NSF) Engineering Research Centers (ERC) Innovations Program grant EEC-0946463, Mathworks, the Air Force Office of Scientific Research (AFOSR)/the Air Force Research Laboratory (AFRL) LRIR 11RI01COR. The authors would like to thank James Brock for developing the OpenCL code examples, and Devon Yablonski for writing the conjugate gradient GPU code and analyzing the performance of GPU computing. This work was completed while both James and Devon were graduate students, and had worked under the supervision of Miriam Leeser at Northeastern University.

References

1. amd.com, Programming guide: AMD accelerated parallel processing openCL. http://goo.gl/t9pkN (2012)
2. anl.gov, The Message Passing Interface (MPI) standard. http://goo.gl/D6zO0
3. apache.org, Apache Hadoop. http://goo.gl/tnkf
4. Bauer, M., Cook, H., Khailany, B.: CudaDMA: optimizing GPU memory bandwidth via warp specialization. In: Proceedings of 2011 International Conference for High Performance Computing, Networking, Storage and Analysis, SC'11, Seattle, pp. 12:1–12:11. ACM, New York (2011). doi:10.1145/2063384.2063400
5. Bordawekar, R., Bondhugula, U., Rao, R.: Believe it or not!: Mult-core CPUs can match GPU performance for a FLOP-intensive application! In: Proceedings of the 19th International Conference on Parallel Architectures and Compilation Techniques, PACT'10, Vienna, pp. 537–538. ACM, New York (2010). doi:10.1145/1854273.1854340
6. Buck, I.: GPU computing with NVIDIA CUDA. In: ACM SIGGRAPH 2007 Courses, SIGGRAPH'07, San Diego. ACM, New York (2007). doi:10.1145/1281500.1281647
7. Chafi, H., Sujeeth, A.K., Brown, K.J., Lee, H., Atreya, A.R., Olukotun, K.: A domain-specific approach to heterogeneous parallelism. In: Proceedings of the 16th ACM Symposium on Principles and Practice of Parallel Programming, PPoPP '11, San Antonio, pp. 35–46. ACM, New York (2011). doi:10.1145/1941553.1941561
8. Chen, L., Huo, X., Agrawal, G.: Accelerating MapReduce on a coupled CPU-GPU architecture. In: Proceedings of the International Conference on High Performance Computing, Networking, Storage and Analysis, SC'12, Salt Lake City, pp. 25:1–25:11. IEEE Computer Society Press, Los Alamitos (2012). doi:10.1109/SC.2012.16
9. cloudsecurityalliance.org, Cloud Security Alliance (CSA). http://goo.gl/u6VtG

10. Dean, J., Ghemawat, S.: MapReduce: simplified data processing on large clusters. Commun. ACM **51**(1), 107–113 (2008). doi:10.1145/1327452.1327492
11. Elteir, M., Lin, H., Feng, W., Scogland, T.: StreamMR: An optimized MapReduce framework for AMD GPUs. In: Proceedings of the 17th IEEE International Conference on Parallel and Distributed Systems, ICPADS'11, Tainan, pp. 364–371. IEEE Computer Society, Washington (2011). doi:10.1109/ICPADS.2011.131
12. Farber, R.: CUDA, Supercomputing for the masses: Part 2. Dr. Dobb's J. http://www.drdobbs.com/parallel/cuda-supercomputing-for-the-masses-part/207402986 (2008)
13. Feng, W., Lin, H., Scogland, T., Zhang, J.: OpenCL and the 13 dwarfs: a work in progress. In: Proceedings of the 3rd Joint WOSP/SIPEW International Conference on Performance Engineering, ICPE'12, Boston, pp. 291–294. ACM, New York (2012). doi:10.1145/2188286.2188341
14. green500.org, The Green 500: Ranking the world's most energy efficient supercomputers. http://goo.gl/WTDeM (2012)
15. Han, T.D., Abdelrahman, T.S.: hiCUDA: a high-level directive-based language for GPU programming. In: Proceedings of 2nd Workshop on General Purpose Processing on Graphics Processing Units, GPGPU'09, Washington, DC, pp. 52–61. ACM, New York (2009). doi:10.1145/1513895.1513902
16. Harris, M.: Many-core GPU computing with NVIDIA CUDA. In: Proceedings of the 22nd Annual International Conference on Supercomputing, ICS'08, Aegean, Sea, pp. 1–1. ACM, New York (2008). doi:10.1145/1375527.1375528
17. He, B., Fang, W., Luo, Q., Govindaraju, N.K., Wang, T.: Mars: a MapReduce framework on graphics processors. In: Proceedings of the 17th International Conference on Parallel Architectures and Compilation Techniques, PACT'08, Toronto, pp. 260–269. ACM, New York (2008). doi:10.1145/1454115.1454152
18. Hong, C., Chen, D., Chen, W., Zheng, W., Lin, H.: MapCG: writing parallel program portable between CPU and GPU. In: Proceedings of the 19th International Conference on Parallel Architectures and Compilation Techniques, PACT'10, Vienna, pp. 217–226. ACM, New York (2010). doi:10.1145/1854273.1854303
19. Hong, S., Kim, H.: An analytical model for a GPU architecture with memory-level and thread-level parallelism awareness. In: Proceedings of the 36th Annual International Symposium on Computer Architecture, ISCA'09, Austin, pp. 152–163. ACM, New York (2009). doi:10.1145/1555754.1555775
20. Huang, N.F., Hung, H.W., Lai, S.H., Chu, Y.M., Tsai, W.Y.: A GPU-based multiple-pattern matching algorithm for network intrusion detection systems. In: Proceedings of the 22nd International Conference on Advanced Information Networking and Applications – Workshops, AINAW'08, Ginowan, pp. 62–67. IEEE Computer Society, Washington, DC (2008). doi:10.1109/WAINA.2008.145
21. Huang, S., Xiao, S., Feng, W.: On the energy efficiency of graphics processing units for scientific computing. In: Proceedings of 2009 IEEE International Symposium on Parallel & Distributed Processing, IPDPS'09, Rome, pp. 1–8. IEEE Computer Society, Washington, DC (2009). doi:10.1109/IPDPS.2009.5160980
22. Isayev, O.: hpcadvisorycouncil.com, Computational chemistry: Toward real–life petascale simulations. http://goo.gl/gbuxi (2011). HPC Advisory Council Stanford Workshop
23. Jablin, T.B., Prabhu, P., Jablin, J.A., Johnson, N.P., Beard, S.R., August, D.I.: Automatic cpu-gpu communication management and optimization. ACM SIGPLAN Notices **47**(6), 142–151 (2011). doi:10.1145/2345156.1993516
24. Jang, K., Han, S., Han, S., Moon, S., Park, K.S.: SSL Shader: Cheap SSL acceleration with commodity processors. In: Proceedings of the 8th USENIX Symposium on Networked Systems Design and Implementation, Boston (2011)
25. Jooybar, H., Fung, W.W., O'Connor, M., Devietti, J., Aamodt, T.M.: GPUDet: A deterministic GPU architecture. In: Proceedings of the 8th International Conference on Architectural Support for Programming Languages and Operating Systems, ASPLOS'13, Houston, pp. 1–12. ACM, New York (2013). doi:10.1145/2451116.2451118

26. khronos.org, OpenCL – The open standard for parallel programming of heterogeneous systems. http://goo.gl/7aM3s
27. Kim, J., Kim, H., Lee, J.H., Lee, J.: Achieving a single compute device image in OpenCL for multiple GPUs. In: Proceedings of the 16th ACM Symposium on Principles and practice of parallel programming, PPoPP'11, San Antonio, pp. 277–288. ACM, New York (2011). doi:10.1145/1941553.1941591
28. Kirk, D.: NVIDIA CUDA software and GPU parallel computing architecture. In: Proceedings of the 6th International Symposium on Memory management, ISMM'07, Montreal, pp. 103–104. ACM, New York (2007). doi:10.1145/1296907.1296909
29. Kirk, D., mei Hwu, W.: Programming Massively Parallel Processors. Morgan Kaufmann, Boston (2010)
30. Kluter, T., Brisk, P., Ienne, P., Charbon, E.: Speculative DMA for architecturally visible storage in instruction set extensions. In: Proceedings of the 6th IEEE/ACM/IFIP International Conference on Hardware/Software Codesign and System Synthesis, CODES+ISSS'08, Atlanta, pp. 243–248. ACM, New York (2008). doi:10.1145/1450135.1450191
31. Lee, J., Kim, J., Seo, S., Kim, S., Park, J., Kim, H., Dao, T.T., Cho, Y., Seo, S.J., Lee, S.H., Cho, S.M., Song, H.J., Suh, S.B., Choi, J.D.: An OpenCL framework for heterogeneous multicores with local memory. In: Proceedings of the 19th International Conference on Parallel Architectures and Compilation Techniques, PACT'10, Vienna, pp. 193–204. ACM, New York (2010). doi:10.1145/1854273.1854301
32. Leeser, M., Yablonski, D., Brooks, D., King, L.S.: The challenges of writing portable, correct and high performance libraries for GPUs. SIGARCH Comput. Archit. News **39**(4), 2–7 (2011). doi:10.1145/2082156.2082158
33. Leung, A., Vasilache, N., Meister, B., Baskaran, M., Wohlford, D., Bastoul, C., Lethin, R.: A mapping path for multi-GPGPU accelerated computers from a portable high level programming abstraction. In: Proceedings of the 3rd Workshop on General-Purpose Computation on Graphics Processing Units, GPGPU'10, Washington, DC, pp. 51–61. ACM, New York (2010). doi:10.1145/1735688.1735698
34. Li, C., Wu, H., Chen, S., Li, X., Guo, D.: Efficient implementation for MD5-RC4 encryption using GPU with CUDA. In: Proceedings of the 3rd International Conference on Anti-Counterfeiting, Security, and Identification in Communication, ASID'09, Hong Kong, pp. 167–170. IEEE, Piscataway (2009)
35. Lin, Y., Wang, W., Gui, K.: OpenGL application live migration with GPU acceleration in personal cloud. In: Proceedings of the 19th ACM International Symposium on High Performance Distributed Computing, HPDC'10, Chicago, pp. 280–283. ACM, New York (2010). doi:10.1145/1851476.1851510
36. Luebke, D.: ucdavis.edu, Siggraph 2008: GPU architecture: implications & trends. http://goo.gl/KHZmo (2008)
37. Luley, R., Usmail, C., Barnell, M.: Energy efficiency evaluation and benchmarking of AFRL's Condor high performance computer. In: Proceedings of the High Performance Computing Modernization Program Users Group Conference, Portland, pp. 1–11 (2011). URL http://goo.gl/RR6X5
38. Marston, S., Li, Z., Bandyopadhyay, S., Zhang, J., Ghalsasi, A.: ssrn.com, Cloud Computing – The business perspective. http://goo.gl/aYM1s (2011)
39. mathworks.com, Parallel computing toolbox. http://goo.gl/cpM2O
40. Mattson, T.G., Sanders, B.A., Massingill, B.L.: Patterns for Parallel Programming. Addison Wesley, Boston (2005)
41. Meng, J., Morozov, V.A., Kumaran, K., Vishwanath, V., Uram, T.D.: GROPHECY: GPU performance projection from CPU code skeletons. In: Proceedings of 2011 International Conference for High Performance Computing, Networking, Storage and Analysis, SC'11, Seattle, pp. 14:1–14:11. ACM, New York (2011). doi:10.1145/2063384.2063402

42. Mistry, P., Gregg, C., Rubin, N., Kaeli, D., Hazelwood, K.: Analyzing program flow within a many-kernel OpenCL application. In: Proceedings of the 4th Workshop on General Purpose Processing on Graphics Processing Units, GPGPU'11, Newport Beach, pp. 10:1–10:8. ACM, New York (2011). doi:10.1145/1964179.1964193
43. nvidia.com, OpenCL programming guide for the CUDA architecture. http://goo.gl/GcWuK (2012)
44. nvidia.com, High performance computing with CUDA tutorial. http://goo.gl/2DNBY
45. openmp.org, OpenMP. http://goo.gl/M5YVS
46. Qiu, Q., Wu, Q., Bishop, M., Pino, R., Linderman, R.: A parallel neuromorphic text recognition system and its implementation on a heterogeneous high performance computing cluster. IEEE Trans. Comput. (**99**), 1 (2012). doi:10.1109/TC.2012.50
47. Rangan, R., Vachharajani, N., Ottoni, G., August, D.I.: Performance scalability of decoupled software pipelining. ACM Trans. Archit. Code Optim. **5**(2), 8:1–8:25 (2008). doi:10.1145/1400112.1400113
48. Rong, H., Tang, Z., Govindarajan, R., Douillet, A., Gao, G.R.: Single-dimension software pipelining for multidimensional loops. ACM Trans. Archit. Code Optim. **4**(1) (2007)
49. Ross, V.W.: 500 TeraFLOPS heterogeneous cluster (Air Force's largest interactive HPC). In: The 7th Mohawk Valley Technology Symposium, Rome (2011)
50. Saidi, S., Tendulkar, P., Lepley, T., Maler, O.: Optimizing explicit data transfers for data parallel applications on the cell architecture. ACM Trans. Archit. Code Optim. **8**(4), 37:1–37:20 (2012). doi:10.1145/2086696.2086716
51. sci.utah.edu, SCIRun: A scientific computing problem solving environment (PSE). http://goo.gl/Li6YX
52. Shewchuk, J.R.: An introduction to the conjugate gradient method without the agonizing pain. Technical Report, Carnegie Mellon University (1994)
53. Shimokawabe, T., Aoki, T., Muroi, C., Ishida, J., Kawano, K., Endo, T., Nukada, A., Maruyama, N., Matsuoka, S.: An 80-Fold speedup, 15.0 TFlops full GPU acceleration of non-hydrostatic weather model ASUCA production code. In: Proceedings of the 2010 ACM/IEEE International Conference for High Performance Computing, Networking, Storage and Analysis, SC'10, New Orleans, pp. 1–11. IEEE Computer Society, Washington, DC (2010). doi:10.1109/SC.2010.9
54. Silberstein, M., Maruyama, N.: An exact algorithm for energy-efficient acceleration of task trees on CPU/GPU architectures. In: Proceedings of the 4th Annual International Conference on Systems and Storage, SYSTOR'11, Haifa, pp. 7:1–7:7. ACM, New York (2011). doi:10.1145/1987816.1987826
55. Sim, J., Dasgupta, A., Kim, H., Vuduc, R.: A performance analysis framework for identifying potential benefits in GPGPU applications. ACM SIGPLAN Not. **47**(8), 11–22 (2012). doi:10.1145/2370036.2145819
56. Song, F., Dongarra, J.: A scalable framework for heterogeneous GPU-based clusters. In: Proceedings of the 24th ACM Symposium on Parallelism in Algorithms and Architectures, SPAA'12, Pittsburgh, pp. 91–100. ACM, New York (2012). doi:10.1145/2312005.2312025
57. Stuart, J.A., Chen, C.K., Ma, K.L., Owens, J.D.: Multi-GPU volume rendering using MapReduce. In: Proceedings of the 19th ACM International Symposium on High Performance Distributed Computing, HPDC'10, Chicago, pp. 841–848. ACM, New York (2010). doi:10.1145/1851476.1851597
58. top500.org, Top 500 Supercomputer sites. http://goo.gl/P4Qaf (2012)
59. ufl.edu, Sparse matrix collection. http://goo.gl/OgeW
60. Vasiliadis, G., Polychronakis, M., Ioannidis, S.: MIDeA: a multi-parallel intrusion detection architecture. In: Proceedings of the 18th ACM Conference on Computer and Communications Security, CCS'11, Chicago, pp. 297–308. ACM, New York (2011). doi:10.1145/2046707.2046741
61. Viswanathan, B.: cloudtweaks.com, cloud auditing Making sure that your cloud works per your expectations. http://goo.gl/rXwpc (2012)
62. Volkov, V.: Better performance at lower occupancy. In: Graphics Technology Conference (GTC) (2010). http://goo.gl/M6H3k

63. Vuduc, R., Chandramowlishwaran, A., Choi, J., Guney, M., Shringarpure, A.: On the limits of GPU acceleration. In: Proceedings of the 2nd USENIX conference on Hot Topics in Parallelism, HotPar'10, Berkeley, pp. 13–13. USENIX Association, Berkeley (2010)
64. Wang, P.H., Collins, J.D., Chinya, G.N., Jiang, H., Tian, X., Girkar, M., Yang, N.Y., Lueh, G.Y., Wang, H.: EXOCHI: Architecture and programming environment for a heterogeneous multi-core multithreaded system. In: Proceedings of the 2007 ACM SIGPLAN Conference on Programming Language Design and Implementation, PLDI'07, San Diego, pp. 156–166. ACM, New York (2007). doi:10.1145/1250734.1250753
65. Yang, C., Wang, F., Du, Y., Chen, J., Liu, J., Yi, H., Lu, K.: Adaptive optimization for petascale heterogeneous CPU/GPU computing. In: Proceedings of 2010 IEEE International Conference on Cluster Computing, CLUSTER'10, Heraklion, pp. 19–28. IEEE Computer Society, Washington, DC (2010). doi:10.1109/CLUSTER.2010.12

CPU-GPU System Designs for High Performance Cloud Computing

Yiran Chen, Jie Guo, and Zhenyu Sun

Abstract Improvement of parallel computing capability will greatly increase the efficiency of high performance cloud computing. By combining the powerful scalar processing on CPU with the efficient parallel processing on GPU, CPU-GPU systems provide a hybrid computing environment that can be dynamically optimized for cloud computing applications. One of the critical issues in CPU-GPU system designs is the so called memory wall, which denotes the design complexity of memory coherence, bandwidth, capacity, and power budget. The optimization of the memory designs can not only improve the run-time performance but also enhance the reliability of the CPU-GPU system. In this chapter, we will introduce the mainstream and emerging memory hierarchy designs in CPU-GPU systems, discuss the techniques that can optimize the data allocation and migration between CPU and GPU for performance and power efficiency improvement, and present the challenges and opportunities of CPU-GPU systems.

1 Introduction

The core of cloud computing requires to have capability of *exascale computing* to process big data applications. General-purpose Programming on Graphics Processing Unit (GPU) architectures [18, 33] have been introduced to process vertices and fragments in parallel, which frequently occur in online data/transaction processing. By leveraging the intensive computing capability of GPGPU and the functional flexibility of Central Processing Unit (CPU), the hybrid CPU-GPGPU architectures [28, 37] have been proposed to satisfy the ever-increasing computing capacity of big data applications.

Y. Chen (✉) • J. Guo • Z. Sun
University of Pittsburgh, Pittsburgh, PA, USA
e-mail: yic52@pitt.edu; jig26@pitt.edu; zhs25@pitt.edu

K.J. Han et al. (eds.), *High Performance Cloud Auditing and Applications*,
DOI 10.1007/978-1-4614-3296-8_11, © Springer Science+Business Media New York 2014

Data management and communication are two critical problems in hybrid CPU-GPGPU computing system design and programming. The technology fusion between CPU and GPGPU dramatically raises the complexity of data management because of the different requirements on memory metrics in CPU and GPGPU as well as the throughput discrepancy. Many memory designs have been proposed to mitigate these issues and optimize the data allocation and migration between CPUs and GPUs. The reliability and power consumption of the memory hierarchy in CPU-GPGPU systems are also major considerations in these designs. All the above problems will be discussed in this chapter.

The rest of the chapter is organized as follows. Section 2 presents CPU and GPU hierarchies, and limitations of CPU and GPU. Section 3 provides an overview of discrete CPU-GPU architecture, CPU-GPU communication, optimization and scalability. Section 4 provides an overview of integrated CPU-GPU architecture, memory hierarchy for CPU-GPU systems and design issues. Finally, Section 5 presents our conclusions.

2 Motivation for CPU-GPU Systems

2.1 CPU Memory Hierarchy

In computer systems, CPU is in charge of processing various applications. The executions of instructions in pipelined CPU designs require low data access latency to memory hierarchy. Figure 1 shows a typical memory hierarchy design [9], which consists of cache, main memory and secondary storage. As the component closest to the microprocessor, cache is usually implemented by static random-access memory

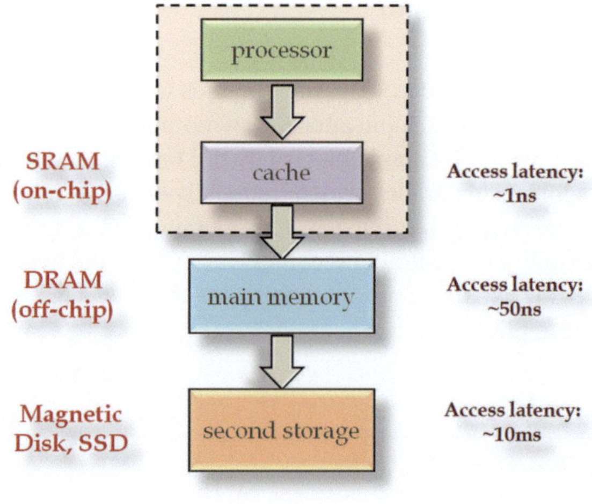

Fig. 1 The conventional memory hierarchy

(SRAM) to offer the fastest access time. However, the high power consumption and large memory cell size limit the cache capacity, which is usually from tens KB (L1 cache) to a few MB (L2 cache and beyond). The spatial locality of the cache access ensures a high cache hit rate (e.g., higher than 90%) [9] to supply the data for most of the accesses without referring to lower-level storage component. Main memory, which is implemented by dynamic random-access memory (DRAM), usually has a much larger capacity than cache but also a longer access latency. Finally, the second storage system (e.g., hard disk drive (HDD)) offers the largest capacity in the memory hierarchy, but also the longest access latency up to milliseconds. Very recently, NAND flash technology is applied as the second storage system (e.g., solid state drive (SSD) in embedded applications.

2.2 GPU Memory Hierarchy

A graphics processing unit (GPU) consists of an array of highly scalable multi-core processors that were specially designed for accelerating the display output generation. Recently, GPU has been also applied in general-purpose parallel computing applications due to its powerful data processing capability. A typical present-day GPU is composed of multiple streaming multiprocessors (SM) which are hardware multi-threaded and each SM contains eight or more streaming processors (SP). Such a highly parallel architecture requires also high data transfer bandwidth.

Compute unified device architecture(CUDA) is a widely adopted parallel programming model mainly designed for NVDIA GPUs. To provide high data throughout required by multi-processors under general computing and visual display applications, CUDA uses a memory subsystem distinct from CPU. The subsystem consists of *cache*, *share memory*, *constant memory*, *texture memory* and *global memory*. Similar to the scenario in CPU designs, cache is on the top of this memory hierarchy and closest to the processors. Instead of building a large memory capacity for hit rate improvement in CPU designs, GPU cache designs focus on promoting the memory bandwidth for highly threaded applications. In GPU memory hierarchy, below the cache, shared memory supports both read and write accesses to each processor in the same SM. Due to requirements on data throughput, both cache and shared memory are implemented by on-chip multi-bank SRAM. Constant memory, texture memory and global memory reside at the lower levels of memory hierarchy.

Both constant memory and texture memory store read-only data and support access from all GPU threads. In particular, constant memory has relatively short access latency for simultaneous requests to the same word while texture memory is designed to support the requests with strong spatial locality. Global memory is for the data transfer between the GPU and the host applications. It is also accessible to all threads of the GPU. In general, constant memory, texture memory, and global memory all reside in the external memory. Therefore, the external memory has to provide enough bandwidth for the multi-processors in the GPU and the data

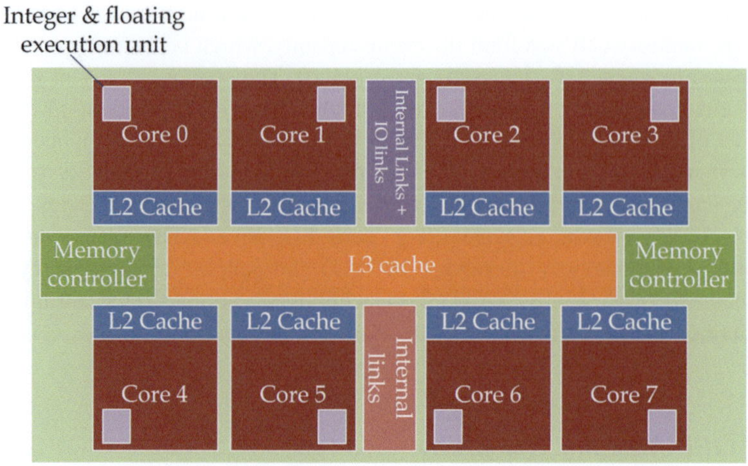

Fig. 2 IBM® POWER7™ processor chip view (Adapted from Ware et al. [29])

migration between the host memory and the global memory to prevent performance degradation. Usually, graphics double data rate (GDDR) memory is employed as the external memory.

2.3 Limitations of CPU and GPU

The high performance of modern CPUs is due to the combination of many technology advancements: high transistor density and performance, pipeline technology, superscalar execution, speculative execution, caching and so on. The evolution of the compiler also greatly reduces the programming difficulty of the modern CPU as the hardware details of micro-architecture is almost invisible to the software developers. However, CPUs are mainly designed for a wide variety of applications with a balanced response time to different tasks. Due to the complex control logic, the integer and floating-point execution units only occupy a very small fraction of the die area in a modern CPU, as shown in Fig. 2. It is no surprise that CPU is relatively inefficient for data-intensive high performance computing applications with high parallelism.

The multi-core structure and high memory bandwidth make GPUs ideal computing platforms for graphic applications with a large degree of data parallelism. Compared to CPUs, the smaller cache capacity and lower cache hit rate (i.e., lower than 90%) and longer memory access latency limit the performance of GPUs for single-threaded applications. Nonetheless, the high tolerance of graphic applications to memory access latency allow GPUs to trade single-thread performance for parallel processing capability. The GPU performance is ultimately limited by the percentage of the scalar section of the program [15].

In modern CPU-GPU systems, CPUs mainly handle the dynamic workloads with short sequences of computational operations and unpredictable control flow while GPUs are mainly in charge of gaming, multimedia encoding and other popular PC applications. Combining the computing flexibility of CPUs and the parallel computing strength of GPUs, the CPU-GPU system has become one of the popular architectures in both PC and enterprise markets. In the next section, we will introduce two kinds of mainstream heterogeneous systems.

3 Discrete CPU-GPU Systems

3.1 Overview of Discrete CPU-GPU Architecture

The architecture of a typical discrete CPU-GPU system is shown in Fig. 3. GPU is connected to a chipset, which is attached to CPU via a $\times 16$ peripheral component interconnect express (PCIe) link. Both CPU and GPU have their own memory subsystems which are disjoint from each other. The data computing in the GPU invokes data movement between the main memory of the host and the memory of the GPU on the PCIe bus. In the existing program models such as CUDA, the host-to-device data transfer is manually and explicitly managed. Since all the data transfer between host and GPU is through the PCIe bus, additional timing and control overhead costs are incurred. As a consequence, the PCI data transfer is often the bottleneck of the GPU performance [2].

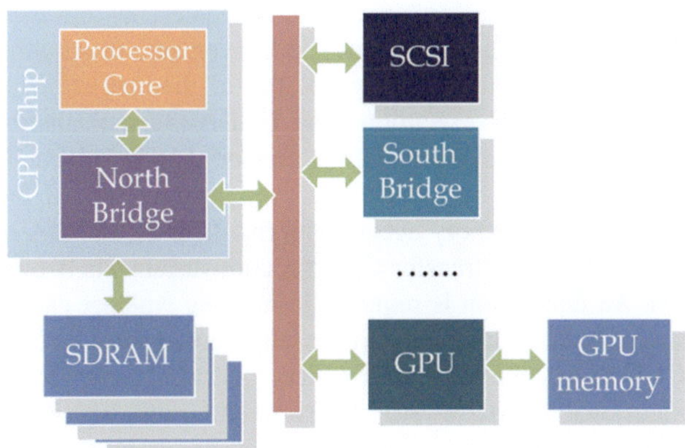

Fig. 3 CPU-GPU architecture

3.2 CPU-GPU Communication and Optimization

If the CPU or GPU needs data outside their memories, the data must be explicitly loaded from other memories into their own memories. The process of copying data between these memories is called *managing communication*. There are two generic types of data communication, *cyclic communication* and *acyclic communication*. Because cyclic communication is often several orders of magnitude slower than acyclic communication, transforming cyclic communication patterns to acyclic patterns is of importance to computing system designs, and is called *optimizing communication*. In CPU-GPU systems, copying data to GPU memory, generating a GPU function, and copying the results back to CPU memory yield typical cyclic communication patterns. Copying data to the GPU in the preheader, generating multiple GPU functions, and copying the result back to CPU memory at the loop exit lead to a typical acyclic communication pattern [10]. An effective communication optimization can minimize the number of access states and prevent the accessing of inconsistent data. However, the communication patterns between CPUs and GPUs could be very complicated because of the mismatch of data rate, access patterns, processing throughput, etc.

A fully automatic CPU-GPU communication management and optimization is proposed in [10]. The proposed scheme consists of a run-time library and a set of compiler transformations working together for the management and optimization of the CPU-GPU communications. This scheme does not rely on the strength of the static compile-time analysis or programmer-supplied annotations. The optimized automatic GPU parallelization yields on average a 5.36$\times$ speedup of the overall system over the best sequential CPU-only execution.

In [27], a CUDA-lite communication scheme translates low-performance, naive CUDA functions into high performance code by coalescing and exploiting GPU shared memory. For execution efficiency improvement, programmers need to provide annotations describing certain properties of the data structures and code regions designated for GPU execution. The CUDA-lite tool analyzes the code with these annotations and determines if the memory bandwidth can be conserved and the access latency can be reduced by utilizing any special memory types and/or by massaging the memory access patterns. In [32], JCUDA uses the Java type system to automatically transfer GPU function arguments between CPU and GPU memories. An annotation is required to indicate whether each parameter is live-in, live-out, or both. If Java implements multidimensional arrays as arrays of references, JCUDA uses the type information to flatten these arrays to Fortran-style multidimensional arrays. In [30], the Portland Group, Inc (PGI) Fortran and C compiler provides a mode for semi-automatic parallelization of GPUs. The PGI compiler cannot parallelize loops containing general pointer arithmetic, while CPU-GPU Communication Manager (CGCM) preserves the semantics of pointer arithmetic. Unlike CGCM, the PGI compiler does not automatically optimize the communication across GPU function invocations.

Acyclic communication can be achieved in the rare instances when dynamic dependence information is reusable. In [24], program annotations are used to prevent the undesired reuse of the dynamic dependence information. In [23], a dynamic check on the relevant program state is adopted to determine if the dependence information is reusable. However, the dynamic check is usually very costly and if the check fails, the system goes back to cyclic communication by default.

In [14], an automatic compiler is proposed to conduct the source-to-source translation from standard OpenMP applications to CUDA-based GPGPU applications. The system automatically transfers the named regions between CPU and GPU using two passes: The first pass copies all named annotated regions to the GPU for each GPU function, and the second cleanup pass removes all the copies that are not live-in. The two passes produce a communication pattern equivalent to an un-optimized CGCM communication.

Recently, message passing interface (MPI) [17] has been incorporated in the CUDA programming model to facilitate the data transfer via the PCIe bus, due to its outstanding inter-node communication capacity. However, the duplicate buffers introduced by the mixed MPI and CUDA programming model lead to the waste of memory space and code complexity. Furthermore, the MPI library is not able to support intra-node communication and the data movement between the host and memory is required under the current mixed MPI-CUDA model. However, such an approach results in longer latency and decreases bandwidth due to the significant amount of message transfers. In [11], the authors propose an address-aware MPI-GPU programming model to eliminate the redundant data copying between intermediate buffers by adding support to the GPU memory buffer in MPICH2 that is a open source of the MPI implementation. Asynchronous direct memory access (DMA) mode is also adopted to further improve the data copying throughput.

3.3 Power Efficiency Optimization

To satisfy the high data throughput requirements in video display applications and high performance computing, GDDR memories are specialized for current CPU-GPU systems. GDDR memories feature high working frequency and extended bus width. However, the increase in working frequency significantly raises power consumption and higher heat sinking cost. Although voltage and frequency scaling techniques can reduce the power consumption to a certain degree, they may also result in significant performance degradation. To improve the power efficiency of CPU-GPU systems, Zhao et al. [35] present a design to integrate GPU processor, GDDR-like graphic memories and their controllers in a single package. The data throughput is dramatically improved by reducing the number of memory package pins, without incurring any performance overheads. To further reduce power consumption, a reconfigurable memory controller is designed to dynamically adjust the bus width and frequency, based on the characteristics of different applications.

3.4 Scalability

To maximize parallel computing performance, multiple GPUs may be integrated in each core (or a node) in some CPU-GPU systems, such as Tianhe-1A, Tsubame, and KIDS. However, such designs impose new challenges for system architects in the design of efficient data transfer algorithms. In addition, increasing the number of GPUs introduces a performance bottleneck to the discrete heterogeneous system under the current topology where each node connects the GPU to the I/O hub via the PCI Express bus. Compared to the memory bandwidth of GPUs, PCIe bus is still quite narrow: it has a maximum of 16 PCI lanes, which can support only a peak bandwidth of 8 GB/s while the peak bandwidth of GPU memory (e.g., Nvidia Tesla M2070) can reach up to 148 GB/s [16]. This great throughput gap has emerged as the major performance bottleneck in modern CPU-GPU systems.

4 Integrated CPU-GPU Systems

4.1 Overview of CPU-GPU Systems

Following the advances of VLSI technologies, the integration of CPUs and GPUs on the same chip has become a hot research area. Such CPU-GPU systems provide the opportunity to leverage both the high computational power from GPUs for regular applications and flexible execution from CPUs for irregular workloads. In this section, we will introduce the latest integrated CPU-GPU architectures, including Intel's Sandy Bridge, AMD's Fusion, and NIVIDIA's Denver. We will also discuss the memory designs in such systems.

4.1.1 Intel Sandy Bridge

Figure 4 shows the architecture of the Sandy Bridge processor [34] that integrates four high performance Intel Architecture cores, one GPU, on-chip cache, memory controller and peripheral component interconnect (PCI) controller on the same die. A 8 MB L3 cache memory is shared by both CPUs and GPU. The data flow is optimized by a high performance on-die interconnect fabric (called "ring"). The ring connects the CPUs, the GPU, the L3 cache and the system agent (SA) unit, which houses a 1,600 MT/s dual-channel DDR3 memory controller, a 20-lane PCIe gen2 controller, 2 parallel pipe display engines, a power management control unit and testability logic. Sandy Bridge also has 6 power planes, 13 PLLs driving independent clock domains, temperature control with 2 types of thermal sensors, independent debug bus, as well as variable power supply for power and performance

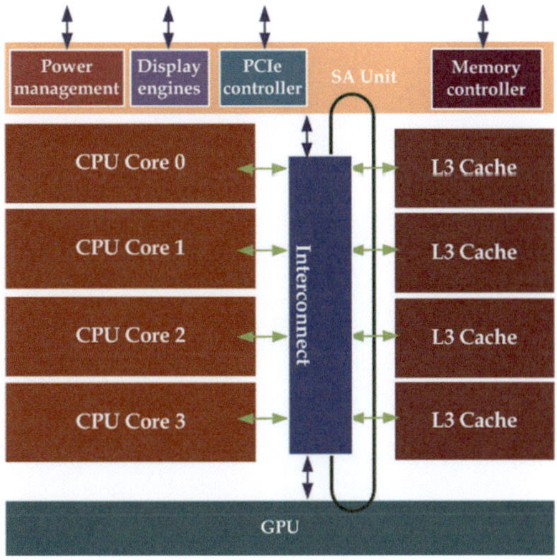

Fig. 4 Block diagram of Sandy Bridge (Adapted from Zhao et al. [35])

optimization. Because the L3 cache capacity of Sandy Bridge is large (3, 4 or 8 MB depending on the configuration), the power dissipated by the cache memory accounts for a big portion of the overall design.

4.1.2 AMD Fusion

An accelerated processing unit (APU) is a system with additional processing capability designated to accelerate one or more types of computations outside a CPU. Some examples are general-purpose computing on graphics processing unit (GPGPU), a field-programmable gate array (FPGA), or similar specialized processing system. In some marketing relevant usage of the term, APU also describes the processing device which integrates a CPU and a GPU on the same die, thus improving the data transfer rates between these components while reducing the total power consumption.

AMD Fusion is the series of such APU designs [1, 7, 8], including the version for *desktop processors*, *mobile processors*, *ultra-mobile processors*, and *embedded processors*. As the most representative version, the *Llano* variant is a complex integration of processor, graphics, and multimedia resources. Here, we use Llano architecture as the example to discuss the CPU-GPU system design in detail.

As shown in Fig. 5, the Llano variant combines four ×86 processor cores, a unified video decoder, an integrated DirectX11 graphics core, and an integrated two-head display controller. The entire integrated complex fits in a die area of 227 mm^2 at 32-nm silicon on insulator (SOI) process. Twenty-four lanes of PCIe Gen2 are also supported for high performance connectivity. Among them, 4 lanes are dedicated to the Llano-to-Fusion controller hub (FCH) link, 4 lanes to general-purpose ports, and the other 16 lanes to the discrete graphics that can expand

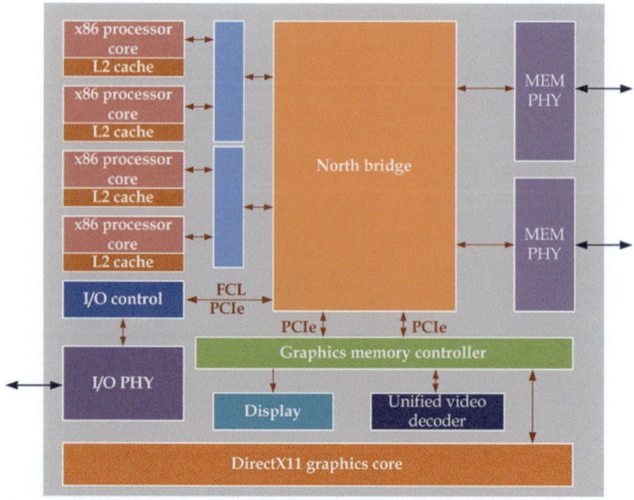

Fig. 5 Block diagram of a fusion APU of AMD (Adapted from Branover et al. [1])

computing and graphics stream capabilities. Two display streams can be directed to any combination of up to six display ports.

The integrated NorthBridge is the node where the processor, graphics, and multimedia devices are jointly managed. It also mainly determines the APU performance and power consumption. In NorthBridge, processor cores, I/O interfaces, graphics and video accelerators are connected to two 64-bit channels of DRAM through a multistage memory controller. Memory operates at data transfer rates up to DDR3-1866 megatransfers per second (MT/s).

Llano uses a unified memory architecture (UMA) in which the processor and graphics share a common memory. A portion of this memory is also dedicated to graphics frame buffer. Graphics, multimedia, and display memory traffic is routed through the graphics memory controller (GMC), which arbitrates between the requestors and issues a well-behaved (from a memory access perspective) stream of memory requests over the Radeon memory bus (RMB) to the NorthBridge. The accesses from GMC to frame buffer memory are non-coherent and do not snoop the processor caches. The RMB supports a 256-bit read data path and a 256-bit write data path for each of the two memory channels.

The coherent accesses from graphics or multimedia to the DRAM are directed over the fusion control link (FCL). The FCL consists of separated 128-bit read and write data paths to serve as the path for the processor accesses to the I/O devices and the dedicated graphics memory, and for the I/O accesses to the DRAM. Also, display traffic imposes a real-time requirement on the whole memory system. The display controller maintains a local frame buffer to store the next few lines to be sent to the display panel; buffer underrun must be avoided because it results in visible tearing effects on the panel. The DRAM controller prioritizes display requests indicating a nearly empty buffer over other traffics. Rather than supplying a continuous request stream to refill the frame buffer, the display controller issues the requests to memory in a burst, followed by a period of no request activities. In an idle system, this lets the memory refresh itself between bursts, thus maximizing power efficiency.

The other graphics-related streams are managed in both the GMC and DRAM controllers to ensure that the average memory latency never exceeds a configurable threshold. The utilizations of processors and I/O traffic are arbitrated in the NorthBridge front-end stage by allocating proper bandwidth to each of them. The front-end output stream is arbitrated with the graphics-related stream in the DRAM controller. The DRAM controller contains a starvation-prevention mechanism that tracks the outstanding activity and ensures that the minimum required bandwidth is allocated to each stream. Starvation timers indicate when a specific traffic class needs to be stalled longer than the dynamically configurable threshold, and allows forward progress. Software drivers or on-die firmware can dynamically adapt the arbitration timers for specific workload scenarios.

4.1.3 NVIDIA Denver

Project Denver was developed by NVIDIA [3]. The innovative features of Denver include an NVIDIA CPU running the ARM instruction set, which will be fully integrated on the same chip as the NVIDIA GPU. By extending the performance range of the ARM instruction-set architecture, project Denver opens a new era for computing that enables the ARM architecture to cover a large portion of the computing space. Coupled with an NVIDIA GPU, it provides the hybrid CPU-GPU computing platform of the future with ultra-high performance and energy efficiency.

4.2 Memory Hierarchy for CPU-GPU Systems

The main bottleneck for most discrete GPU-based applications is the cost of data transfers to and from the GPU over PCIe. The recent introduction of the AMD Fusion provides a novel architecture to eliminate this bottleneck. By placing the CPU on the same die as a SIMD engine, the fused device is able to eliminate most of the costs of data transfer (e.g., PCIe transfers). The typical memory hierarchy for such a fused system is shown in Fig. 6.

Table 1 summarizes the information of three AMD Fusion systems including memory hierarchy configurations [2]. All systems have only one CPU core and the same memory capacity of 2 GB., CPU cache hierarchies of the three processors are all 2-level but with different capacities. The numbers of processing elements for GPU vary from 80 to 1,600. The GPU side has a similar memory hierarchy but a different number of processing units, yielding different throughput. Also, system throughput is determined by not only the memory but also other factors such as

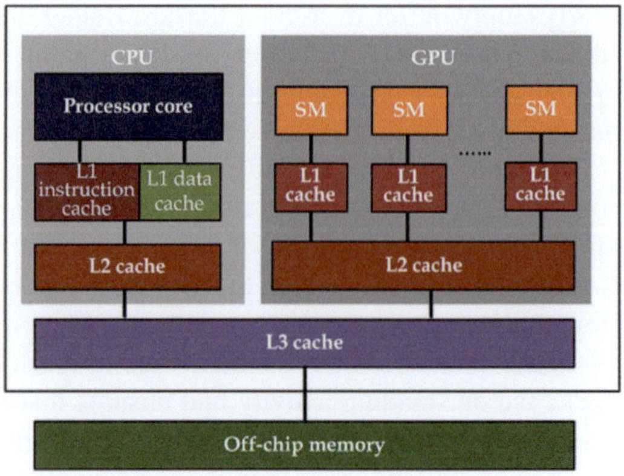

Fig. 6 Memory hierarchy for CPU-GPU systems

Table 1 Information of different fusion systems

Platform	AMD Zacate APU	AMD Radeon HD 5870	MD Radeon HD 5450
Host			
Processor	AMD Engg. sample	Intel Xeon E5405	Intel Celeron 430
Frequency	1.6 GHz	2.0 GHz	1.8 GHz
System memory	2 GB (NA)	2 GB DDR2	2 GB DDR2
Cache	L1: 32K, L2: 512K	L1: 32K, L2: 6M	L1: 32K, L2: 512K
Kernel	Ubuntu 2.6.35.22	Ubuntu 2.6.28.19	Ubuntu 2.6.32.24
Stream processors	80	1,600	80
Compute units	2	20	2
Memory bus type	NA	GDDR5	DDR3
Device memory	192 MB	1,024 MB	512 MB
Local memory	32 KB	32 KB	32 KB
Max. workgroup size	256 threads	256 threads	128 threads
Peak core clock freq.	492 MHz	850 MHz	675 MHz
Peak FLOPS	80 GFlop/s	2,720 GFlop/s	104 GFlop/s

operating frequency, on-chip interconnect, timing and memory controller. Different workloads also generate different system performances on different platforms [2].

4.3 Design Issues in CPU-GPU Systems

As aforementioned, CPUs are designed for a wide range of applications while GPUs are specialized for large scale parallel computing. Latency-sensitive CPUs tend to adopt the memory with large capacity, like the DDR3 in the Llano APU. In contrast, GPUs conventionally use GDDR memory with higher clock frequency and wider buses to provide high data throughout. Compared to DDR3 memory, longer-latency GDDR memory reduces its capacity to minimize the power budget. In integrated CPU-GPU systems, these two diverse cores share the same type of memory, incurring performance degradation for either counterpart due to the mismatch of memory requirements [25]. Furthermore, sharing memory between CPU and GPU invokes access contention. It has been shown that the performance penalty incurred by contention is up to 20% [25].

4.4 Adopting Emerging Nonvolatile Memory in CPU-GPU Systems

Besides software-level data managements and optimizations, many hardware solutions can be also implemented for the performance and power improvements of CPU-GPU memory systems. As an example, emerging nonvolatile memory technologies such as spin-transfer torque random access memory (STT-RAM) and phase change memory (PCM) have been adopted for both cache and main memory designs in CPU memory hierarchy.

Due to its relatively fast access speed and zero-leakage power, STT-RAM has been adopted to replace SRAM as on-chip cache, which accounts for a large portion of energy consumption. However, there are two major obstacles in using STT-RAM for on-chip caches: long write latency and high write energy. When a STT-RAM cell operates in the sub-10 ns region, the resistance switching mechanism of *magnetic tunnel junction* (MTJ), which is the data storage device in STT-RAM, is dominated by *spin precession*. The required switching current rises exponentially as the MTJ switching time reduces. As a consequence, the size of MOS driving transistor increases accordingly, leading to a large memory cell area. The lifetime of the STT-RAM cell also degrades exponentially as the voltage across the oxide barrier of the MTJ increases. Therefore, a 10 ns programming time is widely accepted as the performance limit of STT-RAM designs in mainstream STT-RAM research and development [4, 5, 13, 26, 31]. Several proposals have been made to address the write speed and energy limitations of the STT-RAM. For example, the early write termination scheme [36] mitigates the performance degradation and energy overhead by terminating the unnecessarily long write pulse to STT-RAM cells. The dual write speed scheme [31] improves the average access time of a STT-RAM cache by having a fast and a slow cache partition. A SRAM/STT-RAM hybrid cache hierarchy with 3D stacking structure was proposed in [26] to compensate the performance degradation caused by STT-RAM writes by migrating write intensive data block into SRAM-based cache way.

Due to its small cell size and multi-level cell (MLC) capability, PCM is considered to be the replacement of DRAM in high-density main memory designs. The first issue in PCM is its long programming latency and high programming energy. A DRAM/PCM hybrid memory structure was proposed in [22] to merge frequent writes into the DRAM cache so that the number of write accesses to the PCM devices can be greatly reduced. Write pausing [19] was proposed to allow performance critical read operations to preempt writes. A morphable memory system [20] was proposed to improve the read latency of MLC PCM by converting one MLC PCM page into two SLC pages when there is sufficient memory. Another issue in PCM is its limited programming endurance, since a PCM cell can only be reprogrammed about 10^6 times [6]. A start-gap scheme was used in [21] to efficiently enhance the PCM main memory lifetime by distributing the write accesses among all the PCM cells. In [12], line-level mapping and salvaging are combined to raise the limitation of wear-leveling technique for PCM-based main memory.

5 Conclusion

CPU-GPU systems are widely employed in high performance cloud computing to offer the computing capability in diverse scientific research fields. The major issues in the existing CPU-GPU system designs include data communication performance bottleneck, power efficiency and system scalability. In this chapter, we introduced the currently prevalent CPU-GPU system architectures. The memory hierarchies of these two kinds of computing units are extensively discussed. Based on the architectural characteristics, we conducted a deep analysis on the performance bottleneck of data transferring between CPU and GPU, the power efficiency of the GPU memory, and their architecture limitations on scalability. Finally, we discussed the adoption of emerging non-volatile memories in CPU-GPU systems for performance optimization and power reduction.

Acknowledgements This material is based upon work partially supported by the National Science Foundation (NSF) grant CNS-1116171 and the Air Force Research Laboratory (AFRL) Visiting Faculty Research Program (VFRP) extension grant LRIR 11RI01COR. We are grateful to Prof. Hai (Helen) Li from the University of Pittsburgh Department of Electrical and Computer Engineering for generous help.

References

1. Branover, A., Foley, D., Steinman, M.: AMD fusion APU: Llano. IEEE Micro **32**(2), 28–37 (2012). doi:10.1109/MM.2012.2
2. Daga, M., Aji, A.M., Feng, W.c.: On the efficacy of a fused CPU+GPU processor (or APU) for parallel computing. In: Proceedings of the 2011 Symposium on Application Accelerators in High-Performance Computing, SAAHPC'11, Knoxville, pp. 141–149 (2011). doi:10.1109/SAAHPC.2011.29
3. Dally, B.: nvidia.com, PROJECT DENVER: Processor to usher in new era of computing. http://goo.gl/HepP5 (2011)
4. Desikan, R., Lefurgy, C., Keckler, S., Burger, D.: ibm.com, On-chip MRAM as a high-bandwidth low-latency replacement for DRAM physical memories. http://goo.gl/lyvV2 (2008)

5. Dong, X., Wu, X., Sun, G., Xie, Y., Li, H., Chen, Y.: Circuit and microarchitecture evaluation of 3D stacking magnetic RAM (MRAM) as a universal memory replacement. In: Proceedings of the 45th Annual Design Automation Conference, DAC'08, Anaheim, pp. 554–559. ACM, New York (2008). doi:10.1145/1391469.1391610
6. Ferreira, A.P., Zhou, M., Bock, S., Childers, B., Melhem, R., Mossé, D.: Increasing PCM main memory lifetime. In: Proceedings of the Conference on Design, Automation and Test in Europe, DATE'10, Leuven, pp. 914–919. European Design and Automation Association, Leuven (2010)
7. Foley, D., Bansal, P., Cherepacha, D., Wasmuth, R., Gunasekar, A., Gutta, S., Naini, A.: A low-power integrated x86–64 and graphics processor for mobile computing devices. IEEE J. Solid-State Circuits **47**(1), 220–231 (2012)
8. Gutta, S.R., Foley, D., Naini, A., Wasmuth, R., Cherepacha, D.: A low-power integrated x86-64 and graphics processor for mobile computing devices. In: Proceedings of the 2011 IEEE International Solid-State Circuits Conference (ISSCC), San Francisco, pp. 270–272. IEEE (2011). doi:10.1109/ISSCC.2011.5746314
9. Hennessy, J., Patterson, D.: Computer Architecture: A Quantitative Approach, 4th edn. Morgan Kaufmann, Burlington (2007)
10. Jablin, T.B., Prabhu, P., Jablin, J.A., Johnson, N.P., Beard, S.R., August, D.I.: Automatic CPU-GPU communication management and optimization. SIGPLAN Notice **47**(6), 142–151 (2011). doi:10.1145/2345156.1993516
11. Ji, F., Aji, A.M., Dinan, J., Buntinas, D., Balaji, P., Feng, W.c., Ma, X.: Efficient intranode communication in GPU-accelerated systems. In: Proceedings of the 26th International Parallel and Distributed Processing Symposium Workshops & PhD Forum, IPDPSW'12, Shanghai, pp. 1838–1847 (2012). doi:10.1109/IPDPSW.2012.227
12. Jiang, L., Du, Y., Zhang, Y., Childers, B.R., 0002, J.Y.: LLS: Cooperative integration of wear-leveling and salvaging for PCM main memory. In: Proceedings of the 41st International Conference on Dependable Systems & Networks (DSN), Hong Kong, pp. 221–232. IEEE (2011). doi:10.1109/DSN.2011.5958221
13. Kawahara, T., Takemura, R., Miura, K., Hayakawa, J., Ikeda, S., Lee, Y., Sasaki, R., Goto, Y., Ito, K., Meguro, T.: 2Mb SPRAM (SPin-Transfer Torque RAM) with bit-by-bit bi-directional current write and parallelizing-direction current read. IEEE J. Solid-State Circuit **43**(1), 109–120 (2008)
14. Lee, S., Min, S.J., Eigenmann, R.: OpenMP to GPGPU: a compiler framework for automatic translation and optimization. In: Proceedings of the 14th ACM SIGPLAN Symposium on Principles and Practice of Parallel Programming, PPoPP'09, Raleigh, pp. 101–110. ACM, New York (2009). doi:10.1145/1504176.1504194
15. Lee, V.W., Kim, C., Chhugani, J., Deisher, M., Kim, D., Nguyen, A.D., Satish, N., Smelyanskiy, M., Chennupaty, S., Hammarlund, P., Singhal, R., Dubey, P.: Debunking the 100X GPU vs. CPU myth: an evaluation of throughput computing on CPU and GPU. SIGARCH Comput. Archit. News **38**(3), 451–460 (2010). doi:10.1145/1816038.1816021
16. Meredith, J., Roth, P., Spafford, K., Vetter, J.: Performance implications of nonuniform device topologies in scalable heterogeneous architectures. IEEE Micro **31**(5), 66–75 (2011). doi:10.1109/MM.2011.79
17. mpi-forum.org, MPI: A message-passing interface standard version 2.2. http://goo.gl/SEqm1 (2009)
18. Nere, A., Lipasti, M.: Cortical architectures on a GPGPU. In: Proceedings of the 3rd Workshop on General-Purpose Computation on Graphics Processing Units, GPGPU'10, Pittsburgh, pp. 12–18. ACM, New York (2010). doi:10.1145/1735688.1735693
19. Qureshi, M.K., Franceschini, M., Lastras-Montaño, L.A.: Improving read performance of phase change memories via write cancellation and write pausing. In: Proceedings of the 16th International Symposium on High Performance Computer Architecture (HPCA), Bangalore, pp. 1–11. IEEE Computer Society (2010). doi:10.1109/HPCA.2010.5416645

20. Qureshi, M.K., Franceschini, M.M., Lastras-Montaño, L.A., Karidis, J.P.: Morphable memory system: a robust architecture for exploiting multi-level phase change memories. SIGARCH Comput. Archit. News **38**(3), 153–162 (2010). doi:10.1145/1816038.1815981
21. Qureshi, M.K., Karidis, J., Franceschini, M., Srinivasan, V., Lastras, L., Abali, B.: Enhancing lifetime and security of PCM-based main memory with start-gap wear leveling. In: Proceedings of the 42nd Annual IEEE/ACM International Symposium on Microarchitecture, MICRO 42, New York, pp. 14–23. ACM, New York (2009). doi:10.1145/1669112.1669117
22. Qureshi, M.K., Srinivasan, V., Rivers, J.A.: Scalable high performance main memory system using phase-change memory technology. In: Proceedings of the 36th Annual International Symposium on Computer Architecture, ISCA'09, Austin, pp. 24–33. ACM, New York (2009). doi:10.1145/1555754.1555760
23. Rauchwerger, L., Amato, N., Padua, D.: A scalable method for run-time loop parallelization. Int. J. Parallel Program. **23**(6), 537–576 (1995)
24. Saltz, J., Mirchandaney, R., Crowley, K.: Run-time parallelization and scheduling of loops. IEEE Trans. Comput. **40**(5), 603–612 (1991)
25. Spafford, K.L., Meredith, J.S., Lee, S., Li, D., Roth, P.C., Vetter, J.S.: The tradeoffs of fused memory hierarchies in heterogeneous computing architectures. In: Proceedings of the 9th Conference on Computing Frontiers, CF'12, Caligari, pp. 103–112. ACM, New York (2012). doi:10.1145/2212908.2212924
26. Sun, G., Dong, X., Xie, Y., Li, J., Chen, Y.: A novel architecture of the 3D stacked MRAM L2 cache for CMPs. In: IEEE Symposium on High-Performance Computer Architecture (HPCA), Los Alamitos, pp. 239–249. IEEE Computer Society, Los Alamitos (2009). doi: 10.1109/HPCA.2009.4798259
27. Ueng, S.Z., Lathara, M., Baghsorkhi, S.S., Hwu, W.M.W.: CUDA-Lite: reducing GPU programming complexity. Languages and Compilers for Parallel Computing, pp. 1–15. Springer, Berlin/Heidelberg (2008)
28. Venkatasubramanian, S., Vuduc, R.W.: Tuned and wildly asynchronous stencil kernels for hybrid CPU/GPU systems. In: Proceedings of the 23rd International Conference on Supercomputing, ICS'09, Yorktown Heights, pp. 244–255. ACM, New York (2009). doi:10.1145/1542275.1542312
29. Ware, M., Rajamani, K., Floyd, M., Brock, B., Rubio, J., Rawson, F., Carter, J.: Architecting for power management: The IBM® POWER7™ approach. In: Proceedings of the 16th IEEE International Symposium on High Performance Computer Architecture, HPCA'10, Bangalore, pp. 1–11 (2010). doi:10.1109/HPCA.2010.5416627
30. Wolfe, M.: Implementing the PGI accelerator model. In: Proceedings of the 3rd Workshop on General-Purpose Computation on Graphics Processing Units, GPGPU'10, Pittsburgh, pp. 43–50. ACM, New York (2010). doi:10.1145/1735688.1735697
31. Xu, W., Sun, H., Wang, X., Chen, Y., Zhang, T.: Design of last-level on-chip cache using spin-torque transfer RAM (STT RAM). IEEE Trans. Very Large Scale Integr. (VLSI) Syst. **19**(3), 483–493 (2011). doi:10.1109/TVLSI.2009.2035509
32. Yan, Y., Grossman, M., Sarkar, V.: JCUDA: A programmer-friendly interface for accelerating Java programs with CUDA. In: Proceedings of the 15th International Euro-Par Conference on Parallel Processing, Euro-Par'09, Delft, pp. 887–899. Springer-Verlag, Berlin/Heidelberg (2009). doi:10.1007/978-3-642-03869-3_82
33. Yang, Y., Xiang, P., Kong, J., Mantor, M., Zhou, H.: A unified optimizing compiler framework for different GPGPU architectures. ACM Trans. Archit. Code Optim. **9**(2), 9:1–9:33 (2012). doi:10.1145/2207222.2207225
34. Yuffe, M., Knoll, E., Mehalel, M., Shor, J., Kurts, T.: A fully integrated multi-CPU, GPU and memory controller 32nm processor. In: Proceedings of the 2011 IEEE International Solid-State Circuits Conference (ISSCC), San Francisco, pp. 264–266 (2011). doi:10.1109/ISSCC.2011.5746311

35. Zhao, J., Sun, G., Loh, G., Xie, Y.: Energy-efficient GPU design with configurable package graphic memory. In: ISLPED'12, Redondo Beach, pp. 403–408 (2012)
36. Zhou, P., Zhao, B., Yang, J., Zhang, Y.: Energy reduction for STT-RAM using early write termination. In: Proceedings of the 2009 International Conference on Computer-Aided Design, ICCAD'09, New York, pp. 264–268. ACM, New York (2009). doi:10.1145/1687399.1687448
37. Zidan, M.A., Bonny, T., Salama, K.N.: High performance technique for database applications using a hybrid GPU/CPU platform. In: Proceedings of the 21st Edition of the Great Lakes Symposium on VLSI, GLSVLSI'11, Lausanne, pp. 85–90. ACM, New York (2011). doi:10.1145/1973009.1973027

MapReduce Performance in Federated Cloud Computing Environments

Praveenkumar Kondikoppa, Chui-Hui Chiu, and Seung-Jong Park

Abstract Large scale scientific and engineering applications, and cloud auditing generate huge amounts of data. MapReduce framework coupled with cloud computing is emerging as the viable solution for distributed big data processing. Specifically, if data is generated from distributed sources and computation is also distributed then multiple clouds need to be set up to minimize data transfer, which introduces us to federated distributed or multi-domain clouds. In addition to security concerns of general clouds, distributed clouds expose new challenges to the performance of cloud based applications including cloud auditing and analysis. This book chapter focuses on a method to deploy distributed clouds and evaluates the performance of various cloud based applications over distributed clouds. It also proposes a method to optimize the performance of cloud based applications over high speed networks.

1 Introduction

Scientific research communities amass data at an astounding and accelerating rate, which creates a large data storage, analysis and management problem referred to as *Big Data Analysis* [4, 43, 44]. As technology advances over time, the size of data sets that qualify as big data will also increase and is beyond the ability of traditional database software tools to manage, store and analyze. Companies like Google and Microsoft routinely processes petabytes of data. A simple search query [10] will be processed against hundreds of gigabytes of data and no doubt this hunger to share, store and analyze data will continue to grow. This leads to the problem of establishing an efficient computational model and a software framework to enable distributed computing. Cloud computing [40] provides a unique solution to

P. Kondikoppa (✉) • C.-H. Chiu • S.-J. Park
Louisiana State University, Baton Rouge, LA, USA
e-mail: pkondi1@lsu.edu; cchiu1@lsu.edu; sjpark@lsu.edu

K.J. Han et al. (eds.), *High Performance Cloud Auditing and Applications*,
DOI 10.1007/978-1-4614-3296-8_12,

storage and computing power for big data analysis [43, 44]. Emerging service cloud infrastructures such as Amazon EC2, OpenStack [33], Rackspace [35], OpenCirrus, Eucalyptus [11] and IBMSmartCloud [17] are a big leap forward towards providing computing power over the internet. On demand provisioning [29], ease of use, seamless computing and storage, and the fact that computing power is provided through low cost commodity clusters draws huge attention both from academia and industry. Many of these commercial clouds provide unlimited computing power. In addition to those commercial clouds, many universities and research laboratories have deployed their own private clouds for research and education.

MapReduce [5, 8] is a distributed programming [30] model for large scale data processing over a commodity cluster. MapReduce framework [9] makes automatic parallelization of the data while hiding the complex issues of parallelization from the user. The programmer breaks down his computations into Map and Reduce tasks. The issues of how to parallelize the computations, distribute the data, and handle failures are managed by the framework. There are very limited implementations of MapReduce programming model [3, 16]: Google MapReduce [23], Apache Hadoop [2] and Cloudera Hadoop [6]. Apache Hadoop is a widely used open source implementation of MapReduce. Cloud computing coupled with MapReduce programming model [3] have made huge strides in big data analysis. MapReduce is extensively used on compute clouds at Amazon, Facebook and Yahoo. Recently, the Yahoo Grid team took a moderate sized Hadoop cluster of 910 nodes and won the TeraSort benchmark [32]. Amazon's Elastic Compute cloud enables Hadoop to be run as a separate service. This has led many of the organizations to set up MapReduce based private clusters for internal data processing.

Conventional MapReduce jobs are supposed to be launched over a single cluster which connects computing nodes within local networks. Therefore, it is hard to deploy MapReduce jobs over distributed clusters because nodes at remote clusters cannot communicate with each other. Although they can communicate with other nodes at remote clusters, there are several issues on performance of MapReduce [1, 13, 20, 28, 41, 42]. For example, network latency and bandwidth between a master node and slave nodes reduce the performance of MapReduce because MapReduce configuration requires a direct communication between a master node and slave nodes. Additionally, scheduling [27, 39] tasks among slave nodes which are scattered over remote clusters becomes an important factor on performance because of data locality [15]. Geographically separated clouds expose new challenges to cloud based applications. Apart from increased computing power, federated clouds would be helpful for sharing the data between the clouds. At times, a data set resides on one cloud and a map task is started at another cloud and the data needs to be transferred.

The rest of the chapter is organized as follows. Section 2 covers an overview of MapReduce. Section 3 describes MapReduce performance evaluation on federated cloud. Section 4 presents MapReduce performance evaluation in high-speed networks. Conclusions are presented in Section 5.

2 MapReduce in Cloud Computing

MapReduce has been widely used as a powerful parallel data processing tool for data-intensive and compute-intensive applications. Most cloud providers (e.g., Google, Yahoo!, Facebook, etc.) adopt MapReduce to process big data in multi-tenant cloud computing environments. An overview of MapReduce is introduced in this section, and MapReduce performance evaluations of data-intensive and compute-intensive computing will be presented in Section 3.

2.1 MapReduce Framework

MapReduce paradigm is a distributed computing framework for data-intensive and compute-intensive applications [25]. It is inspired by the Map and Reduce constructs of the functional programming model. The program is split into Map and Reduce parts. Input and output are expressed in terms of (key,value) pairs. Map takes (key, value) as input and produces intermediate (key1, value1) output pairs. Reducer takes (key1,value1) as input and produces (key2,value2) as output. Map functions run in parallel. The output of the mapper will be sent to the Reduce function as input, and the reducer will sort the input key-value pairs according to the key and combine the value of key-value pairs into a list which have the same key. Lastly, the reducers, which are also run in parallel, generate the output key-value pairs.

The MapReduce framework abstracts programmers from the complexities of input data distribution, parallelization [21, 45], fault tolerance and synchronization of various constructs. The open source implementation of MapReduce framework consists of computing environment called Hadoop and distributed filesystem called Hadoop Distributed File System (HDFS) [31]. Hadoop acts like runtime environment for the Map and Reduce functions provided by programmers. HDFS is designed to provide replication, integrity and fault tolerance of the data. The architecture of Hadoop has ***master services*** namely JobTracker and NameNode and ***client services*** namely TaskTrackers and DataNodes. The JobTracker receives the job submitted by the user and splits it into map and reduce tasks. It assigns tasks to TaskTrackers, monitors task completion status, and when all the tasks for a job are complete, it reports back to the user. TaskTrackers are configured with fixed number of map and reduce slots which determine the number of tasks to run simultaneously. Scheduling [22] in Hadoop is done by a JobTracker which manages the number of TaskTrackers and a default scheduler is the First In, First Out (FIFO) scheduler.

2.1.1 MapReduce Execution Flow

The execution of MapReduce job involves the steps below:

1. Client submits the MapReduce job to the JobTracker.

2. JobTracker co-ordinates the job run. It splits the input data into chunks and each chunk is submitted to a map task.
3. TaskTracker operates on the chunk of the data assigned to it. It can independently access the filesystem.
4. HDFS is used for sharing the file system with other entities.

2.2 *MapReduce Scheduling*

In this subsection, we give an overview of MapReduce scheduling methodologies in [22]. FIFO, fair and capacity schedulers are briefly discussed.

2.2.1 FIFO Scheduler

Scheduling in Hadoop is centralized. Instead of JobTracker assigning the tasks, TaskTrackers pull the tasks from the job which needs to be completed. The default scheduling in Hadoop is through FIFO scheduler. The communication between the master node and slave nodes happens through heartbeat packets. When a slave node with empty map slot sends a heartbeat packet to the JobTracker, the scheduler checks the head of a job queue for the tasks. If the job has tasks whose input splits are located on the slave node, then the current task is assigned to the TaskTracker. If scheduler does not find a local map task, then it assigns only one non local map task to this TaskTracker.

2.2.2 Fair Scheduler

Fair scheduling ensures that every job gets an equal share of resources over time. If there is a single job running, then the entire cluster is allocated for the job. If new jobs are submitted, then task slots which free up are allocated to the newly submitted jobs, so that each job gets roughly equal share of cpu time. FAIR scheduler organizes jobs into pools, and inside each pool FIFO or fair queuing of the jobs can be configured. By default, each user gets assigned to a pool. Fair sharing can also use job priorities for scheduling. The priorities are used as weights to determine the fraction of total compute time that each job should get.

In addition to providing fair sharing, the fair Scheduler allows the assignment of guaranteed minimum shares to pools, which is useful for ensuring that certain users, groups or production applications always get sufficient resources. When a pool contains jobs, it gets at least its minimum share, but when the pool does not need its full guaranteed share, the excess is split among other running jobs. This lets the scheduler guarantee capacity for pools while utilizing resources efficiently when these pools do not contain jobs.

The fair Scheduler can limit the number of concurrent running jobs per user and per pool. This can be useful when a user must submit hundreds of jobs at once, or for ensuring that intermediate data does not fill up disk space on a cluster when too many concurrent jobs are running. Setting job limits causes jobs submitted beyond the limit to wait until some of the user/pool's earlier jobs finish. Jobs to run from each user/pool are chosen first in order of priority and then submit time.

2.2.3 Capacity Scheduler

Capacity scheduler supports multiple queues and each of the submitted jobs will be assigned to a queue. Each queue gets a certain portion of the available resources. The scheduler picks a queue to schedule a job based on priority, and inside the queue a job is selected based on job priority. Free resources can be allocated to any queue beyond its capacity. However, once a job is running, it will not be preempted for a higher priority job although new tasks from the higher priority job will be preferentially scheduled. When there is demand for these resources from queues running below capacity at a future point in time, as tasks scheduled on these resources complete, they will be assigned to jobs on queues running below capacity. In order to prevent one or more users from monopolizing its resources, each queue enforces a limit on the percentage of resources allocated to a user at any given time.

2.3 Deploying MapReduce over Federated Clouds

2.3.1 Hierarchical MapReduce

In this setup [27], MapReduce operates in two layers. The top layer consists of job scheduler, data manager and work load manager. The bottom layer consists of distributed clusters running local MapReduce. In each of the clusters, a master process runs on a node and remaining computing nodes run client processes. When a job is submitted, it is split into sub jobs, which are assigned to distributed clusters. JobTracker at each of the clusters reports the job progress and sends the result back to a global controller where merging or reduction takes place.

The disadvantage of this approach is that it is suitable for map-intensive or map-mostly types of applications. If the intermediate data generated from the map phase is huge and needs to be transferred to the reducer, the data transfer process becomes a bottleneck for the performance of Hadoop. The majority of the data intensive applications involve significant reduce phase computations. Also users need to add additional programming construct to perform global reduction at the global controller.

2.3.2 MapReduce on Virtual Machines

Hadoop MapReduce cannot be directly deployed over distributed clusters to form a single MapReduce cluster since master node services require direct communication between slave node services. But the internal nodes of a cluster are not accessible from outside. We address this issue of internal nodes accessible from outside by introducing a virtualization on top of all the physical clusters. Cloud computing software like Eucalyptus [11] or Openstack [33] is installed on each of the underlying physical clusters. Computing units are provided as virtual machines (VMs) from any of the distributed clusters. A virtual layer unifies the resources provided by all the distributed clusters into a single infrastructure layer of VMs. MapReduce cluster is set up on top of these VMs where a single JobTracker takes the responsibility of scheduling the jobs and managing the TaskTrackers. The advantage of this approach is that it is easy to manage distributed clusters through cloud computing software. However, the critical aspect in this approach is to make a scheduler be topology-aware in order to accomplish data local computation of map tasks.

The proposed work focuses on improving data locality while scheduling map tasks for global MapReduce so that all types of data intensive applications (such as map-only, map-mostly and map-reduce application types) are supported. In the sections below, we describe the data locality issue and propose network aware scheduling [22] to make the global MapReduce scheduler data aware.

2.4 MapReduce Optimization

There are several ways of optimizing MapReduce over distributed clusters [19, 26, 36]. The first way is to set up independent MapReduce clusters at each site via a centralized job scheduler to manage job submission. The second approach is to aggregate underlying physical clusters into a single virtual cluster and run MapReduce tasks on top of this virtual cluster. The third approach is to make the MapReduce framework work directly with multiple clusters without additional virtual clusters. We deploy global MapReduce over federated clusters to utilize the computing power effectively. However, deploying MapReduce framework directly on federated clouds degrades the performance because of the failure to exploit data locality [14, 34].

Achieving data locality is the biggest challenge of deploying MapReduce over federated clouds. Current scheduling schemes are not optimized to exploit data locality. We propose to improve the data locality in MapReduce scheduler as follows. The object of task scheduling in MapReduce is to move computation towards data. If this objective can't be met for a given task, then we propose to schedule the task on a node which is requesting for a task, causing data to be transferred to a compute node for processing.

In the case of federated clouds, clients from any of the underlying clouds might request a map task. A technique to make MapReduce scheduler aware of network topology [24] is to extend the rack aware feature of the existing scheduler to provide one more level of caching. Provision is made so that a master node has information about the physical location of the requesting client node.

3 MapReduce Performance Evaluation over Federated Clouds

3.1 Factors Impacting Performance of MapReduce

It is important to understand where performance bottlenecks may reside in the dataflow of a MapReduce job over federated clouds.

Once the job starts, the individual Map tasks are usually executed on machines which have already stored the HDFS data block corresponding to the Map task; these are normal, block-local tasks. This stage does not rely on network bandwidth or latency. If a node becomes idle during the Map phase, it may be assigned a Map task for which it does not have the data block locally stored; it would then need to download that block from another HDFS data location, which could be costly depending on which data source it chooses for the download. Finally, and most importantly, is the Reduce phase.

The Reduce operation is an all-to-all transmission of intermediate data between the Map task output data and Reduce tasks. If there is a significant amount of intermediate data, this all-to-all communication could be costly depending on the bandwidth of each end-to-end link.

Figure 1 shows the data flow in MapReduce job.

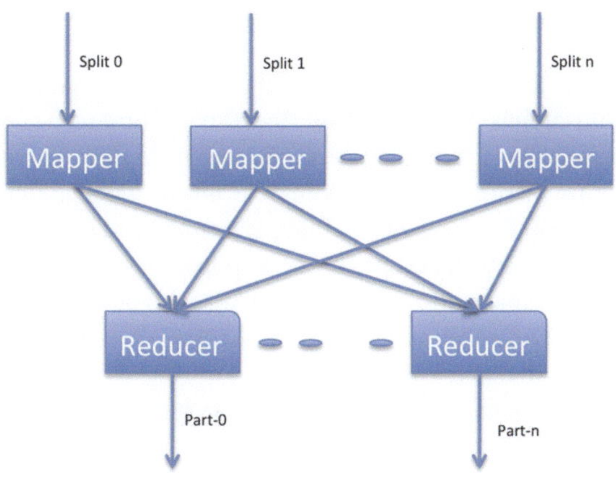

Fig. 1 Data flow in MapReduce job

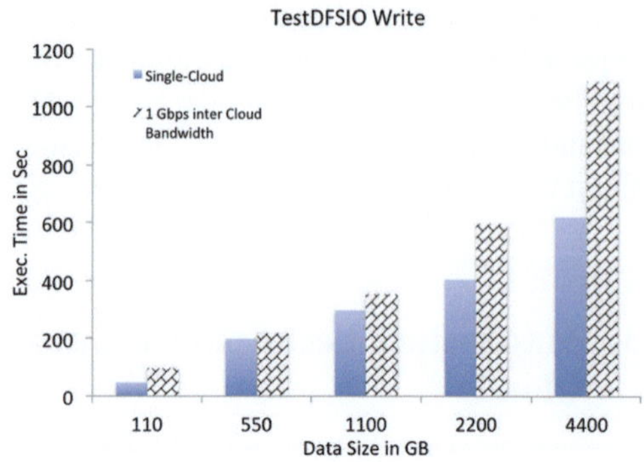

Fig. 2 Performance of distributed file system for *write* over federated clouds

3.2 Evaluation of Data-Intensive Applications

We evaluate the performance of MapReduce on federated clouds by executing various benchmark applications. We compare the performance of data intensive benchmark applications such as TestDFSIO [32] and Sort in both single cloud and multicloud scenarios.

3.2.1 TestDFSIO

TestDFSIO is a benchmark application for evaluating performance of distributed file system I/O. This benchmark has two parts: (1) TestDFSIO-Write writes a file of given size to the file system, and (2) TestDFSIO-Read reads the data from the distributed file system. Because of the replication factor, the write test does more I/O and generates substantial networking traffic.

This benchmark takes the number of files to write and file size as constraints. For each file, a map task is initiated and all the writes by clients happen in parallel. In our experiments, we run one map task per each node to write/read a file to/from the file system, with file size varying from 100 MB to 4 GB per map task.

Figure 2 shows the performances of distributed file system over single cloud and multi-clouds. For small data size, the performance is comparable but for larger data sets there is a notable latency in the write operations. Performance degradation is due to writing to a node which is not in the same cloud as the one executing the map task. Figure 3 shows that performance of read is significantly better than write operation. For read operation, map task first chooses a data block which is located on that same node. If it does not find the data block, it searches for a node which is located in the same rack. In this way, data block search takes place from node locality to cluster locality.

Performance of distributed file system operations read and write with respect to throughput is shown in Figs. 4 and 5. The throughput of write operation is considerably less than read operation because write operation involves writing to replicated locations.

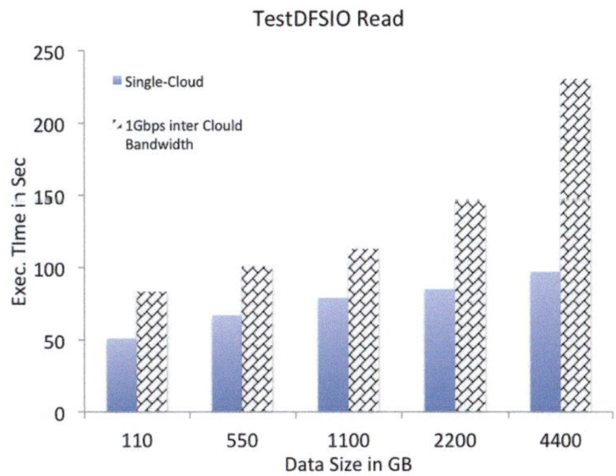

Fig. 3 Performance of distributed file system for *read* over federated clouds

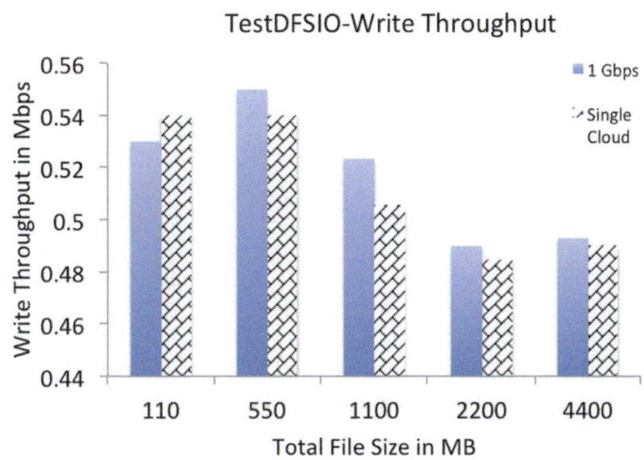

Fig. 4 Throughput for TestDFSIO *write* over federated clouds

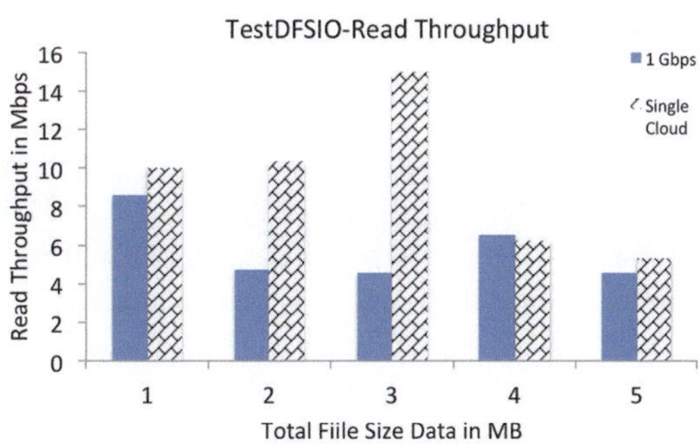

Fig. 5 Throughput for TestDFSIO *read* over federated clouds

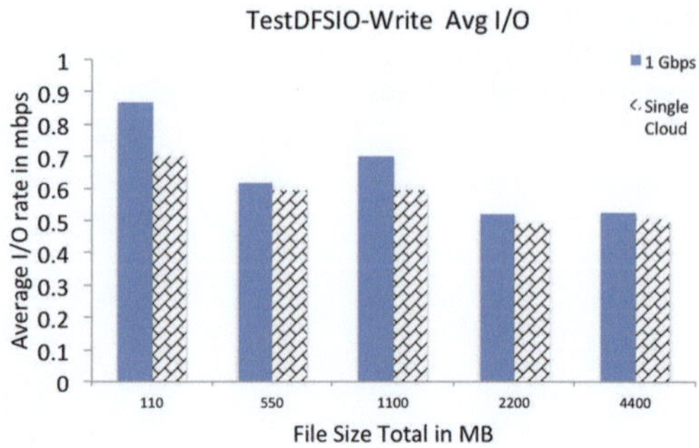

Fig. 6 Average I/O for TestDFSIO for *write* over federated clouds

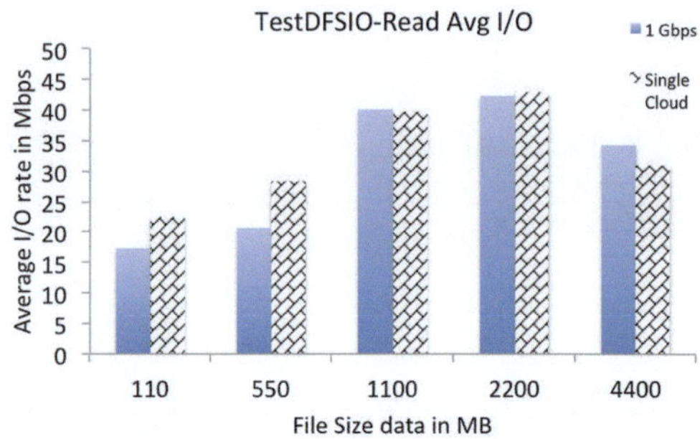

Fig. 7 Average I/O for TestDFSIO for *read* over federated clouds

3.2.2 Sort

Sort is an important benchmark application from the perspective of both MapReduce and distributed file system. It consists of considerable computation, networking, and storage I/O and it is often considered to be representative of real MapReduce-based cloud applications. An application goes through map, Shuffle and Reduce, the important constructs of MapReduce programming framework. In our experiments, the input data of Sort is generated using the RandomWriter. Each of the map tasks reads data from the local data-node and performs sorting in parallel. The result of the map is written to local disk. In the shuffle phase, the data will be transferred to the node performing the reduce task.

For replication needs, the data blocks will be written on other nodes also. Thus sort constitutes both disk I/O bound and network I/O bound and expects reasonable disk and network performance. The I/O rate for read indicates the rate at which the file is read by the mapper from HDFS. Since I/O rate for local map tasks is high in general, average I/O rate for read is high. However, write operation not only involves writing to the local disk but also to other replicated nodes and if a replicated node is located at a remote cluster, then average I/O write will be high.

The results from Figs. 6 and 7 indicate this aspect of HDFS. A reduce phase follows the sorting phase to merge the locally-sorted data chunks from all data nodes and writes them to a new file.

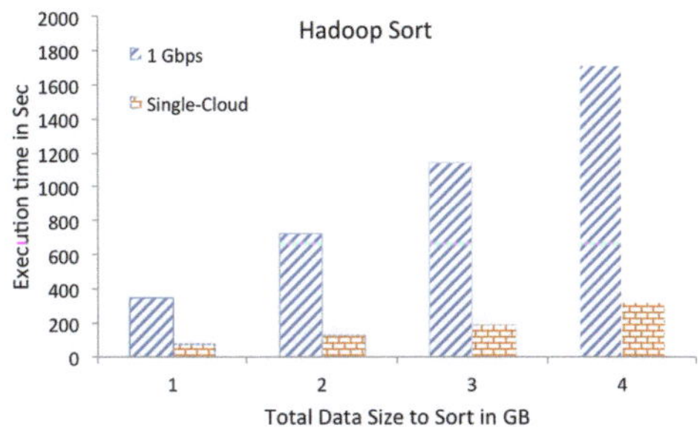

Fig. 8 Performance of sort application over federated clouds

From Fig. 8, it is clear that the performance of MapReduce is greatly degraded over distributed clouds for high data intensive applications because of big data transfer from map phase to reduce phase.

3.3 Evaluation of Compute-Intensive Applications

We experimented using MapReduce-based compute intensive applications such as CloudBurst [37] and Autodock Vina [38].

3.3.1 CloudBurst

CloudBurst [37] is a bioinformatics application used for aligning short reads [37]. It uses the Hadoop implementation of MapReduce to efficiently execute in parallel on multiple compute nodes, thus making it feasible to perform highly sensitive alignments on large read sets. It uses well- known seed-and-extend algorithms to map reads to a reference genome. In the map phase, the map function emits for a given *k* all *k*-mers from the reference sequences, and all non-overlapping *k*-mers from the reads. In the reduce phase, the seeds are extended into end-to-end alignments. In our experiment, 100,000 short reads data-sets are aligned against Streptococcus suis genome. This sample data set is given with the CloudBurst application. The execution time of CloudBurst scales linearly with the number of processors. Experimental result of CloudBurst is shown in Fig. 9.

3.3.2 AutoDock Vina

AutoDock Vina [38] has been used for molecular docking in the field of structure-based drug design. This application is originally developed to run on a single computer. We implement the MapReduce version of this application to run on the cloud using Hadoop. Each map task receives a file name as input to process. The actual file will be located on the local disk of the map task. The output will

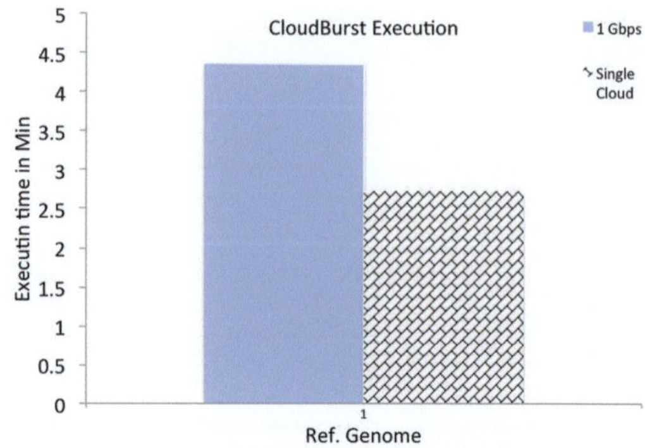

Fig. 9 Performance of CloudBurst, a moderately compute intensive application over federated clouds

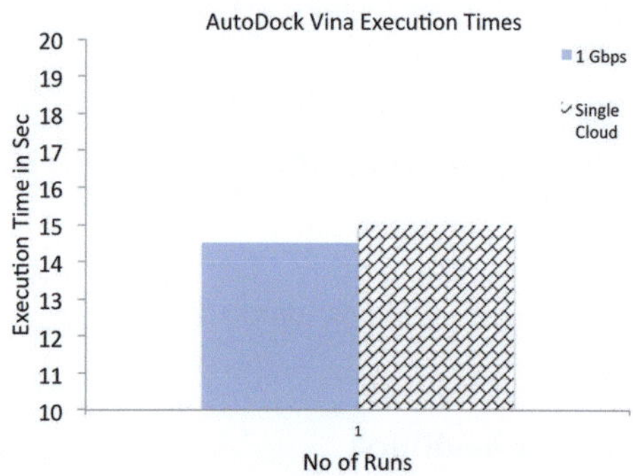

Fig. 10 Performance of AutoDockVina, a compute intensive application over federated clouds

be written on the local disk of the map task. This application does not have a reduce phase, thus making it a purely compute-intensive application. We chose this application to obtain insights about the performance of compute-intensive applications over federated clouds (see Fig. 10). The results from compute-intensive applications, which have very minimal data movement between the clouds, show that MapReduce-based cloud auditing analysis tools can be deployed over federated clouds with minimal architectural changes.

4 MapReduce Performance Evaluation in High Speed Networks

To evaluate the performance of network aware Hadoop in a controlled testbed environment, we use an emulated network testbed. After two virtual clouds were created and connected by emulated network links, network parameters for the emulated links, such as bandwidth and delay, were varied. To provide the emulated network infrastructure for cloud computing experiments, we use a Cyberinfrastructure of Reconfigurable Optical Networking (CRON) [7] testbed. CRON provides

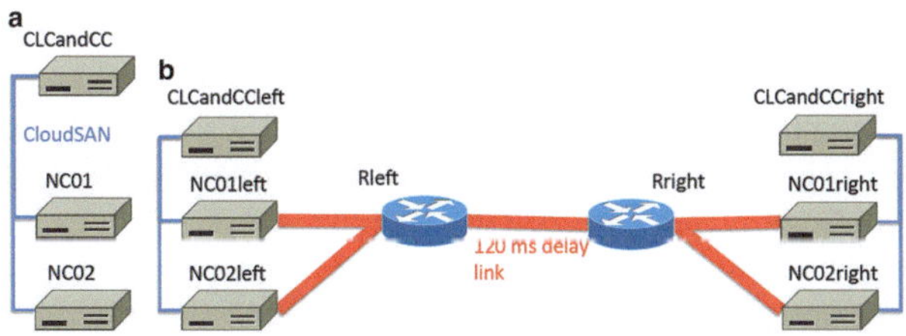

Fig. 11 Cloud computing on the CRON testbed. (**a**) Single Eucalyptus cloud. (**b**) Inter-cloud communication

multiple emulation testbeds operating up to 10 Gbps bandwidth. It consists of two main components: (a) hardware (H/W) components, including a switch, optical fibers, network emulators, and the workstations required to physically compose optical paths or functions at the ends of these paths; and (b) software (S/W) components, creating an automatic configuration server that will integrate all the H/W components to create virtual network environments based on the users' requirements. All components are connected with 100/1,000 Mbps Ethernet links for control. Each workstation is connected with 10 Gbps optical fibers for data movement. The CRON Testbed is an ideal platform for cloud computing and cloud auditing research in high speed networks. It automatically provides ready-to-use and easy-to-modify computer clouds to users. Meanwhile, it provides a computer cloud specifically enhanced for federated cloud communication.

Cloud Computing introduces a flexible method for managing computing resources and a revolutionary manner for providing computing as a service. For the service providers, the conventional computer clusters are regarded as a pool of computing and storage resources. When a request arrives, the Cloud Computing software, on demand, picks up the resources corresponding to the request with the help of the Hypervisors, wraps them into Virtual Machines (VMs) or virtual disk volumes, and returns the access information to accommodate the request. The users only pay for the certain amount of remote computing facilities for a certain period of time; they do not have to own and maintain the facilities.

The CRON Testbed (see Fig. 11) is capable of automatically setting up the building blocks which enable the cloud computing computer clusters, Hypervisors, and cloud computing software, as follows.

- For computer clusters, the testbed allocates nodes and connects them in the form of computer clusters as constructing a regular experimental network topology.
- For Hypervisors, the testbed deploys an OS image/indexOS image with the Kernel-based Virtual Machine (KVM) that is a hardware-accelerated Linux kernel module to make multiple OSs run as processes on a hosting OS on a physical machine.
- For Cloud computing software, on the basis of the OS image with KVM, the testbed adds the Eucalyptus and shell scripts for configuring it.

The OS image has the binaries for the Eucalyptus cloud computing software which can be configured to any of its five components, cloud controller (CLC), Walrus, cluster controller (CC), storage controller (SC), and node controller (NC). After the image is loaded on a machine, shell scripts for automatic Eucalyptus post-installation configuration are launched to make the system operate.

Figure 11a illustrates a single Eucalyptus cloud scenario on the testbed. In order to preserve more nodes for computing tasks, the node "CLCandCC", by default, runs the CLC, Walrus, CC, and SC. The actual computing nodes "NC01" and "NC02" only run the NC. All nodes connect to each other by the "CloudSAN". Additionally, the Eucalyptus comes with source code, and users can easily modify the system and test it. Along with the standard Eucalyptus cloud, the testbed deploys the NEuca patch [12] to satisfy the surging desires for inter-cloud communication. The function for the NEuca patch is to add additional network interface cards (NICs) to each VM. With the enhancement, users can emulate high speed channels which connect VMs in different clouds, such as the Internet2 ION Circuit on the CRON Testbed. Figure 11b shows the scenario for federated cloud communication between NEuca-patched Eucalyptus clouds.

In our experiment, we used Eucalyptus as the cloud computing software. We set up two NEuca-patched Eucalyptus Clouds to construct the distributed cloud scenario. We have two computing nodes in each Cloud. For improved execution of the benchmark application, each computing node accommodates three VMs so that each VM is allocated with 2 physical CPU cores, 2 GB of physical memory, and 10 GB of local hard disk space. All VMs connect to each other through the additional virtual NICs to form a single virtual cluster. The VirtIO interface is adopted to serve the virtual NIC so that the transmission rate is as high as the Hypervisor can provide. Then, we deploy our proposed Hadoop on the virtual cluster to examine its performance in the complicated network environment. The scenario in Fig. 11b is used for conducting the experiments.

4.1 MapReduce Optimization in CRON Testbed

In this section we compare the performance of native Hadoop (i.e., original Hadoop from Apache software foundation which is not optimized) and network aware Hadoop which is optimized for federated clouds to improve data locality. Table 1 shows hardware and software equipments for experiments. To evaluate performance in a practical test environment, we prepare different jobs consisting of map tasks with varying execution times. Multiple jobs having varying map execution times are needed to evaluate the performance of the fair scheduler since a fair scheduler tries to schedule multiple jobs to improve the response time of each job.

Table 1 Hardware and software equipments for performance tests

Nodes	Quantity	Hardware and Hadoop
Master node	1	2 CPU core 2 GB RAM 10 Gbps NIC JobTracker and NameNode
Slave nodes	12	2 CPU core 2 GB RAM 10 Gbps NIC TaskTracker and DataNode Hadoop-0.20.203.0 2 map and 2 reduce tasks

4.2 Native Hadoop

In this set of experiments, we ran Hadoop over a distributed cluster without network awareness. One of the nodes was a master node running JobTracker and NameNode services. The remaining nodes from both clusters ran the slave services TaskTrackers and DataNodes. We used the word count application for evaluation. The input file size for the word count application was varied so that the average execution time for the map task differed.

4.3 Optimized Hadoop

In this set of experiments Hadoop with network awareness was configured over the federated cluster. One JobTracker managed the scheduling over all the clusters. TaskTrackers from different clusters requested map tasks as map slots became free. JobTracker used cluster awareness to schedule tasks on these TaskTrackers, thereby improving data locality. We performed experiments with varying inter-cloud bandwidth and delay, and measured the performance of Hadoop with network awareness [22].

4.4 Preliminary Results and Analysis

We compare the execution times of Hadoop native scheduling with network-aware Hadoop for varying delay values. Figure 12 shows the execution times for various numbers of map tasks, with the FIFO scheduler with network awareness identified as FIFO+NA. The results show a reduction in execution time of about 12% on average and maximum reduction of 15% when the number of map tasks is higher compared

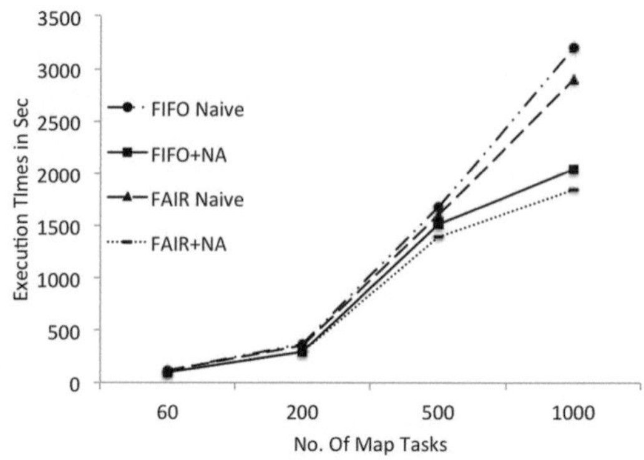

Fig. 12 Execution times for different schedulers for 1 Gbps, 30 ms inter-cloud bandwidth and delay

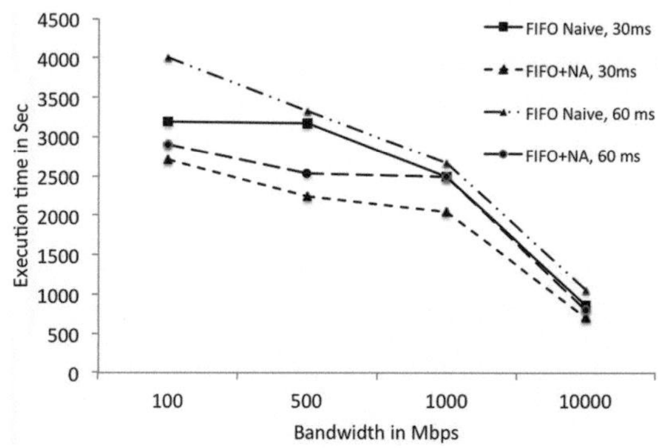

Fig. 13 Execution times for varying inter-cloud bandwidth

to Hadoop Naive scheduler (identified as FIFO Naive). Fair scheduler with Network Awareness (identified as Fair+NA) exhibits similar results. Fair scheduler in general takes less execution time than FIFO, the reason being more effective utilization of the cluster. Network aware Hadoop shows maximum performance improvement for map tasks between 500 and 1,000. As the map task number increases, the data splits associated with the map tasks will be spread over the distributed clusters. When JobTracker has to schedule a map task over a TaskTracker, it checks if the TaskTracker is located in the same cluster as the DataNode having the data to process. If this condition is met, then this task will be assigned to the TaskTracker; otherwise, it will be delayed as specified in the delay scheduling.

Bandwidth is a significant factor for federated clusters. The performance of Hadoop greatly depends on the bandwidth between the clusters. Figure 13 illustrates various inter-cloud bandwidths with fixed delay. We emulate the varying bandwidth cases on our experimental set up using dummynet [18]. At 100 Mbps bandwidth between the clouds, there is significant difference in execution times for processing 10 GB data. If the tasks are not executed either with node local or rack local, then large amounts of data will be moved from one cluster to another cluster for map task execution. With 100 Mbps of bandwidth there will be less available bandwidth

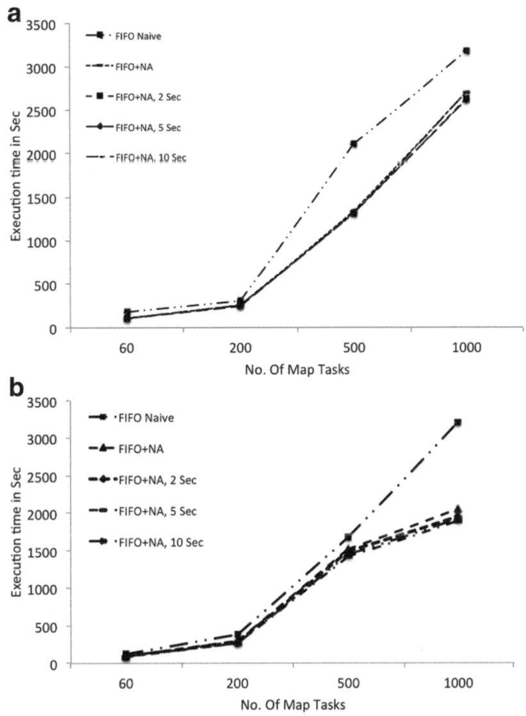

Fig. 14 Execution times for 30 ms inter-cloud delay. (**a**) 100 Mbps bandwidth and 30 ms delay between clusters. (**b**) 1 Gbps bandwidth and 30 ms delay between clusters

between nodes. As the bandwidth provisioned between the clusters is increased, overall execution time decreases. However, for processing terabytes of data, there should be sufficient bandwidth provisioned between the clusters. We performed experiments with different network conditions.

In all the cases significant reduction in execution time is observed between the Native Hadoop scheduler and the Network-aware scheduler as the number of map tasks increases. It is evident that for higher inter-cloud bandwidth the performance of Hadoop on the network is greatly increased compared to the performance of native Hadoop scheduling.

Figure 14 shows the results for 30 ms inter-cloud delay for varying bandwidths of 100 Mbps and 1 Gbps. For the same number of map tasks, the execution times for a job is less for 1 Gbps bandwidth. In case of 100 Mbps bandwidth, the gap between curves for different scheduling schemes is less significant. The less available bandwidth between the clusters causes a bottleneck for data split movement for tasks scheduled on distributed clusters. In Fig. 14a, for 500–1,000 tasks, a significant difference is observed in the execution times for native Hadoop and different variants for Network aware Hadoop, though the inter-cluster bandwidth is less. Figure 14b shows experimental results for 1 Gbps inter cloud bandwidth and 30 ms delay. Since higher bandwidth is provisioned between the clusters, the overall

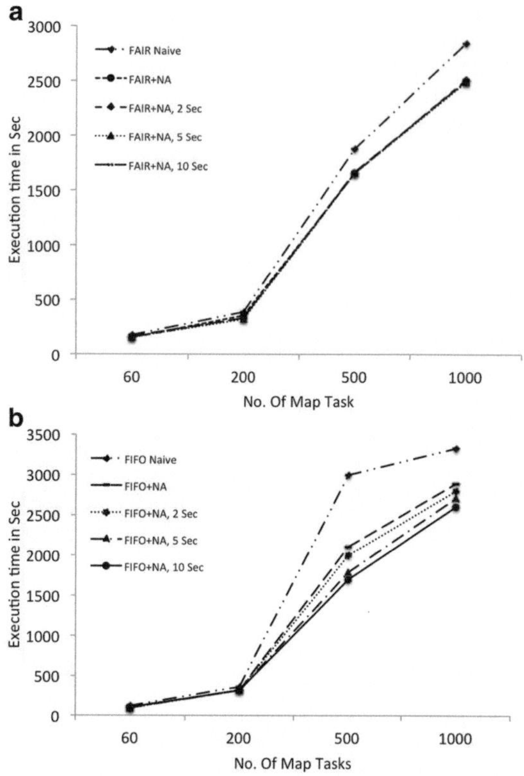

Fig. 15 Execution times for 60 ms inter-cloud delay. (**a**) 100 Mbps bandwidth and 60 ms delay between clusters. (**b**) 1 Gbps bandwidth and 60 ms delay between clusters

execution times for the jobs compared to Fig. 15a is less. Furthermore, for jobs with higher number of tasks (i.e., tasks ranging from 500 to 1,000 map tasks), the performance of Network aware Hadoop is significantly improved. This shows that the performance of Hadoop could be increased with provisioning higher bandwidth, and less delay coupled with locality awareness. Similar results were observed for experiments conducted with the FAIR scheduler.

Figure 15 shows 60 ms inter-cloud delay and bandwidths of 100 Mbps and 1 Gbps. There is significant reduction in the overall job execution times between 100 Mbps and 1 Gbps scenarios. However, Fig. 15a shows results for 100 Mps and 60 ms inter cloud network parameters. Because of the higher delay between the clusters, there is not significant improvement between the native Hadoop and the different variants of network aware Hadoop. Figure 15b is for 1 Gbps inter-cloud bandwidth and 60 ms delay. In this case also, because of higher delay between the federated clusters, less performance improvement is observed (but more of an improvement than in the 100 Mbps scenario).

5 Conclusion

As the amount of data to analyze increases, the demand on federation among multiple distributed clusters is growing. However, most software frameworks on cloud computing such as MapReduce and Hadoop are not developed for distributed high speed network environments. Due to the unawareness of distributed network environments, the performance of Hadoop can be degraded. Therefore, optimizing the performance of Hadoop over integrated and distributed cloud computing resources is critical. In this book chapter, we showed the performance degradation of Hadoop over federated cloud computing clusters connected with high speed networks. Based on the preliminary results, we modified the scheduler inside Hadoop so that the network aware scheduler can distribute tasks based on the locality of data. We provided network awareness to the FIFO and fair schedulers in Hadoop. Performance showed an improvement of 12–15 % in both FIFO and fair schedulers. Customized Hadoop implementation and optimization is required to perform compute-intensive cloud auditing analysis. In the future, it would be desirable to extend network awareness at the reduce phase because the overhead of reduce phase in Hadoop contributes significantly to the overall execution time.

Acknowledgements This material is based upon work partially supported by the National Science Foundation (NSF) GENI-grant and grant MRI-0821741 (CRON project), the Department of Defense Experimental Program to Stimulate Competitive Research (DEPSCoR) N0014-08-1-0856, and the Air Force Research Laboratory (AFRL) Visiting Faculty Research Program (VFRP) extension grant LRIR 11RI01COR. Any opinions, findings, and conclusions or recommendations expressed in this publication are those of the authors and do not necessarily reflect the views of the funding agency.

References

1. Ahmad, F., Chakradhar, S.T., Raghunathan, A., Vijaykumar, T.N.: Tarazu: Optimizing MapReduce on heterogeneous clusters. In: Proceedings of the 7th International Conference on Architectural Support for Programming Languages and Operating Systems, ASPLOS'12, London, pp. 61–74. ACM, New York (2012). doi:10.1145/2150976.2150984
2. apache.org, Apache Hadoop. http://goo.gl/tnkf
3. Babu, S.: Towards automatic optimization of mapreduce programs. In: Proceedings of the 1st ACM Symposium on Cloud Computing, SoCC'10, Indianapolis, pp. 137–142. ACM, New York (2010). doi:10.1145/1807128.1807150
4. Bajda-Pawlikowski, K., Abadi, D.J., Silberschatz, A., Paulson, E.: Efficient processing of data warehousing queries in a split execution environment. In: Proceedings of the 2011 ACM SIGMOD International Conference on Management of Data, SIGMOD'11, Athens, pp. 1165–1176. ACM, New York (2011). doi:10.1145/1989323.1989447
5. Cardosa, M., Wang, C., Nangia, A., Chandra, A., Weissman, J.: Exploring mapreduce efficiency with highly-distributed data. In: Proceedings of the 2nd International Workshop on MapReduce and its Applications, MapReduce'11, San Jose, pp. 27–34. ACM, New York (2011). doi:10.1145/1996092.1996100
6. cloudera.com, Hadoop resources. http://goo.gl/uW8nr

7. cron.loni.org, CRON project: Cyberinfrastructure for reconfigurable optical networking environment. http://goo.gl/yNRrn (2011)
8. Dean, J., Ghemawat, S.: MapReduce: simplified data processing on large clusters. Commun. ACM **51**(1), 107–113 (2008). doi:10.1145/1327452.1327492
9. Dou, A., Kalogeraki, V., Gunopulos, D., Mielikainen, T., Tuulos, V.H.: Misco: a MapReduce framework for mobile systems. In: Proceedings of the 3rd International Conference on PErvasive Technologies Related to Assistive Environments, PETRA'10, Samos, pp. 32:1–32:8. ACM, New York (2010). doi:10.1145/1839294.1839332
10. Eastman, C.M., Jansen, B.J.: Coverage, relevance, and ranking: the impact of query operators on Web search engine results. ACM Trans. Inf. Syst. **21**(4), 383–411 (2003). doi:10.1145/944012.944015
11. eucalyptus.com, Open source software for building AWS-compatible private and hybrid clouds. http://goo.gl/FpdG5
12. geni-orca.renci.org, NEuca patch for Eucalyptus cloud computing software. http://goo.gl/UbQDG
13. GonzáLez-VéLez, H., Kontagora, M.: Performance evaluation of MapReduce using full virtualisation on a departmental cloud. Int. J. Appl. Math. Comput. Sci. **21**(2), 275–284 (2011). doi:10.2478/v10006-011-0020-3
14. Guo, Z., Fox, G., Zhou, M.: Investigation of data locality and fairness in mapreduce. In: Proceedings of 3rd International Workshop on MapReduce and its Applications, MapReduce'12, Ottawa, pp. 25–32. ACM, New York (2012). doi:10.1145/2287016.2287022
15. Guo, Z., Fox, G., Zhou, M.: Investigation of data locality in MapReduce. In: Proceedings of the 12th IEEE/ACM International Symposium on Cluster, Cloud and Grid Computing, CCGRID'12, Delft, pp. 419–426. IEEE Computer Society, Washington, DC (2012). doi:10.1109/CCGrid.2012.42
16. Horey, J.: A programming framework for integrating web-based spatiotemporal sensor data with mapreduce capabilities. In: Proceedings of the ACM SIGSPATIAL International Workshop on GeoStreaming, IWGS'10, San Jose, pp. 51–58. ACM, New York (2010). doi:10.1145/1878500.1878511
17. ibm.com, IBM SmartCloud services. http://goo.gl/BMhy2
18. iet.unipi.it, The dummynet project. http://goo.gl/smxg5
19. Jahani, E., Cafarella, M.J., Ré, C.: Automatic optimization for MapReduce programs. Proc. VLDB Endow. **4**(6), 385–396 (2011)
20. Jiang, D., Ooi, B.C., Shi, L., Wu, S.: The performance of MapReduce: an in-depth study. Proc. VLDB Endow. **3**(1–2), 472–483 (2010)
21. Karloff, H., Suri, S., Vassilvitskii, S.: A model of computation for MapReduce. In: Proceedings of the 21st Annual ACM-SIAM Symposium on Discrete Algorithms, SODA'10, Austin, pp. 938–948. Society for Industrial and Applied Mathematics, Philadelphia (2010)
22. Kondikoppa, P., Chiu, C.H., Cui, C., Xue, L., Park, S.J.: Network-aware scheduling of MapReduce framework ondistributed clusters over high speed networks. In: Proceedings of the 2012 Workshop on Cloud Services, Federation, and the 8th Open Cirrus Summit, FederatedClouds'12, San Jose, pp. 39–44. ACM, New York (2012). doi:10.1145/2378975.2378985
23. Lämmel, R.: Google's MapReduce programming model – Revisited. Sci. Comput. Program. **70**(1), 1–30 (2008). doi:10.1016/j.scico.2007.07.001
24. Li, M., Subhraveti, D., Butt, A.R., Khasymski, A., Sarkar, P.: CAM: a topology aware minimum cost flow based resource manager for MapReduce applications in the cloud. In: Proceedings of the 21st International Symposium on High-Performance Parallel and Distributed Computing, HPDC'12, Delft, pp. 211–222. ACM, New York (2012). doi:10.1145/2287076.2287110
25. Lin, M.Y., Lee, P.Y., Hsueh, S.C.: Apriori-based frequent itemset mining algorithms on MapReduce. In: Proceedings of the 6th International Conference on Ubiquitous Information Management and Communication, ICUIMC'12, Kuala Lumpur, pp. 76:1–76:8. ACM, New York (2012). doi:10.1145/2184751.2184842

26. Liu, J., Ravi, N., Chakradhar, S., Kandemir, M.: Panacea: towards holistic optimization of MapReduce applications. In: Proceedings of the 10th International Symposium on Code Generation and Optimization, CGO'12, San Jose, pp. 33–43. ACM, New York (2012). doi: 10.1145/2259016.2259022
27. Luo, Y., Guo, Z., Sun, Y., Plale, B., Qiu, J., Li, W.W.: A hierarchical framework for cross-domain MapReduce execution. In: Proceedings of the 2nd International Workshop on Emerging Computational Methods for the Life Sciences, ECMLS'11, San Jose, pp. 15–22. ACM, New York (2011). doi:10.1145/1996023.1996026
28. Luo, Y., Plale, B.: Hierarchical mapreduce programming model and scheduling algorithms. In: Proceedings of the 12th IEEE/ACM International Symposium on Cluster, Cloud and Grid Computing, CCGRID'12, Ottawa, pp. 769–774. IEEE Computer Society, Washington, DC (2012). doi:10.1109/CCGrid.2012.132
29. Mandal, A., Xin, Y., Baldine, I., Ruth, P., Heerman, C., Chase, J., Orlikowski, V., Yumerefendi, A.: Provisioning and evaluating multi-domain networked clouds for Hadoop-based applications. In: Proceedings of the 3rd IEEE International Conference on Cloud Computing Technology and Science, CLOUDCOM'11, Athens, pp. 690–697. IEEE Computer Society, Washington, DC (2011). doi:10.1109/CloudCom.2011.107
30. Mantha, P.K., Luckow, A., Jha, S.: Pilot-mapreduce: an extensible and flexible mapreduce implementation for distributed data. In: Proceedings of 3rd International Workshop on MapReduce and its Applications Date, MapReduce'12, Delft, pp. 17–24. ACM, New York (2012). doi:10.1145/2287016.2287020
31. Molina-Estolano, E., Gokhale, M., Maltzahn, C., May, J., Bent, J., Brandt, S.: Mixing Hadoop and HPC workloads on parallel filesystems. In: Proceedings of the 4th Annual Workshop on Petascale Data Storage, PDSW'09, Portland, pp. 1–5. ACM, New York (2009). doi:10.1145/1713072.1713074
32. Noll, M.G.: michael-noll.com, Benchmarking and stress testing an Hadoop cluster with TeraSort, TestDFSIO. http://goo.gl/zhxD5 (2011)
33. OpenStack: Open source software for building private and public clouds. http://goo.gl/sWpx
34. Palanisamy, B., Singh, A., Liu, L., Jain, B.: Purlieus: locality-aware resource allocation for mapreduce in a cloud. In: Proceedings of 2011 International Conference for High Performance Computing, Networking, Storage and Analysis, SC'11, Seattle, pp. 58:1–58:11. ACM, New York (2011). doi:10.1145/2063384.2063462
35. rackspace.com, The Rackspace cloud. http://goo.gl/5b2iU
36. Sandholm, T., Lai, K.: MapReduce optimization using regulated dynamic prioritization. In: Proceedings of the 11th International Joint Conference on Measurement and Modeling of Computer Systems, SIGMETRICS'09, Seattle, pp. 299–310. ACM, New York (2009). doi: 10.1145/1555349.1555384
37. Schatz, M.C.: CloudBurst: highly sensitive read mapping with MapReduce. Bioinformatics **25**(11), 1363–1369 (2009). doi:10.1093/bioinformatics/btp236
38. scripps.edu, Auto dock vina from scripps lab. http://goo.gl/JM2Pw
39. Tan, J., Meng, X., Zhang, L.: Delay tails in MapReduce scheduling. SIGMETRICS Perform. Evaluation Rev. **40**(1), 5–16 (2012). doi:10.1145/2318857.2254761
40. Vaquero, L.M., Rodero-Merino, L., Buyya, R.: Dynamically scaling applications in the cloud. SIGCOMM Comput. Commun. Rev. **41**(1), 45–52 (2011). doi:10.1145/1925861.1925869
41. Verma, A., Cherkasova, L., Campbell, R.H.: Resource provisioning framework for MapReduce jobs with performance goals. In: Proceedings of the 12th International Middleware Conference, Middleware'11, Lisbon, pp. 160–179. International Federation for Information Processing, Laxenburg (2011)
42. Wang, G., Butt, A.R., Pandey, P., Gupta, K.: Using realistic simulation for performance analysis of mapreduce setups. In: Proceedings of the 1st ACM Workshop on Large-Scale System and Application Performance, LSAP'09, Garching, pp. 19–26. ACM, New York (2009). doi:10.1145/1552272.1552278

43. Xu, H., Li, Z., Guo, S., Chen, K.: CloudVista: interactive and economical visual cluster analysis for big data in the cloud. Proc. VLDB Endow. (PVLDB) **5**(12), 1886–1889 (2012)
44. Zhang, J., Li, T., Pan, Y.: Parallel rough set based knowledge acquisition using mapreduce from big data. In: Proceedings of the 1st International Workshop on Big Data, Streams and Heterogeneous Source Mining: Algorithms, Systems, Programming Models and Applications, BigMine '12, Beijing, pp. 20–27. ACM, New York (2012). doi:10.1145/2351316.2351320
45. Zhou, C.: Fast parallelization of differential evolution algorithm using MapReduce. In: Proceedings of the 12th Annual Conference on Genetic and Evolutionary Computation, GECCO'10, Portland, pp. 1113–1114. ACM, New York (2010). doi:10.1145/1830483.1830689

Improving Cloud Performance with Router-Based Filtering

Chin-Tser Huang, Heath Carroll, and James Perretta

Abstract Our goal in this chapter is to introduce a router-based filtering technology aimed at enhancing the availability and performance of cloud computing. When this technology is integrated with cloud auditing methods, it can make use of cloud auditing information to detect malicious intrusion and traffic anomalies, and to define appropriate filtering rules that can be exchanged between routers in the network, for filtering malicious traffic early and rerouting excessive legitimate requests to other suitable replicated servers. We first give an overview of the specification and generation of filtering rules used by routers. Then we present a theoretical model to find the best locations for hardware routers in a network to block malicious traffic, and discuss how to integrate this theoretical model with cloud auditing techniques. Finally, we present results of experiments that validate our router-based filtering approach.

1 Introduction

Cloud computing has attracted much attention because it realizes the long-desired goal of on-demand network access to a scalable shared pool of computing resources, such as servers, storage, applications, platforms, and networks. However, along with

"Approved for Public Release; Distribution Unlimited: 88ABW-2013-0140, 18-Jan-2013"

C.-T. Huang (✉) • H. Carroll
University of South Carolina, Columbia, SC, USA
e-mail: huangct@cec.sc.edu; carrollh@cec.sc.edu

J. Perretta
Air Force Research Laboratory, Rome, NY, USA
e-mail: James.Perretta@us.af.mil

K.J. Han et al. (eds.), *High Performance Cloud Auditing and Applications*,
DOI 10.1007/978-1-4614-3296-8_13,

cloud computing also come concerns about security issues, such as authentication, integrity, privacy and availability. In this chapter, our main focus is to address the availability issues.

Availability in cloud computing [8, 19, 39] can be considered from two aspects: communication channel availability and cloud resource availability. Communication channel availability is gaining more significance as more and more services are migrating to the cloud. Since clients entrust cloud servers to meet their storage and computation needs, it is important that clients can obtain service when needed. In particular, many real-time cloud applications have emerged in the past few years, such as teleconferencing, remote surgery, massively multi-player online games, and so on. Real-time delivery of the service is paramount for these time-sensitive applications [3, 12]. If the cloud in question is a public cloud, the clients and the servers are not in the same domain; in this case the Internet is required to deliver the service. However, the reliability of the Internet is often impacted by rampant denial of service (DoS) and distributed denial of service (DDoS) attacks [29]. If the unwanted packets from these attacks are allowed to traverse the routers and links in transit to their destinations, they can consume a huge amount of router bandwidth and link bandwidth, and disrupt the timely delivery of legitimate packets. On the other hand, cloud resource availability is no less important than communication channel availability. Cloud service providers try their best to analyze and predict the pattern of client needs, and allocate the cloud resources accordingly. However, the level of service requests from clients is always dynamic. To ensure that cloud resources can continue to meet client needs, the cloud service provider must be able to quickly readjust the allocation of cloud resources [24, 30] when facing an excessive level of service requests, or reroute service requests to replicated servers [32, 44] that provide backup services [43].

This chapter introduces a router-based filtering technology [18, 21, 34] aimed at enhancing the security [7] and performance [42] of cloud computing. When this technology is integrated with cloud auditing methods, it can make use of cloud auditing information to detect malicious intrusion and traffic anomalies, and define appropriate filtering rules that can be exchanged between routers in the network for filtering malicious traffic early and rerouting excessive legitimate requests to other suitable replicated servers.

This chapter is organized as follows. Section 2 provides an overview of the specification and generation of filtering rules used by routers. Section 3 presents a theoretical model to find the best locations for hardware routers in a network to block malicious traffic, and introduces how to integrate this theoretical model with cloud auditing techniques. Section 4 discusses the results of experiments conducted to evaluate the efficacy of our theoretical model based approach to router-based filtering and rerouting in cloud computing. Our experimental results validate the effectiveness and benefits of our approach. Finally, Section 5 presents conclusions of the router-based early filtering technology.

2 Router-Based Filtering and BGP Flow-Spec Rules

Routers are originally designed to forward packets toward their specified destination according to the routing table, which contains entries specifying the routing information they exchange with adjacent routers. However, it is also suitable to use routers for access control and filtering [18] since packets need to pass through routers on their way to the destination. Compared to using traditional firewalls [6, 25, 41] installed at the end host to filter malicious unwanted traffic, using routers for filtering has the potential to filter the malicious unwanted traffic at a location closer to the source, therefore saving the bandwidth that would otherwise be consumed by the malicious unwanted traffic [41]. An additional benefit is that using routers for filtering allows for the realization of multiple instantiations of dynamic filtering rules [1, 2, 23] because routers frequently communicate with neighboring peer routers to advertise new routes or route changes and update their routing tables. Routers can disseminate filtering rules along with routing information update messages. If their routing tables are appropriately updated, routers can be instructed to discard malicious packets immediately, limit the rate of malicious packets, redirect legitimate packets to a different legitimate server, or forward them to a security center for further analysis and diagnosis. Figure 1 illustrates the operation of this router-based filtering and redirection technology [17].

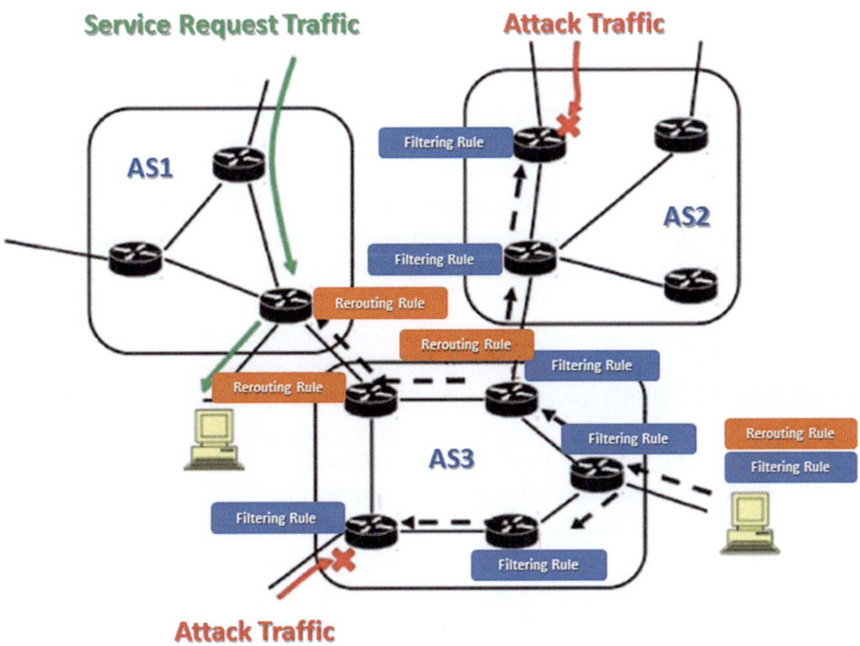

Fig. 1 Router-based filtering and redirection

In order to coordinate the routers in the Internet to perform the task of early filtering of malicious unwanted traffic, we need a means for routers to exchange filtering rules with adjacent routers. The flow specification (Flow-Spec) rules defined for the Border Gateway Protocol (BGP) [26, 31] provide a way of accomplishing this goal. If filtering rules are defined as Flow-Spec rules, they can be passed using BGP UPDATE messages. Since the purpose is to filter the malicious unwanted traffic closer to the source, and BGP is the de facto inter-domain routing protocol, BGP UPDATE messages are the appropriate messages to use to forward the filtering rules upstream toward malicious traffic sources.

BGP is used to exchange routing information across different domains in the Internet and is the protocol used between Internet service providers (ISPs). A domain, or autonomous system (AS), is a network or group of networks under a common administration and with common routing policies. After two peer BGP routers establish a connection, they will use an OPEN message to set up initial parameters, and exchange their routing table [16]. The two BGP routers will send periodical KEEPALIVE messages if there is no change in the routes, and use UPDATE messages to convey any route change information to the other router. BGP UPDATE messages contain a field called Network Layer Reachability Information (NLRI). The Flow-Spec rule is defined as one new type of NLRI, so that it can be carried by BGP UPDATE messages. Formally, a Flow-Spec rule is a vector consisting of several matching criteria that can be applied to IP traffic. The matching criteria of a Flow-Spec are specified by encoding various characteristics of the offending traffic, including destination prefix, source prefix, internet protocol (IP), port, destination port, source port, Internet Control Message Protocol (ICMP) type, ICMP code, TCP flags, packet length, differentiated services code point (DSCP), and fragment. If a given packet satisfies the conjunction of all the specified criteria of a defined Flow-Spec, it is considered a match to the Flow-Spec and should be filtered.

A Flow-Spec received from an external domain needs to be validated against unicast routing before being accepted and installed. The underlying principle of validation is to check that a BGP router [15, 16, 22] from a neighboring AS that advertises a best unicast route for a destination can only advertise a Flow-Spec rule which contains a more or equally specific destination prefix. Moreover, neighboring domains need to establish trust that they will only advertise valid reachability information and will not advertise an NLRI on behalf of a prefix for which they do not provide service; otherwise a domain may cause a DoS attack on that given prefix. More details about the validation of the Flow-Specs and the establishment of trust between neighboring domains are discussed in [26].

As of now, BGP Flow-Spec has received support from multiple major vendors such as Juniper [20]. Cisco has also announced plans to incorporate BGP Flow-Spec into their routers [11]. Although direct Flow-Spec support is not yet available in Cisco routers, it is possible to use Cisco's existing Access Control List (ACL) to emulate a subset of Flow-Spec's features. In our own research projects, we have successfully implemented BGP Flow-Spec rules on a routing software suite called

Table 1 RFC5575 extended community values [26]

Type	Extended community	Encoding
$0x8006$	Traffic-rate	2-byte AS number, 4-byte float
$0x8007$	Traffic-action	bitmask
$0x8008$	Redirect	6-byte route target
$0x8009$	Traffic-marking	DSCP value

Note: Diffserv Code Point (DSCP) and Autonomous System (AS)

Quagga [28], and have also developed Tool command language (Tcl) scripts which allow Cisco routers to exchange access control list (ACL)rules with other routers and install received rules.

In the following subsections, we continue to introduce the filtering actions associated with Flow-Spec rules and the factors associated with the router-based early filtering approach.

2.1 Filtering Actions

A router continues to compare the received traffic with the Flow-Spec rules in its rule base according to their order in the rule base, until a matching rule is found or the end of the rule base is reached with no matching rule being found. When the router finds a matching Flow-Spec rule, i.e. the traffic matches all the criteria of the matching rule, the router will apply the action specified by the rule on the traffic. The BGP Flow-Spec standard [26] specifies a minimum set of typical filtering actions beyond the default action in which the traffic matching the Flow-Spec rule in question is accepted. Four types of BGP extended community values are defined to standardize the minimum set of filtering actions (see Table 1) and are described as follows:

- *Traffic-rate extended community* is often used for policing applications. The encoding is 6 bytes. The first 2 bytes stand for an AS number, while the remaining 4 bytes carry rate information in IEEE floating point format, with the units expressed in bytes per second. The traffic in question is totally filtered if the rate is specified as 0.
- *Traffic-action extended community* consists of 6 bytes but currently only the 2 least significant bits of the 6th byte (from left to right) are defined. When the Terminal Action bit (bit 47) is set, subsequent filtering rules will be applied by the filtering engine if the current rule is not a match. If this bit is not set, subsequent filtering rules will not be applied no matter whether the current rule is a match or not. When the Sampling bit (bit 46) is set, traffic sampling and logging for this flow is enabled.
- *Redirect extended community* allows the traffic to be redirected to a Virtual Routing and Forwarding (VRF) routing instance. Its 6-byte value encodes the route target.

- *Traffic-marking extended community* instructs a system to modify the DSCP bits of a transiting IP packet to the corresponding value. This extended community value is encoded as a sequence of 5 zero bytes followed by the DSCP value encoded in the 6 least significant bits of the 6th byte.

Although the above extended community values have been defined, the standard specification leaves it open for each unique implementation to use arbitrary extended community values for various additional filtering actions, as heterogeneous networks and routing devices make it difficult to standardize on all possible filtering actions.

2.2 Factors of Router-Based Filtering

It is expected that the router-based early filtering approach can filter malicious unwanted traffic early and so can benefit the legitimate traffic with reduced latency and increased throughput. However, the benefits are not always guaranteed along with the deployment. This is mainly due to the cost of filtering: a router's action of filtering packets against a rule will incur extra packet processing overhead, and all the packets, no matter whether unwanted packets or legitimate packets, need to be matched against the installed rules. In this subsection, we discuss the factors that need to be considered to ensure the effectiveness and benefits of the router-based early filtering approach.

2.2.1 Degree of Rule Dissemination

Many unwanted packets can be attributed to DDoS attacks [13, 27] launched by botnets [10, 40], in which the sources are highly distributed [10]. In this case, if a corresponding filtering rule is disseminated to more routers, then there is a better chance to filter more unwanted traffic at an earlier time. We regard the number of hops for which a rule is disseminated as its degree of dissemination. Depending on the type of attack, the rule can be disseminated at each hop to only one specific upstream router (if the offending source is already known) or to every neighboring router except the one from which the rule is received. However, a higher degree of rule dissemination also increases the overall filtering overhead.

2.2.2 Filtering Overhead

Although filtering unwanted traffic early can save bandwidth for more efficient network utilization by legitimate traffic, this benefit does not come for free. The action of filtering will consume the computation resource of a router, which will slow down the router's packet processing to some degree. Therefore, to ensure that there is a considerable net gain, it is necessary to consider the tradeoffs between the bandwidth saved from the filtered unwanted traffic and the overhead resulting from early filtering. Assume that the size of the attacking botnet (i.e., the number of participating bots in the botnet) is N, and each participating bot contributes traffic of average bandwidth u to the victim. If the filtering rule is disseminated outbound for h hops in a breadth-first manner, the overall bandwidth we can save by filtering the attack traffic is roughly $h \cdot N \cdot u$ (assuming the sources are at least h hops away from the victim). On the other hand, the use of filtering rules will lead to a slowdown in packet processing. Assume that the rule corresponding to the attack is installed at i routers, and the overhead of filtering against one rule will result in an average reduction of th in throughput. Thus, the cost for achieving the above saving is estimated at $i \cdot th$. The goal is that the saving is larger than the cost, which is very likely considering the large size of botnets and the ever growing processing speed and capabilities of routers.

2.2.3 Size of Rule Cache and Replacement of an Old Rule

It is desirable to limit the filtering overhead on each router to an acceptable range. Besides limiting the degree of rule dissemination, we can also specify the maximum number of rules each router installs. The installed rules can be maintained in a rule cache. When a new filtering rule is received and the rule cache is full, the router has to make a decision about whether to drop the newly received rule or to replace an old rule with the newly received rule.

2.2.4 Revocation of Obsolete Rules

A filtering rule will not stay valid forever because most offending unwanted traffic will eventually stop. It is desirable to revoke obsolete rules in order to save filtering overhead. We can revoke obsolete rules with two methods. First, when the victim of the unwanted traffic detects that the attack has stopped, it can send out a revocation rule to notify the routers to revoke the corresponding obsolete rule. The revocation rule must be authenticated to prevent an attacker from spoofing a request to revoke an effective filtering rule. Second, it is also possible to include a lifetime or expiration time in the filtering rule, so that the routers know when to revoke a rule.

2.2.5 Location of Legitimate Sources

For a router, installing a filtering rule spells for more processing overhead. Therefore, the dissemination of filtering rules should not continue without a bound. When considering how far a filtering rule should be disseminated, one important observation is that at any given time a host only communicates with a certain number of legitimate hosts. Assume that the victim host in question is communicating with n other legitimate hosts, and the number of hops from the victim to each of these hosts is $h_1, h_2, \ldots, h_n$. We shall call the largest value among $h_1, h_2, \ldots, h_n$ as the communication radius for the victim. If the filtering rule is disseminated beyond the victim's radius of communication, the legitimate sources will not get additional benefit, because the links over which the bandwidth is saved due to the extra dissemination are not on the paths from the legitimate sources to the victim. If a victim knows that it will only communicate with legitimate hosts within communication radius r, then the victim can include r in the filtering rule it sends out, so that the routers will not to disseminate the rule beyond r hops from the victim.

3 Selecting Appropriate Routers to Install Filtering Rules

One important problem that remains to be addressed is how to determine which set of routers in the network should be selected to install a given filtering rule of malicious traffic in order to maximize the benefit of early filtering. There are at least two reasons why this problem is important. First, in our previous work [21], we showed that more router bandwidth can be saved by forwarding filtering rules to the routers closer to the attack sources. This is straightforward to realize if all the malicious traffic converges on the same single path, because moving away from the victim on the path means moving toward attack sources. However, in today's Internet the attacks and threats often originate from very distributed sources, which largely complicate this problem. Second, for consideration of router efficiency it is not possible to install an unlimited number of filtering rules in a router because the overhead of matching every received packet against every filtering rule will be too large. To meet this constraint, it is desirable to install a filtering rule only in a chosen subset of routers instead of in all routers in the network.

In this section, we formally formulate the problem of automatic selection of routers for filter placement, and present an analytical solution to this problem which was also introduced in [18]. We start with the simple case of a tree topology in which there is only one path from each node to the root, and then generalize it to a cyclic topology. We also discuss how to apply our early filtering approach to quality of service (QoS) based routing with traffic distribution in the networks.

In our network model, we assume that the routing paths from the distributed attackers to the target server (the victim) form an acyclic tree topology, in which there is only one path from each attacker to the victim. We begin our discussion with the tree topology for the sake of simplicity. In a later section, we will discuss an extension of our model and algorithm to cover cyclic topologies. The root of the tree is the victim, and other nodes on the tree represent intermediate routers. Therefore, the network can be modeled as a tree $T = (V, E)$, where V is a set of routers and E is a set of links connecting adjacent routers in V. There may be a number of hosts, legitimate or malicious, connected to each router. Malicious hosts can send malicious traffic toward the victim. In our network model, our concern

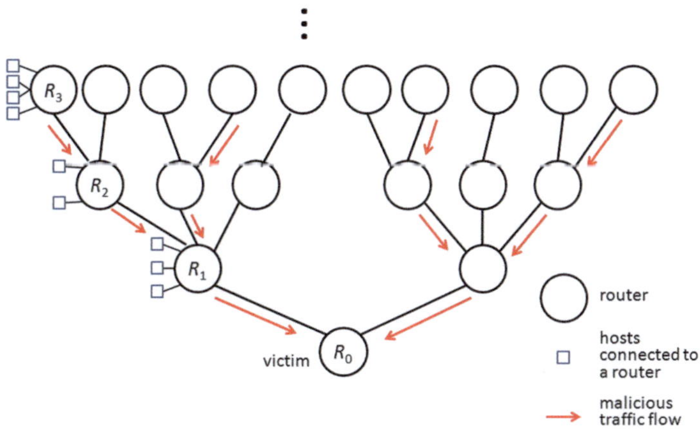

Fig. 2 The network model (Reproduced from Huang et al. [18])

is focused on the routers, not the hosts, because all the malicious traffic flows are forwarded by the routers toward the victim. Without loss of generality, we consider one routing path [5, 37] at a time.

An example of our network model is shown in Fig. 2. In this example, if router R_3 has malicious traffic destined to the victim R_0, then R_3 will forward the malicious traffic to R_2, which is the next router on the path from R_3 to R_0. Next, R_2 will forward the malicious traffic received from R_3, along with the malicious traffic received from malicious hosts connected to R_2 (if any), to router R_1. Finally, R_1 will forward the malicious traffic received from R_2, along with the malicious traffic received from malicious hosts connected to R_1 (if any), to R_0. The goal of our approach is to install filtering rules at a set of appropriate routers in the network, such that the malicious traffic flows are blocked early, soon after their packets enter the network. This approach has at least two significant benefits. First, it can save bandwidth of the routers and links from being occupied by malicious traffic, and make the saved bandwidth available for the use of legitimate traffic. Second, it can confine other possible damage that could be caused by the malicious traffic (e.g., slowing down the hosts or routers, modifying or deleting files in the hosts or routers) to local regions.

Figure 3 illustrates the case in which some malicious traffic flows are blocked at routers where the corresponding filtering rule is installed (represented by a gray circle). In this example, since router R_3 installs the corresponding filtering rule, R_3 will block the malicious traffic it receives. As such, there will be no malicious traffic flowing from router R_3 to router R_2. Router R_2 does not install the corresponding filtering rule, so R_2 will forward the malicious traffic it receives toward router R_1. Router R_1 installs the corresponding filtering rule, so R_1 will block the malicious traffic it receives, and there will be no malicious traffic flowing from router R_1 to the victim R_0.

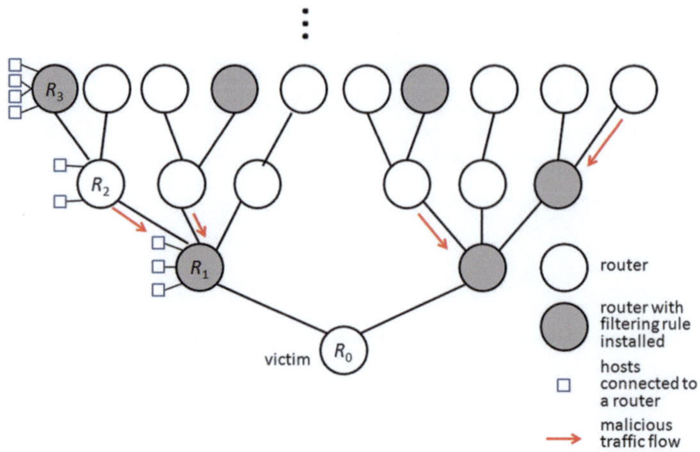

Fig. 3 Malicious traffic flows are blocked at some routers where the corresponding filtering rule is installed (Reproduced from Huang et al. [18])

To formulate the problem, we have to first define the saving and cost resulting from filtering a malicious traffic flow at a router u. It is apparent to see that there is a miss penalty if one specific malicious traffic flow is not filtered at a given router u. The miss penalty includes at least the following two components: (1) the bandwidth (of the link to the next router on the routing path to the victim) that will be occupied by the malicious flow if not filtered, (2) the cost of the threat or damage caused by the malicious flow if not filtered. The first component is apparently a numeric value which could be normalized according to the capability of the router and the capacity of the link, while the second component needs to be quantified using some predefined metrics. Let $m(u, f_i)$ be the miss penalty of allowing one unit of malicious traffic flow f_i (e.g., one individual packet of the malicious flow, or a set of consecutive packets that can cause damage when allowed to pass) at router u. We can define $m(u, f_i)$ as

$$m(u, f_i) = \sum_{(u_1,u_2)\in PATH(u,u')} c(u_1, u_2, f_i)$$

where u' is the nearest node on the path from u to the root that installs the filtering rule on f_i, u_1, u_2 are adjacent nodes on the path from u to u', $(u_1, u_2) \in E$, and $c(u_1, u_2, f_i)$ is the penalty of not filtering one unit of f_i on the link from u_1 to u_2.

Next, we analyze the tradeoff between the saving and cost when early filtering is exerted. The bandwidth saved at a router due to filtering is related to the packet rate of the malicious traffic flow arriving at the router. If the rate of a malicious traffic flow is higher, then the bandwidth saved by filtering this flow is higher. On the other hand, installing a filtering rule at a router imposes processing overhead on the router because the router will have to match the filtering rule against every received packet to determine whether the packet should be allowed or blocked. (Even if a received packet is legitimate, it will be matched against every filtering rule so that it can be determined to be legitimate.) This overhead will negatively impact the throughput at the router. Let $r(u, f_i)$ be the rate of malicious flow f_i observed at router u, and $o(u, f_i)$ be the amount of reduced throughput due to the overhead of installing the filtering rule on f_i at router u. Then the saving of filtering f_i at router u can be derived as

$$r(u, f_i) \cdot m(u, f_i) - o(u, f_i)$$

In addition, there is one possible source of performance loss in the throughput. We note that due to the overhead of filtering there should be an upper bound on the number of installed filtering rules at each router, because if a router installs an unlimited number of filtering rules then it may have to spend too much of its computation resources on filtering, thus leading to long delay and low throughput. If a router wants to install a new rule but the number of installed filtering rules at the router has reached the upper bound, then one existing rule should be removed in order to make space for installing this new rule. We assume that there is a heuristic algorithm available for a router which replaces the rule that generates the least saving. In particular, let $l(u, f_i)$ be the performance loss in the throughput due to replacing the least-saving rule at router u, then $l(u, f_i)$ can be derived as

$$l(u, f_i) = min_{every\ f_j\ currently\ filtered\ at\ u}\{r(u, f_j) \cdot m(u, f_j)\}$$

It is clear that the rule selected to be replaced also imposes processing overhead on router u, whose value is the same as $o(u, f_i)$ because this new rule also needs to be matched against every received packet. Therefore, the net saving of filtering f_i at router u can be derived as

$$r(u, f_i) \cdot m(u, f_i) - o(u, f_i) - (l(u, f_i) - o(u, f_i)) = r(u, f_i) \cdot m(u, f_i) - l(u, f_i)$$

Now we can formulate the problem. Without loss of generality, we limit our consideration to only one routing path in the network and one type of malicious traffic. (Note that other routing paths in the network and other types of malicious traffic can be formulated in the same way.) Suppose one routing path of the malicious traffic contains n routers on the path, R_1 through R_n, and R_0 is the victim of the malicious traffic. The filtering rule corresponding to the given malicious traffic can be installed in up to k routers, where $0 \leq k \leq n$, on the routing path. This leads to an optimization problem on figuring out which routers (up to k) should install the filtering rule in order to maximize the saving.

Let r_i be the rate of the malicious traffic observed at router R_i, and m_i be the miss penalty of the malicious traffic at router R_i. Since every intermediate router may possibly be connected to malicious hosts that will inject more malicious traffic toward the victim R_0, we have $r_1 \geq r_2 \geq \ldots \geq r_n$, which means a router closer to R_0 will observe higher rate of the malicious traffic. Assume $x \leq k$ intermediate routers $R_{v_1}, R_{v_2}, \ldots, R_{v_x}$, where $v_1, v_2, \ldots, v_x$ are integers and $1 \leq v_1 \leq v_2 \leq \ldots \leq v_x \leq k$, are chosen to install the filtering rule. Since R_{v_i+1} would have filtered rate r_{v_i+1} of the malicious traffic, the rate of malicious traffic to be observed and filtered by R_{v_i} is only $r_{v_i} - r_{v_i+1}$. Also, let l_i be the performance loss due to replacing an existing filtering rule at router R_i. Thus the overall saving achieved by filtering at the aforementioned x routers $R_{v_1}, R_{v_2}, \ldots, R_{v_x}$ is given by the following summation:

$$\sum_{i=1}^{x}((r_{v_i} - r_{v_i+1}) \cdot m_{v_i} - l_{v_i})$$

Therefore, the problem can be formally defined as follows.

Formulation of filter placement k-optimization problem

Given $R_0, R_1, R_2, \ldots, R_n$, which are nodes on a tree $T = (V, E)$ and R_i is the parent of R_{i+1} on the tree T, and associated parameters $r_1, r_2, \ldots, r_n, r_{n+1}$ (where $r_1 \geq r_2 \geq \ldots \geq r_n \geq r_{n+1} = 0$), $m_1, m_2, \ldots, m_n, l_1, l_2, \ldots, l_n$, and integer k such that $0 \leq k \leq n$. The filter placement problem is to find integer x satisfying $0 \leq x \leq k$, and integers $v_1, v_2, \ldots, v_x$ where $1 \leq v_1 \leq v_2 \leq \ldots \leq v_x \leq k$ (and $v_0 = 0$), such that

$$\sum_{i=1}^{x}((r_{v_i} \cdot \sum_{j=v_{i-1}+1}^{v_i} m_j) - l_{v_i})$$

is maximized. The maximum value is the net saving achieved by filtering when up to k of the routers R_1 through R_n can install filtering. It is straightforward to use dynamic programming to solve this maximization problem. Regarding this model as a function of $(x, v_1, v_2, \cdots, v_x)$, filter placement via maximization of objective function over $(x, v_1, v_2, \cdots, v_x)$ is our approach to filter placement based on the theoretical model. □

3.1 *Resolving Cycles in Network Topology*

In the previous discussion, it is assumed that the network topology is tree-shaped for the sake of simplicity. However, in reality a network topology often contains one or more redundant paths between two routers, for several reasons such as reliability and load balancing. The undirected edges representing these paths form cycles in the topology.

There is one incompatibility between cyclic topology and our algorithm. When the malicious traffic reaches one router on the cycle, the malicious traffic may split into two portions, each following a different path toward the victim. This violates our basic assumption that a router closer to the root of the tree (in this case the victim) will observe a higher rate of the malicious traffic due to convergence.

Our algorithm can be extended with the following operation to resolve the cycles in the topology. If a cycle is found in the topology, we can find the router at the split point where the malicious traffic splits into two paths, and install the filtering rule in that router. Therefore, no malicious traffic will flow from the router at the split

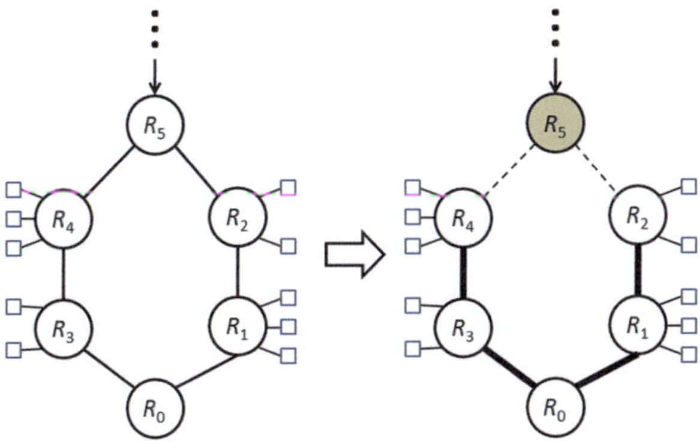

Fig. 4 Configuration of testbed network (Reproduced from Huang et al. [18])

point to the next router on each of the two paths. In this case, we can temporarily remove the two edges connecting the router at the split point and the next router on each of the two paths, and divide the topology into two subgraphs: one subgraph is rooted at the victim and the other is rooted at the router at the split point. With this operation, the assumption that a router closer to the root will observe a higher rate of the malicious traffic is again satisfied. This operation can be applied in multiple instances until all the cycles in the topology are resolved. Then, our algorithm can be applied on each subgraph separately.

We can use the example topology in Fig. 4 to illustrate this extension. Suppose R_0 is the victim and the sequence (R_0, R_3, R_4, R_5, R_2, R_1, R_0) forms a cycle. When the malicious traffic reaches R_5, it will split into two paths R_5 R_4 R_3 R_0 and R_5 R_2 R_1 R_0. If the filtering rule is installed at R_5, then there will be no malicious traffic flowing from R_5 to R_4 and from R_5 to R_2. Therefore, we can divide the topology into two subgraphs, one rooted at R_0 and the other one rooted at R_5, and each subgraph satisfies the basic assumption.

3.2 Integrating Router-Based Early Filtering and Cloud Auditing with QoS Routing Algorithms

It has been shown in [21] that our router-based early filtering approach can reduce the average transmission latency and increase the average throughput of legitimate traffic. The filter placement approach presented earlier can find an appropriate set of routers automatically to install the filtering rule to maximize the benefit of early filtering. These improved values of transmission latency and average throughput can be used as feedback to QoS routing algorithms. QoS routing [36,38] is a well studied topic and was proposed to provide guaranteed services by finding a path satisfying the QoS requirements of the traffic. There are existing QoS routing algorithms based on dynamic programming [9], graph theory [14], and probability [33].

In general, QoS routing algorithms depend on the link state information, such as available bandwidth, transmission latency, loss probability, and cost function, to make routing decisions that satisfy QoS requirements of specific traffic. Since the network environment changes as time goes by, the link state information which was given as the initial input to the QoS routing algorithm may not persist.

The router-based early filtering approach and cloud auditing techniques can be combined to improve the performance of QoS routing algorithms. Cloud auditing techniques are usually used to monitor the service level and the usage of cloud resources, and analyze the interaction between cloud servers and clients. If the early filtering approach is applied, the shorter transmission latency and higher average throughput resulting from the saved bandwidth will be observed and recorded by cloud auditing tools. When the improved values of transmission latency and average throughput are provided to QoS algorithms, they can make more precise routing decisions that utilize the benefits generated by the early filtering approach.

4 Experimental Procedures and Preliminary Results

We design a set of experiments to evaluate the performance and benefits of our approach introduced in Section 3. The steps we follow to conduct the experiments are summarized as follows:

- *Step 1*: Set up a testbed network which is composed of a single path of routers, one host (playing the role of the distributed attacker) connected to each router, and the destination (the victim).
- *Step 2*: Define the parameters for the experiment, including r_i, m_i, and l_i at router R_i.
- *Step 3*: Inject the target attacking traffic and the secondary attacking traffic whose volume meets the defined parameters.
- *Step 4*: By default, install in each router the filtering rule corresponding to the secondary attacking traffic that directly comes into each router instead of being forwarded by another router.
- *Step 5*: Do the following for each possible combination of routers: for each router in the chosen combination, replace the filtering rule on the secondary attacking traffic with the filtering rule on the target attacking traffic; generate the legitimate traffic toward the destination and measure the throughput of the legitimate traffic.

Next, we describe each step of our simulation in detail.

4.1 Testbed Setup

With our currently available equipment, we set up a testbed as illustrated in Fig. 5. The configuration of our testbed is composed of a single path of 4 Cisco 2911

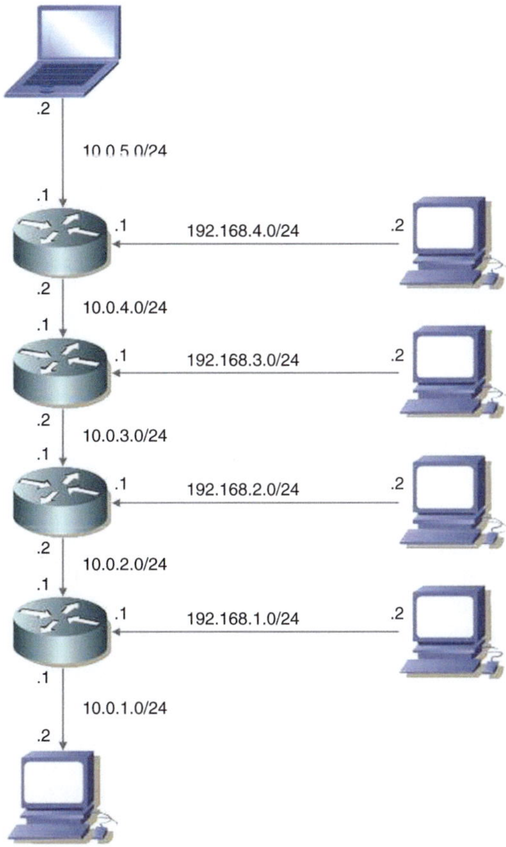

Fig. 5 Configuration of testbed network

routers, with a host connected to each router to simulate distributed attackers. The attackers are from the addresses 192.168.1.2, 192.168.2.2, 192.168.3.2, and 192.168.4.2 as shown in the figure. All the malicious traffic flows are destined to the victim host located at the address 192.168.5.2. To measure the performance of our algorithm, we also connect a host with the address 192.168.6.2 to the last router to send legitimate traffic to the victim host.

4.2 Parameter Selection

We define the parameters for the experiment. Since there are four routers on the path in the testbed, we call the four routers R_4, R_3, R_2, R_1 respectively, while R_0 represents the destination. We define the relevant parameters r_4, r_3, r_2, r_1, m_4, m_3, m_2, m_1, l_4, l_3, l_2, l_1, for the simulation example as shown in Fig. 6. Note that the value of each r_i must be at least equal to the value of $r_i + 1$; this is because router R_i may receive the attacking traffic from $R_i + 1$ and also from the attacker directly

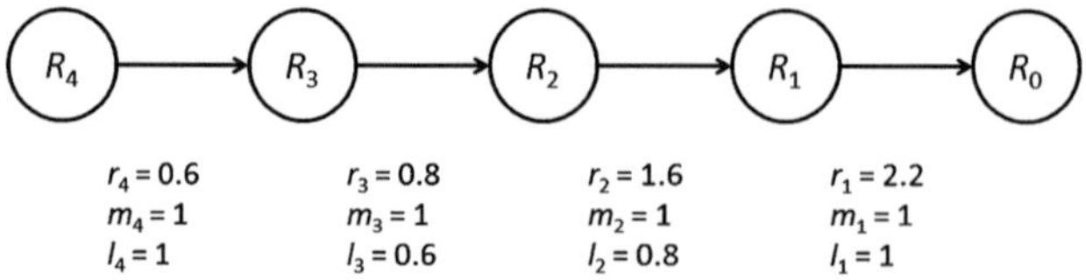

Fig. 6 Parameters used in the simulation experiments (Reproduced from Huang et al. [18])

connected to R_i itself. Also note that we normalize the value of all m_i to 1 in this particular simulation because all the attacking traffic is ping packets and all the routers used in our testbed are of the same model.

4.3 *Traffic Injection*

We need to inject the target attacking traffic and the secondary attacking traffic which meet the parameters defined in Step 2. The target attacking traffic simulates the traffic which corresponds to the new filtering rule, therefore the volume of the target attacking traffic observed by router R_i should be in proportion to the r_i parameter. The secondary attacking traffic simulates the traffic filtered by another filtering rule which will be selected to be replaced by the new filtering rule; therefore the volume of the secondary attacking traffic should be proportional to the l_i parameter. The target attacking traffic is simulated by flooding pings from each attacking host toward the victim host with multiple threads of the following command:

```
ping -f -s 1024 192.168.5.2 \&
```

Note that the ping command is executed in the background (by specifying the "&" at the end), so that we can execute different number of threads of ping at different attacking hosts to simulate different volume of attacking traffic coming into each router.

Moreover, in order to simulate l_i, namely the performance loss due to replacing an existing filtering rule at router R_i, we need to simulate secondary attacking traffic coming into each router and being filtered by a different rule. When one router installs the filtering rule on the target attacking traffic, the other rule filtering the secondary attacking traffic is disabled, so as to simulate the performance loss due to installing the new rule. It would be ideal if we can connect a second host to each router to generate the secondary attacking traffic, so that the filtering of the secondary attacking traffic will be based on source address, and the comparison of the impact on legitimate traffic throughput due to the target attacking traffic and the impact due to the secondary attacking traffic will be fair. However, because of the constraint of our currently available equipment, we have to use an alternative approach in which we distinguish the target attacking traffic from the secondary attacking traffic by using the Type of Service (TOS) value [4], which can be specified as one optional field in the ping packet and can also be specified in a filtering rule. For example, to flood ping packets with TOS value 4 toward the victim 192.168.5.2, we can execute multiple threads of the following command:

```
ping -f  s 1024 Q 4 192.168.5.2 \&
```

We generated the target attacking traffic which traverses all 4 routers as type 0 (the normal type), the secondary traffic coming into R_1 as type 4, the secondary traffic coming into R_2 as type 8, the secondary traffic coming into R_3 as type 12, and the secondary traffic coming into R_4 as type 16.

Based on the parameters defined in Fig. 6, we execute 60 threads of ping at the host 192.168.4.2, 20 threads of ping at the host 192.168.3.2, 80 threads of ping at the host 192.168.2.2, and 60 threads of ping at the host 192.168.1.2, so that the volume of target attacking traffic injected into each router is proportional to r_4, r_3, r_2, and r_1. We also execute an appropriate number of ping threads with TOS assigned so that the volume of secondary attacking traffic injected into each router is proportional to l_4, l_3, l_2, and l_1, respectively.

4.4 Filtering Rule Installation

By default, each router installed the following set of ACL rules to filter the secondary attacking traffic:

```
access-list 100 deny icmp any any TOS X
access-list 100 permit icmp any any
access-list 100 permit ip host 192.168.6.2 any
```

Note: X is a chosen TOS value. If a router is chosen to install the filtering rule on the target attacking traffic in Step 5, the router will disable the above rules and install the following set of ACL rules:

```
access-list 101 deny icmp any any TOS normal
access-list 101 permit icmp any any
access-list 101 permit ip host 192.168.6.2 any
```

4.5 Traffic Generation and Test Results

We generate legitimate traffic and test it with each possible combination of routers installing the filtering rule in order to measure the throughput of legitimate traffic of each case. The legitimate traffic is generated by executing iperf [35] client procedure on the host 192.168.6.2 which is connected to router R_4, and executing

Table 2 Net saving achieved by filtering and measured throughput

Combinations of routers	Net saving	Throughput (MB/s)
No filtering	0	195.2
Filtering at R_1	1.2	270.6
Filtering at R_2	2.4	303.6
Filtering at R_3	1.8	250.6
Filtering at R_4	1.4	222.2
Filtering at R_1, R_2	2	251.9
Filtering at R_1, R_3	2.2	277.2
Filtering at R_1, R_4	2	277.2
Filtering at R_2, R_3	2.6	306.7
Filtering at R_2, R_4	2.6	304.8
Filtering at R_3, R_4	1.4	247.0
Filtering at R_1, R_2, R_3	2.2	248.3
Filtering at R_1, R_2, R_4	2.2	247.8
Filtering at R_1, R_3, R_4	1.8	275.9
Filtering at R_2, R_3, R_4	2.2	300.4
Filtering at R_1, R_2, R_3, R_4	1.8	243.2

Note: second column (net saving calculated by algorithm) and third column (throughput measured by simulations)

Iperf server procedure on the victim host 192.168.5.2. The Iperf tool can measure the amount of data transferred between the client and the server during a given period of time, and calculate the throughput of the data path accordingly. We tested all possible combinations of routers installing the filtering rule to see what the net savings achieved by filtering would be versus the throughput of legitimate traffic as measured via the simulations. The results are given in Table 2. In particular, Table 2 shows that the two cases leading to highest net savings are installing the filtering rule at routers R_2 and R_3 or installing at routers R_2 and R_4. These choices also yield the two highest throughput values. At least for this testbed, the savings achieved by filtering are consistent with the measured throughput, lending credence to the validity of our approach.

5 Conclusion

In this chapter, we presented a router-based filtering technology to address the availability issues in cloud computing. We first introduced the BGP Flow-Spec rules, which provide a handy means for routers to exchange information about early filtering of overwhelming malicious traffic that could impact the availability of cloud services. Then, we presented a theoretical model via which to find the best routers in a network to install the filtering rule against given malicious traffic flow, and discussed how to integrate this algorithm with cloud auditing techniques. Finally, we used experimental results to validate the effectiveness and benefits of our theoretical model and algorithm. Although we have demonstrated the effectiveness of the router-based early filtering approach, how to ensure the responsiveness of this approach remains an open challenge. It may take some time to detect an ongoing malicious attack and generate the corresponding filtering rule, and during this period of time the availability of cloud services could be seriously impacted. It is worth investigating how to enhance the responsiveness of the approach so that the attack can be mitigated shortly after its outbreak. Moreover, with the fast proliferation of mobile devices that can access the cloud services, it is also important to manage the scalability of the proposed approach.

Acknowledgements This material is based upon work partially supported by the National Science Foundation (NSF) grants CNS-0916857, the Air Force Office of Scientific Research (AFOSR) Summer Faculty Fellow Program (SFFP), the Air Force Research Laboratory (AFRL) Visiting Faculty Research Program (VFRP), and AFOSR/AFRL LRIR 11RI01COR.

References

1. Abbes, T., Bouhoula, A., Rusinowitch, M.: An inference system for detecting firewall filtering rules anomalies. In: Proceedings of the 2008 ACM Symposium on Applied Computing, SAC'08, Fortaleza, pp. 2122–2128. ACM, New York (2008). doi:10.1145/1363686.1364197
2. Abramov, E., Mordvin, D., Makarevich, O.: Automated method for constructing of network traffic filtering rules. In: Proceedings of the 3rd International Conference on Security of Information and Networks, SIN'10, Taganrog, Rostov-on-Don, pp. 203–211. ACM, New York (2010). doi:10.1145/1854099.1854141
3. Agapi, A., Birman, K., Broberg, R.M., Cotton, C., Kielmann, T., Millnert, M., Payne, R., Surton, R., Renesse, R.v.: Routers for the cloud: can the Internet achieve 5-nines availability? IEEE Internet Comput. **15**(5), 72–77 (2011). doi:10.1109/MIC.2011.122
4. Almquist, P.: ietf.org, RFC 1349 – type of service in the internet protocol suite. http://goo.gl/6O5jQ (1992)
5. Andersen, D.G., Snoeren, A.C., Balakrishnan, H.: Best-path vs. multi-path overlay routing. In: Proceedings of the 3rd ACM SIGCOMM Conference on Internet Measurement, IMC'03, Miami Beach, pp. 91–100. ACM, New York (2003). doi:10.1145/948205.948218
6. Bartal, Y., Mayer, A., Nissim, K., Wool, A.: Firmato: a novel firewall management toolkit. ACM Trans. Comput. Syst. **22**(4), 381–420 (2004). doi:10.1145/1035582.1035583
7. Bleikertz, S., Schunter, M., Probst, C.W., Pendarakis, D., Eriksson, K.: Security audits of multi-tier virtual infrastructures in public infrastructure clouds. In: Proceedings of the 2010 ACM Workshop on Cloud Computing Security Workshop, CCSW'10, Chicago, pp. 93–102. ACM, New York (2010). doi:10.1145/1866835.1866853
8. Casalicchio, E., Silvestri, L.: An inter-cloud outsourcing model to scale performance, availability and security. In: Proceedings of the 5th IEEE/ACM International Conference on Utility and Cloud Computing, UCC'12, Chicago, pp. 151–158. IEEE Computer Society, Washington, DC (2012). doi:10.1109/UCC.2012.16
9. Chen, S., Nahrstedt, K.: An overview of quality of service routing for next-generation high-speed networks: problems and solutions. Netw. Manage. Global Internetwork. **12**(6), 64–79 (1998). doi:10.1109/65.752646
10. Dainotti, A., King, A., Claffy, K., Papale, F., Pescapè, A.: Analysis of a "/0" stealth scan from a botnet. In: Proceedings of the 2012 ACM Conference on Internet Measurement Conference, IMC'12, Boston, pp. 1–14. ACM, New York (2012). doi:10.1145/2398776.2398778
11. Duvivier, B.: cisco.com, Cisco BGP security enhancements. http://goo.gl/7NDpc (2011)

12. Esposito, C., Cotroneo, D., Gokhale, A.: Reliable publish/subscribe middleware for time-sensitive internet-scale applications. In: Proceedings of the 3rd ACM International Conference on Distributed Event-Based Systems, DEBS'09, Nashville, pp. 16:1–16:12. ACM, New York (2009). doi:10.1145/1619258.1619280
13. François, J., Aib, I., Boutaba, R.: FireCol: a collaborative protection network for the detection of flooding DDoS attacks. ACM Trans. Comput. Syst. **20**(6), 1828–1841 (2012). doi:10.1109/TNET.2012.2194508
14. Ghosh, D., Sarangan, V., Acharya, R.: Quality-of-service routing in IP networks. IEEE Trans. Multimed. **3**(2), 200–208 (2001). doi:10.1109/6046.923819
15. Gill, P., Schapira, M., Goldberg, S.: Let the market drive deployment: a strategy for transitioning to bgp security. In: Proceedings of the 2011 ACM SIGCOMM Conference, SIGCOMM'11, Toronto, pp. 14–25. ACM, New York (2011). doi:10.1145/2018436.2018439
16. Gregori, E., Improta, A., Lenzini, L., Rossi, L., Sani, L.: On the incompleteness of the AS-level graph: a novel methodology for BGP route collector placement. In: Proceedings of the 2012 ACM Conference on Internet Measurement Conference, IMC'12, Boston, pp. 253–264. ACM, New York (2012). doi:10.1145/2398776.2398803
17. Houmansadr, A., Nguyen, G.T., Caesar, M., Borisov, N.: Cirripede: circumvention infrastructure using router redirection with plausible deniability. In: Proceedings of the 18th ACM Conference on Computer and Communications Security, CCS'11, Chicago, pp. 187–200. ACM, New York (2011). doi:10.1145/2046707.2046730
18. Huang, C.T., Han, K.J., Perretta, J.: Automatic selection of routers for placing early filters of malicious traffic. In: Proceedings of IEEE GLOBECOM 2011, Houston, pp. 1–5. IEEE (2011). doi:10.1109/GLOCOM.2011.6133640
19. Juels, A., Oprea, A.: New approaches to security and availability for cloud data. Commun. ACM **56**(2), 64–73 (2013). doi:10.1145/2408776.2408793
20. juniper.net, Enabling BGP to carry flow-specification routes. http://goo.gl/zxv8C (2010)
21. Kalakota, P., Huang, C.T.: On the benefits of early filtering of botnet unwanted traffic. In: Proceedings of the 18th International Conference on Computer Communications and Networks, ICCCN'09, San Francisco, pp. 1–6. IEEE Computer Society, Washington, DC (2009). doi:10.1109/ICCCN.2009.5235325
22. Li, J., Guidero, M., Wu, Z., Purpus, E., Ehrenkranz, T.: BGP routing dynamics revisited. SIGCOMM Comput. Commun. Rev. **37**(2), 5–16 (2007). doi:10.1145/1232919.1232921
23. Li, Y., Liljenstam, M., Liu, J.: Real-time security exercises on a realistic interdomain routing experiment platform. In: Proceedings of the 23rd ACM/IEEE/SCS Workshop on Principles of Advanced and Distributed Simulation, PADS'09, Lake Placid, pp. 54–63. IEEE Computer Society, Washington, DC (2009). doi:10.1109/PADS.2009.12
24. Li, C., Li, L.: An efficient resource allocation for maximizing benefit of users and resource providers in Ad Hoc grid environment. Inf. Syst. Front. **14**(5), 987–998 (2012). doi:10.1007/s10796-011-9310-4
25. Mansmann, F., Göbel, T., Cheswick, W.: Visual analysis of complex firewall configurations. In: Proceedings of the 9th International Symposium on Visualization for Cyber Security, VizSec'12, Seattle, pp. 1–8. ACM, New York (2012). doi:10.1145/2379690.2379691
26. Marques, P., Sheth, N., Raszuk, R., Greene, B., Mauch, J., McPherson, D.: ietf.org, RFC 5575 – Dissemination of flow specification rules. http://goo.gl/y8XVq (2009)
27. Nazario, J.: DDoS: DDoS attack evolution. Netw. Secur. **2008**(7), 7–10 (2008). doi:10.1016/S1353-4858(08)70086-2
28. nongnu.org, Quagga routing software suite. http://goo.gl/FqSJE
29. Oreku, G.S., Li, J., Pazynyuk, T.: An application-driven perspective on wireless devises security: the case of distributed denial-of-service (DDOS). In: Proceedings of the 2nd ACM Workshop on Performance Monitoring and Measurement of Heterogeneous Wireless and Wired networks, PM2HW2N'07, Chania, pp. 81–83. ACM, New York (2007). doi:10.1145/1298275.1298292

30. Rai, A., Bhagwan, R., Guha, S.: Generalized resource allocation for the cloud. In: Proceedings of the 3rd ACM Symposium on Cloud Computing, SoCC'12, San Jose, pp. 15:1–15:12. ACM, New York (2012). doi:10.1145/2391229.2391244
31. Rekhter, Y., Li, T., Hares, S.: RFC 4271 – a border gateway protocol 4 (BGP-4). http://goo.gl/7NDpc (2006)
32. Rodolakis, G., Siachalou, S., Georgiadis, L.: Replicated server placement with QoS constraints. In: Proceedings of the 3rd International Conference on Quality of Service in Multiservice IP Networks, QoS-IP'05, Catania, pp. 207–220. Springer, Berlin/Heidelberg (2005). doi:10.1007/978-3-540-30573-6_16
33. Siachalou, S., Georgiadis, L.: Efficient QoS routing. Comput. Netw. **43**(3), 351–367 (2003). doi:10.1016/S1389-1286(03)00286-X
34. Soldo, F., Argyraki, K., Markopoulou, A.: Optimal source-based filtering of malicious traffic. IEEE/ACM Trans. Netw. **20**(2), 381–395 (2012). doi:10.1109/TNET.2011.2161615
35. sourceforge.net, Iperf. http://goo.gl/4ubV5 (2008)
36. Sun, H., Hughes, H.D.: Adaptive QoS routing by cross-layer cooperation in Ad Hoc networks. EURASIP J. Wirel. Commun. Netw. **2005**(5), 661–671 (2005). doi:10.1155/WCN.2005.661
37. Tague, P., Nabar, S., Ritcey, J.A., Poovendran, R.: Jamming-aware traffic allocation for multiple-path routing using portfolio selection. IEEE/ACM Trans. Netw. **19**(1), 184–194 (2011). doi:10.1109/TNET.2010.2057515
38. Uludag, S., Lui, K.S., Nahrstedt, K., Brewster, G.: Analysis of topology aggregation techniques for QoS routing. ACM Comput. Surv. **39**(3) (2007). doi:10.1145/1267070.1267071
39. Undheim, A., Chilwan, A., Heegaard, P.: Differentiated availability in cloud computing SLAs. In: Proceedings of the 12th IEEE/ACM International Conference on Grid Computing, GRID'11, Lyon, pp. 129–136. IEEE Computer Society, Washington, DC (2011). doi:10.1109/Grid.2011.25
40. West, A.G., Lee, I.: Towards the effective temporal association mining of spam blacklists. In: Proceedings of the 8th Annual Collaboration, Electronic Messaging, Anti-Abuse and Spam Conference, CEAS'11, Perth, pp. 73–82. ACM, New York (2011). doi:10.1145/2030376.2030385
41. Wiebelitz, J., Brenner, M., Kunz, C., Smith, M.: Early defense: enabling attribute-based authorization in grid firewalls. In: Proceedings of the 19th ACM International Symposium on High Performance Distributed Computing, HPDC'10, Chicago, pp. 336–339. ACM, New York (2010). doi:10.1145/1851476.1851524
42. Xia, L., Cui, Z., Lange, J.R., Tang, Y., Dinda, P.A., Bridges, P.G.: VNET/P: bridging the cloud and high performance computing through fast overlay networking. In: Proceedings of the 21st International Symposium on High-Performance Parallel and Distributed Computing, HPDC'12, Delft, pp. 259–270. ACM, New York (2012). doi:10.1145/2287076.2287116
43. Xu, L., Hu, J., Mkandawire, S., Jiang, H.: SHHC: a scalable hybrid hash cluster for cloud backup services in data centers. In: Proceedings of the 31st International Conference on Distributed Computing Systems Workshops, ICDCSW'11, Minneapolis, pp. 61–65. IEEE Computer Society, Washington, DC (2011). doi:10.1109/ICDCSW.2011.31, http://dx.doi.org/10.1109/ICDCSW.2011.31
44. Zegura, E.W., Ammar, M.H., Fei, Z., Bhattacharjee, S.: Application-layer anycasting: a server selection architecture and use in a replicated web service. IEEE/ACM Trans. Netw. **8**(4), 455–466 (2000). doi:10.1109/90.865074

Index

K.J. Han et al. (eds.), *High Performance Cloud Auditing and Applications*,
DOI 10.1007/978-1-4614-3296-8, © Springer Science+Business Media New York 2014

N

O

P

MIX
Papier aus verantwortungsvollen Quellen
Paper from responsible sources
FSC® C105338

If you have any concerns about our products,
you can contact us on
ProductSafety@springernature.com

In case Publisher is established outside the EU,
the EU authorized representative is:
Springer Nature Customer Service Center GmbH
Europaplatz 3, 69115 Heidelberg, Germany

Printed by Libri Plureos GmbH
in Hamburg, Germany